THE REPRODUCTIVE BIOLOGY OF AMPHIBIANS

THE REPRODUCTIVE BIOLOGY OF AMPHIBIANS

Edited by

Douglas H. Taylor
and Sheldon I. Guttman

Miami University
Oxford, Ohio

Springer Science+Business Media, LLC

Library of Congress Cataloging in Publication Data

Society for the Study of Amphibians and Reptiles.
The reproductive biology of amphibians.

"Proceedings of a Symposium held at the nineteenth annual meeting of the Society for the Study of Amphibians and Reptiles, Miami University, Oxford, Ohio, August 9-10, 1976."

Includes bibliographies and indexes.

1. Amphibians – Reproduction – Congresses. 2. Amphibians – Behavior – Congresses. 3. Courtship of animals – Congresses. 4. Sexual behavior in animals – Congresses. I. Taylor, Douglas H. II. Guttman, Sheldon I. III. Title.

QL669.3.S63 1977 597'.6'0456 77-11004

ISBN 978-1-4757-6783-4 ISBN 978-1-4757-6781-0 (eBook)
DOI 10.1007/978-1-4757-6781-0

Proceedings of a Symposium held at the Nineteenth Annual Meeting of the Society for the Study of Amphibians and Reptiles, Miami University, Oxford, Ohio, August 9-10, 1976

Originally published by Plenum Press, New York in 1977
Softcover reprint of the hardcover 1st edition 1977

Preface

During the course of the development of most areas of scientific research certain events occur that serve to bring a particular area into focus. We feel that this symposium volume serves that role with regard to the study of the reproductive biology of amphibians. The collection of works contained in this volume provides an excellent review of a diversity of past contributions to this field of investigation and brings together in one volume the major approaches that have been utilized to study amphibian reproductive biology.

The work emphasizes recently-completed studies and therefore can be considered to be an up-to-date account of the field. Equally significant, however, the contributors to this volume have excelled in pointing out directions for future research that will be of immense value in integrating the component parts of reproductive biology, i.e., genetics, developmental biology, physiology, behavior, and ecology, into an evolutionary framework.

As can be seen from the bibliographies of each contribution, studies of reproductive biology are increasing annually. The reasons for this are many: amphibians can be easily collected in relatively large numbers; intrageneric crosses between most anurans and some salamanders are feasible; they have relatively large chromosomes which can be readily karyotyped; sophisticated biochemical and physiological techniques have been developed which can be utilized to study evolutionary relationships and sensory capabilities; in particular, an increasing awareness of the importance of behavioral studies in the development of general theories of evolutionary relationships and processes has resulted in a renewal of interest in amphibian behavior, a heretofore largely neglected aspect of amphibian biology. Additionally, since, in vertebrate evolution, amphibians are relatively primitive, extrapolations can possibly be made to reproductive mechanisms in the advanced vertebrates.

As we were planning for the 19th annual meeting of the Society for the Study of Amphibians and Reptiles to be hosted by Miami University, we decided that research on the reproductive biology of amphibians had progressed so rapidly that a symposium devoted solely to this topic was in order. We surveyed the recent reproductive biology literature and assembled a list of individuals who possessed either an excellent current publication record in the area or were relatively new to the field but were making significant contributions. The composition of the list was discussed with a variety of other people but most notably with Kraig Adler (Cornell University) and Thomas Uzzell (Philadelphia Academy of Sciences) and a consensus of opinion on potential participants was reached.

Reactions to the symposium itself, held at Miami University from August 9 to August 10, 1976, were extremely positive and went beyond our greatest expectations. This was due in large part to the excellent contributions of the invited participants. We wish formally to thank the participants for their contributions at the symposium sessions and for their promptly-delivered and readily-editable manuscripts. Thomas Uzzell spent many days helping edit each of the manuscripts and this volume owes much of its quality to his fine efforts. We felt that outside peer review of the manuscripts would enhance the quality of the volume and would like to thank James Bogart, Martha Crump, H. Carl Gerhardt, Robert Inger, Roy McDiarmid, Craig Nelson, Ronald Nussbaum, and Dennis Ralin for their assistance in this regard.

During the preparation and presentation of the symposium many people assisted in making the program a success. The following students assisted greatly in the running of the symposium: Jill Auburn, Lynne Lohmeier, Melinda Moore, Debra Christein, Kevin Crofton, Sara Woolpy, Michael McLeod, and Karen Blackburn. Jill Auburn also gave invaluable editorial assistance. Ruth McLeod provided her usual expertise with the typing of manuscripts. Dean Clarence K. Williamson of the College of Arts and Science and Chairman Charles M. Vaughn of the Zoology Department assisted in funding certain aspects of the meeting and the typing of manuscripts. We thank the officers and members of the Society for the Study of Amphibians and Reptiles for their personal and financial assistance. We also wish to acknowledge the Herpetologists' League for their contributions.

Vice President for Development and Alumni Affairs John E. Dolibois supported the symposium by subsidizing, through the generosity of the Miami University Alumni Fund, the travel and housing expenses of the foreign participants. We would not have been able to produce a symposium, and, therefore, a volume, with such breadth of participation without his strong and continued support.

To all of the above, and others not mentioned individually, we wish to express our sincere gratitude. We hope this volume will provide a contemporary review and synthesis of the present state of knowledge in the field of amphibian reproductive biology.

Douglas H. Taylor

Sheldon I. Guttman

Miami University
Oxford, Ohio
June 29, 1977

Contents

STUDIES OF LIFE HISTORIES AND REPRODUCTION IN NORTH AMERICAN PLETHODONTID SALAMANDERS

Stephen G. Tilley

Department of the Biological Sciences, Smith College
Northampton, Massachusetts

Plethodontid salamanders are among the most abundant vertebrates of many North American ecological communities. Their evident ecological importance (Burton and Likens 1975a, 1975b) and unique associations of ubiquity, obscurity, and beauty have popularized plethodontids as subjects of evolutionary and ecological studies. Their reproductive ecologies and life histories have received particular attention in recent years, as studies have progressed from scattered anecdotes to detailed investigations. Salthe and Mecham (1974) reviewed the literature through the 1960s. Since then, studies of several species have appeared: Gyrinophilus porphyriticus (Bruce 1969, 1972), Pseudotriton montanus (Bruce 1969, 1974, 1975), Stereochilus marginatus (Bruce 1971), Eurycea quadridigitata (Harrison 1973), Eurycea neotenes (Bruce 1976), Eurycea nana (Tupa and Davis 1976), Desmognathus fuscus (Danstedt 1975), Desmognathus ochrophaeus (Fitzpatrick 1973; Forester 1977; Huheey and Brandon 1973; Tilley 1972, 1973a, 1973b, 1974; Tilley and Hausman 1976), and Plethodon wehrlei (Hall and Stafford 1972). Houck reviews certain aspects of life histories of terrestrial plethodontids in this volume. Much, however, remains to be learned.

Our expanding knowledge of plethodontid life histories has accompanied a growing interest in life histories in general and in their adaptive significance. Stearns (1976) and Giesel (1976) recently reviewed theories of the evolution of life history patterns, and several authors (Bruce 1969, 1972, 1975; Tilley 1973a; Danstedt 1975) have attempted to reconcile information on plethodontids with these theories.

A major strength of life history studies is that they bear heavily on the phenomenon of adaptation by directly confronting the components of Darwinian fitness: age-specific fertility and mortality. Heritability of variation in these parameters, however, is nearly always assumed without evidence. Life history and reproductive data are also vital to an understanding of a population's role in community metabolism, and the roles of plethodontid populations in many North American ecosystems must be far from trivial. Burton and Likens (1975a, 1975b) estimated that the *Plethodon cinereus* population of the Hubbard Brook Experimental Forest in New Hampshire exceeds the biomass of the entire avian community during the latter's peak breeding season, and fixes about five times as much energy. These conclusions were reached by assuming stationary populations and without detailed knowledge of reproduction and life histories. A detailed picture of the role of a population in ecosystem dynamics requires more information on dispersal, growth, mortality, and fertility than is presently available.

Reproductive and life history patterns have profound implications for the genetic structure and differentiation of populations. It is now possible to quantify levels of genetic variation and divergence among local populations for that subset of the genome revealed by electrophoretic techniques. Investigators in this area are confronted with a problem exactly the opposite of that faced by students of reproductive strategies: whereas the heritability of observed variation is well understood, its relevance to Darwinian fitness is obscure. The ensuing debate about the relative roles of stochastic and deterministic factors in evolution (see reviews in Lewontin 1974; Nei 1975; and Ayala 1976) has underscored our lack of knowledge of breeding structures and effective breeding sizes of populations, important determinants of levels of genetic variation and divergence. Effective breeding sizes are influenced by dispersal rates, reproductive cycles, and survival, and future studies of plethodontids will undoubtedly focus on these problems.

The strength with which a life history study bears on problems of general interest depends on obtaining solid data on age at first reproduction, age-specific survival, and age-specific fertility. This is best done in the field by following individuals of known age through time via some sort of capture-recapture program. Few plethodontids lend themselves to such studies. They are so numerous and distributed over such extensive areas that it is usually possible to capture only a small percentage of a population, doing considerable violence to the habitat in the process. It may be difficult or impossible to locate certain classes of individuals, such as brooding females, or any individuals at all during much of the year. The detection of recaptured animals is frustrated by their ability to regenerate toes and limbs rapidly. Mortality and dispersal rates in some age classes, such as larvae and small juveniles, may be so great that enormous numbers must be marked to yield meaningful numbers of recaptures, yet occasional adults may live so long as to necessitate very long term studies. Capture-recapture studies are thus feasible only in special cases: where large populations can be adequately and repeatedly sampled without damaging their habitat, where individuals can be regularly recognized and recaptured over long periods, and where the investigator is willing to devote several years to the project. It is not surprising that studies of plethodontid life histories have taken another approach: the collection, preservation, and analysis of chronological samples of individuals. This "vertical" approach has yielded information on a number of species, but a number of difficulties and questions remain. Let us first consider the available information, its strengths, and its weaknesses. We will then consider the data emerging from capture-recapture studies of one species, Desmognathus ochrophaeus, that bear on some critical problems.

LIFE HISTORY PARAMETERS OF NORTH AMERICAN PLETHODONTIDS

Age at First Reproduction

In the "vertical" approach, ages at transformation and sexual maturity are estimated by visualizing size and hopefully age classes among larvae and juveniles. These attempts may be frustrated by numerous factors. Tilley (1973a, 1974) found that juveniles of some Desmognathus ochrophaeus populations fell into distinct size classes while those of other populations did not. The number of size classes in one population varied seasonally, leading him initially (1973a) to underestimate age at maturity and incorrectly postulate that it differed between two adjacent populations.

The available data indicate that age at first reproduction varies considerably among North American plethodontids, but that most species attain maturity in 2-3 years. On the early-maturing side are Eurycea quadridigitata, which matures in the autumn of its first year (Harrison 1973); E. neotenes matures during its second year (Bruce 1976); E. multiplicata transforms and matures 5-8 months after hatching (Ireland 1976). Several species appear to require 4-5 years: the primitive hemidactyliines Gyrinophilus porphyriticus, Pseudotriton montanus (females), P. ruber, and Stereochilus marginatus (Bruce 1969, 1971, 1975), Plethodon wehrlei (Hall and Stafford 1972), P. richmondi (Angle 1969). Females apparently first reproduce one or more years later than males in many species: Desmognathus fuscus (Danstedt 1975), Leurognathus marmoratus (Martof 1962), Eurycea longicauda (Anderson and Martino 1966), Pseudotriton montanus (Bruce 1975), Stereochilus marginatus (Bruce 1971), Aneides ferreus (McKenzie 1970), Plethodon glutinosus (Highton 1956), and P. wehrlei (Hall and Stafford 1972).

Age at maturity has been shown to be delayed in high elevation populations of Desmognathus ochrophaeus (Tilley 1973a) relative to low elevation populations. This results in greater body sizes at maturity and throughout life at high elevations; these larger females can produce larger clutches of eggs. Tilley also found evidence for age structure instability in low elevation D. ochrophaeus populations. Tilley (1973a) and Mertz (1971) discussed ways that early maturity might be favored by selection in unstable environments, in which populations might be increasing most of the time even while fluctuating about mean densities. Stearns (1976) is skeptical of this explanation of early maturity. Bruce related his findings to the well-known theory of r- and K- selection (MacArthur and Wilson 1967, Pianka 1970), hypothesizing that low elevation populations might experience more frequent local extinctions and recolonizations. It is indeed tempting to interpret variation in age at first reproduction, a parameter intimately related to Darwinian fitness (Cole 1954, Lewontin 1965, Stearns 1976), in terms of responses to selection in different environments. Unfortunately, it is difficult to demonstrate heritability for this variation.

Fecundity

Plethodontids typically produce relatively modest numbers of large, yolky eggs (Salthe 1969). Clutch sizes are often positively correlated with body sizes (Tilley 1968, Salthe 1969, Bruce 1969, Salthe and Mecham 1974). The form of the relationship between these two parameters varies inter- and intraspecifically. Nearly all investigators to date have attempted to quantify the relationship with least-squares regression techniques. These techniques are inappropriate where both variates are subject to sampling error (Sokal and Rohlf 1969).

Tilley (1968) found that clutch size-body size regressions were similar in the five species of Desmognathus inhabiting southwestern Virginia's Whitetop Mtn.-Mt. Rogers region, but that D. fuscus tended to be more and D. monticola less fecund than would have been predicted from size-fecundity relationships in other species. Female D. fuscus from southwestern Virginia also seemed to be more fecund than females of similar size from Ohio. Danstedt (1975) found variation in size-fecundity relationships between populations of D. fuscus in the Blue Ridge and Piedmont provinces of Maryland, with the Blue Ridge females of a given body size producing larger clutches. Bruce (1969, 1972, 1975) investigated size-fecundity relationships in the large hemidactyliine species Gyrinophilus porphyriticus, Pseudotriton montanus, and P. ruber. He found that fecundity in P. montanus was much greater than in the other two species, even though body sizes are similar among the three. Large clutch size in P. montanus is associated with a biennial (Bruce 1969) or irregular (Bruce 1975) oviposition cycle, whereas G. porphyriticus and P. ruber females appear to oviposit annually. Bruce (1972) also found that female G. porphyriticus from some populations produced more eggs at given body sizes than those from other localities where larger clutches were attained via larger average body sizes. Bruce, Danstedt, and Tilley have all related their findings to the hypothesis that reproductive efforts should be higher where adult life expectancies are short (Williams 1966).

The relationship between age and fecundity has never been established for any plethodontid, since females cannot be aged. It is not known whether females grow enough after maturation for fecundity to increase with age, or whether clutch sizes depend primarily on growth before and size at maturation. Variance in body sizes of female plethodontids leaves much variance in fecundity unexplained (Tilley 1968, 1973a, Bruce 1969, 1972), and age may affect fecundity independently of its effects on body size.

There has been some controversy about annual vs. biennial and biannual oogenic cycles in plethodontids. This stems directly from problems associated with deciphering reproductive cycles from vertical samples rather than from marked individuals. The evidence for biennial oviposition in some species has been the presence in spring, summer, and fall samples of both reproductive (gravid or brooding) and nonreproductive females. The nonreproductive females are thought to have reproduced the previous year, and thus to require an entire year for oogenesis. Biennial oviposition has thus been hypothesized for southern Appalachian Desmognathus fuscus, D. monticola, D. ochrophaeus, D. quadramaculatus, and D. wrighti (Organ 1961), Leurognathus marmoratus (Martof 1962), Plethodon cinereus in Maryland (Sayler 1966), P. richmondi (Angle 1969), northern populations of P. glutinosus (Highton 1962), P. wehrlei in Pennsylvania (Hall and Stafford 1972), and Aneides ferreus (McKenzie 1970). The occurrence of biennial oviposition in Desmognathus ochrophaeus has been questioned by several workers. Tilley and Tinkle (1968) and Tilley (1973b) showed that winter oviposition occurs in that species, and could account for the presence of spent females in the spring and early summer. The possibility of a biannual cycle was introduced by Tilley and Tinkle, but is now discounted (Tilley 1973b). Huheey and Brandon (1973) did not find that adult females fall into two reproductive classes in populations of D. ochrophaeus in southwestern North Carolina, and postulated an annual cycle. Fitzpatrick (1973) found that female D. ochrophaeus are highly efficient at converting ingested energy into ovarian follicles, and postulated an annual cycle for the species in Ohio. He, like Organ (1961) and Tilley (1973b), found spent females in the spring. His interpretation was that these had last oviposited in the fall or in early spring before emergence. Forester (1977) confirmed the species' capacity for annual oviposition by recapturing females brooding their eggs in successive summers.

Biennial oviposition has also been questioned for some populations of D. fuscus. Spight (1967) and Danstedt (1975) failed to find in North Carolina and Maryland populations the bimodality in follicle size distributions noted by Organ. Similarly, Werner (1971) found that Plethodon cinereus populations in Michigan contained much smaller percentages of non-reproductive females during the summer than were found by Sayler (1966) in Maryland. Werner felt that the few non-reproductive females he encountered might have been out of phase with a primarily annual cycle, perhaps because of late maturation. Annual cycles have also been proposed for Desmognathus aeneus (Harrison 1967), Gyrinophilus porphyriticus (Bruce 1969), Stereochilus marginatus (Bruce 1971), Florida populations of Plethodon glutinosus (Highton 1956), and West Virginia populations of P. wehrlei (Hall and Stafford 1972). Bruce (1975) proposed an irregular cycle for Pseudotriton montanus in South Carolina, finding that populations always contained females in varying stages of oogenesis. Tilley (1973b) concluded that the oogenic cycle of Desmognathus ochrophaeus might be more irregular than previously thought, with females occasionally skipping a year's oviposition. This phenomenon might be responsible for the varying proportions of non-reproductive females noted in several of the above studies. Lacking information on the reproductive conditions of individual females in successive years, workers have tended to impose an annual vs. biennial dichotomy on their data. The possibility that individual females may fit neither pattern during their lifetimes has been largely ignored.

Survival

Survivorship has been investigated chiefly in male plethodontids in which peculiarities of the spermatogenic cycle result in lobed testes. Among plethodontids, lobed testes occur in the desmognathines, bolitoglossines, and in some hemidactyliines of the genera Typhlomolge (Brandon 1971), Typhlotriton (Brandon 1971) and Eurycea: E. neotenes (Bruce 1976, Sever 1974), E. multiplicata (Ireland 1971), E. lucifuga (Sever 1974), E. nana (Sever 1974, Tupa and Davis 1976), E. latitans (Brandon 1971), and E. troglodites (Brandon 1971). The adaptive significance of lobed testes is unknown, and the mechanics of lobe formation may differ among salamander taxa. In plethodontid species, numbers of testis lobes increase with body size, presumably as a function of the number of spermatogenic cycles an individual has undergone. Only a limited number of lobes may form along a spermatogenic cord, however, complicating the relationship between number of testis lobes and age. This relationship might be further complicated by the failure of some individuals to undergo annual spermatogenesis. Humphrey (1922) proposed, without direct evidence, that in Desmognathus three years are required to produce the first two complete lobes, and that additional lobes are completed at two-year intervals. Bruce (1976) proposed that new lobes develop annually rather than biennially in Eurycea neotenes. Testis lobes have been used to assign individuals to age classes in Desmognathus fuscus (Organ 1961, Spight 1967, Danstedt 1975), D. monticola (Organ 1961, Tilley 1968), D. ochrophaeus (Organ 1961, Martof and Rose 1963, Tilley 1973a, 1974, Huheey and Brandon 1973), D. quadramaculatus (Organ 1961), Eurycea multiplicata (Ireland 1971) and E. neotenes (Bruce 1976). Organ (1961), Tilley (1968), Spight (1967), and Danstedt (1975) used testis lobation to estimate age-specific survivorship distributions in Desmognathus, and all report individuals with three or more lobes per testis. Whatever the absolute relationship between age and testis structure, testis lobation has encouraged comparisons of survivorship and age structures among species and populations (Organ 1961, Tilley 1968, 1973a, 1974, Danstedt 1975), and allows us to conclude that male Desmognathus occasionally reach ripe old ages for small vertebrates. Life spans of 10 years or more are probably not unusual for plethodontids.

The fact that a few individuals attain considerable ages has implications for the effective breeding sizes of populations. Old male D. ochrophaeus, for instance, are much larger than females and younger males, and are equipped with hooked jaws and enlarged premaxillary teeth; these were illustrated by Noble (1927). Organ (1961) reports aggressive encounters between male Desmognathus, and I have observed them in the field on two occasions. On the first (1 June 1967, near the old Neals Cr. fish hatchery below Mt. Mitchell, Yancey Co., N. C.), two large males were found during the day under a rock, with one's head in the other's mouth. On the second (12 June 1976, near Beauty Spot Gap in the Unaka Mtns., Unicoi Co., Tenn.), two large males were observed at night in a shallow ditch along a road, with their jaws locked together and twisting their bodies in a spiral fashion. Both observations were made during the spring-early summer courtship period, when females are sometimes found with spermatophores in their vents and are actively courted by males in the laboratory. If older, larger males achieve reproductive dominance through such encounters, their genetic contributions might be disproportionate to their numbers. Tilley and Hausman (1976) have shown that multiple inseminations occur in D. ochrophaeus, and older males may thus frequently inseminate females that have already mated with younger ones. The greater numbers of testis lobes in old males suggest that such individuals may thereby produce more sperm, either by producing more fertile spermatophores or more of them. If large old males thus impregnate females with more sperm than younger males, their reproductive dominance would be further enhanced. Selection might also favor females that mate preferentially with large old males which, by virtue of their sizes and/or ages have demonstrated high fitnesses (Trivers 1972). As pointed out by Williams (1974), this would require a high, perhaps unrealistically high, heritability for fitness. These problems lend themselves to experimental analyses, and should be considered by future investigators. The great densities of some *Desmognathus* and other plethodontid populations may belie much more modest effective breeding sizes. This has implications for the abilities of populations to adapt to the arrays of environments exploited by many species, and relates to the evidence for variation in life history and reproductive patterns among populations.

The main feature of the plethodontid literature is still its superficiality. We lack fundamental life history information on many species, both common and rare, and capture-recapture estimates of age at first reproduction, age-specific fecundity, and age-specific mortality are still unavailable for any plethodontid. In the absence of real evidence, we hope that size classes and testis lobe classes approximate age categories. The remainder of this paper will attempt to relieve some of these ambiguities by presenting data from long-term capture-recapture studies of two populations of *Desmognathus ochrophaeus*. Information will be provided on growth before and after maturity, variation in age at maturity between populations, oviposition cycles, and on the relationship between age and testis lobation.

CAPTURE-RECAPTURE STUDIES OF TWO *DESMOGNATHUS OCHROPHAEUS* POPULATIONS

In the summer of 1970, capture-recapture studies were begun on two populations of *Desmognathus ochrophaeus* inhabiting wet rockfaces near Highlands, Macon and Jackson Cos., N. C. The ecology of rockface populations of *D. ochrophaeus* has been investigated by Huheey and Brandon (1973) and Tilley (1973a, 1974). These populations are well-suited for capture-recapture studies. They exploit geometrically simple habitats, and can be effectively sampled while individuals are abroad on warm summer nights. Females brood their egg clutches in rock crevasses, and larvae inhabit water films on vertical surfaces. Larger salamanders that feed on *D. ochrophaeus*, such as *Gyrinophilus porphyriticus*, *Desmognathus monticola*, and *D. quadramaculatus*, are much less adept at climbing. The rockface environment thus provides *D. ochrophaeus* with all its ecological requirements and refuge from some of its major predators, and wet rockfaces throughout the southern Appalachians support thriving populations. Huheey and Brandon (1973) estimated densities of 25 individuals per m^2 on rockfaces near Highlands. Data on densities in woodland habitats are unavailable but are probably considerably lower. Huheey and Brandon (1973) and Tilley (1973a, 1974) found that members of rockface populations were smaller at sexual maturity and as adults than members of woodland populations. Tilley (1973a) initially believed this to be due to earlier maturity in rockface populations, but later (1974) concluded that lower juvenile growth rates were responsible. This suggests that the great densities of rockface populations result in lower food availabilities, although Huheey and Brandon thought that food items, chiefly small arthropods, are sufficiently abundant on rockfaces to provide adequate food supplies.

Study Areas and Methods

The two populations selected for study are separated by 9.3 km and 424 m in elevation. The high elevation population is located on Whiteside Mtn., Jackson Co., N. C., at 1448 m above sea level. It lies at the upper, northern, end of a long, continuous rockface along the abandoned toll road to the summit of the mountain. Much of this rockface was created when the road was constructed, but sections of it are continuous with extensive natural wet rock surfaces above the road. Sampling has been confined to the upper 12.5 m of the artificial rockface, 0.63 km from the junction of the toll road with the Wildcat Ridge Rd. This section was chosen because it remains wet during the driest periods of summer. The rockface is about 3 m high at this point, so that the area sampled embraces about 38 m^2. Much of this area is solid, vertical rock surface, but there are numerous small horizontal surfaces and deep crevasses. The lower, southern, boundary was arbitrarily fixed, and animals are able to move freely across it. The study area supports populations of Gyrinophilus porphyriticus and D. monticola, but neither species is common. Adult Gryinophilus are sometimes seen on the rockface itself and the relatively small body sizes of these animals in the Highlands area (Bruce 1972) may enhance their climbing abilities. Very few D. monticola have been observed on the rockface, but large adults are often seen at its base.

The low elevation population occupies a small rockface at 1024 m above sea level near Dry Falls in the gorge of the Cullasaja R., Macon Co., N. C. The entire rockface, measuring 5.7 x 1.7 m (10 m^2) is sampled. It supports a population of D. monticola as well as D. ochrophaeus. There are numerous horizontal surfaces and deep crevasses and, in contrast with the Whiteside site, adult D. monticola are often seen on the rockface itself as well as at its base. Juvenile D. monticola are also quite abundant.

The two sites have been visited in July and late August, 1970; early July, mid-July, and late August, 1971; and early June and late August or early September, 1972 through 1976. The early June and late August visits have been timed to obtain adult females before and after the summer brooding season. Sampling periods consist of four consecutive nights, the same nights for both populations. Each night 2-4 persons spend 20 minutes at the Whiteside Mtn. site and 15 minutes at the Dry Falls site collecting every D. ochrophaeus seen except larvae and brooding females. By the fourth night's visit very few specimens can be located, indicating that considerable proportions of the animals present are ultimately captured. The specimens taken on a given night are brought to the Highlands Biological Station and refrigerated. The next day they are anesthetized in tricaine methanesulfonate (Finquel, Ayerst Laboratories) and photographed from the dorsal aspect, 5-7 specimens to a 35 mm slide. Specimens are photographed in the anesthetizing solution to eliminate glare. The following information is then recorded for each specimen: snout-vent length (tip of snout to posterior margin of vent), sex (if ovarian follicles or pigmented vasa deferentia are visible through the ventral abdominal wall), tail base and length regenerated if tail regeneration is evident, and diameter and (since 1974) number of ovarian follicles. The animals are then rinsed, allowed to recover from anesthetic, and returned to the refrigerator. All the specimens taken during the four consecutive nights are released the day after the fourth night's visit. Since individuals cannot be returned to their precise points of capture, this procedure may increase mortality rates by displacing animals from their home ranges. Homing behavior has been demonstrated in D. ochrophaeus on rockfaces, but the effect of displacement on survivorship has not been investigated. Whatever the magnitude of this effect, it does not affect the analyses presented here.

The great variability in the dorsal pattern of D. ochrophaeus (Martof and Rose 1963, Tilley 1969, 1973c) permits the recognition of individual animals in different samples. This is done by matching photographs of specimens taken on different occasions. This process is facilitated by assigning each individual a dorsal pattern code based on the spacing of its larval spots. The positions of these spots are apparent in nearly all individuals because they influence the configuration of the dorsolateral melanophore pattern (Tilley 1969, 1973c). The larval spots are paired, and the members of a given pair may be opposite one another or the left or right spot may be anterior to its partner. Each of the first six pairs of spots is thus assigned to one of three categories: "left ahead," "right ahead," or "opposite," and these designations are recorded as punches in the margins of Key Sort Cards (Litton Industries). It is thus possible to find, among the hundreds of cards corresponding to individuals seen on previous occasions, a limited number where sex and larval spot code correspond to those of an animal in question. By individually examining photographs of each member of this subset of specimens it is easy to confirm recaptures. This method is still time-consuming and tedious, but highly effective. It should prove useful in population studies of other plethodontids with variable color patterns.

The major goals of these studies have been the determination of population densities, ages at maturity, and age-specific growth, survival, fecundity, and dispersal rates in the two populations. Other studies (Tilley 1973a) have shown that D. ochrophaeus populations have delayed maturity, larger adult body sizes, and greater fecundities at high elevations, but the evidence for delayed maturity has been indirect. It was hoped that these studies would resolve the questions of whether D. ochrophaeus females oviposit annually, biennially, or irregularly, and whether age affects fecundity independently of body size. Finally, from the summer of 1973 through the present the two populations have been manipulated. All the D. monticola seen each visit at the Dry Falls site have been removed to see what effect this has on survival and density of D. ochrophaeus. At the Whiteside site, all the adult male D. ochrophaeus seen each visit have been removed to see what effect this has on male recruitment. The data pertinent to these questions have not yet been analyzed, but the preservation of the males removed from the Whiteside population permits investigation of another important phenomenon: the relationship between age and testis lobation. Many of the males preserved had been captured before, and the ages of some of them can be estimated fairly accurately.

The results discussed here are derived from samples taken only through 1974. This provides a five-year record of recaptures, and it is possible to draw some firm conclusions about growth rates, ages at maturity, oviposition cycles, and the age-testis lobation relationship.

Growth Rates, Sizes at Maturity, and Fecundities

The relationships among growth rates, body size, sex, season, and elevation can be seen in figures 1 and 2. These figures show growth increments over two time intervals: early through late summer of the same year, and late summer through early summer of the following year. Negative growth increments reflect measurement errors. The regression lines were fitted to the data using Bartlett's 3-group method, since both variates are subject to sampling error (Sokal and Rohlf 1969).

The sizes at which individuals cease growing in the two populations can be estimated from the intersections of the regression lines with the dotted horizontal lines, drawn parallel to the abscissas at the zero growth increments. Males cease growing at about 45 mm at Whiteside and 37 mm at Dry Falls. These body lengths closely approximate the maximum body lengths of males in the two populations. Females cease growth at about 35 mm at Whiteside and 30 mm at Dry Falls. Both sexes become externally sexable at about 30 mm at Whiteside and 25 mm at Dry Falls. Males thus continue to

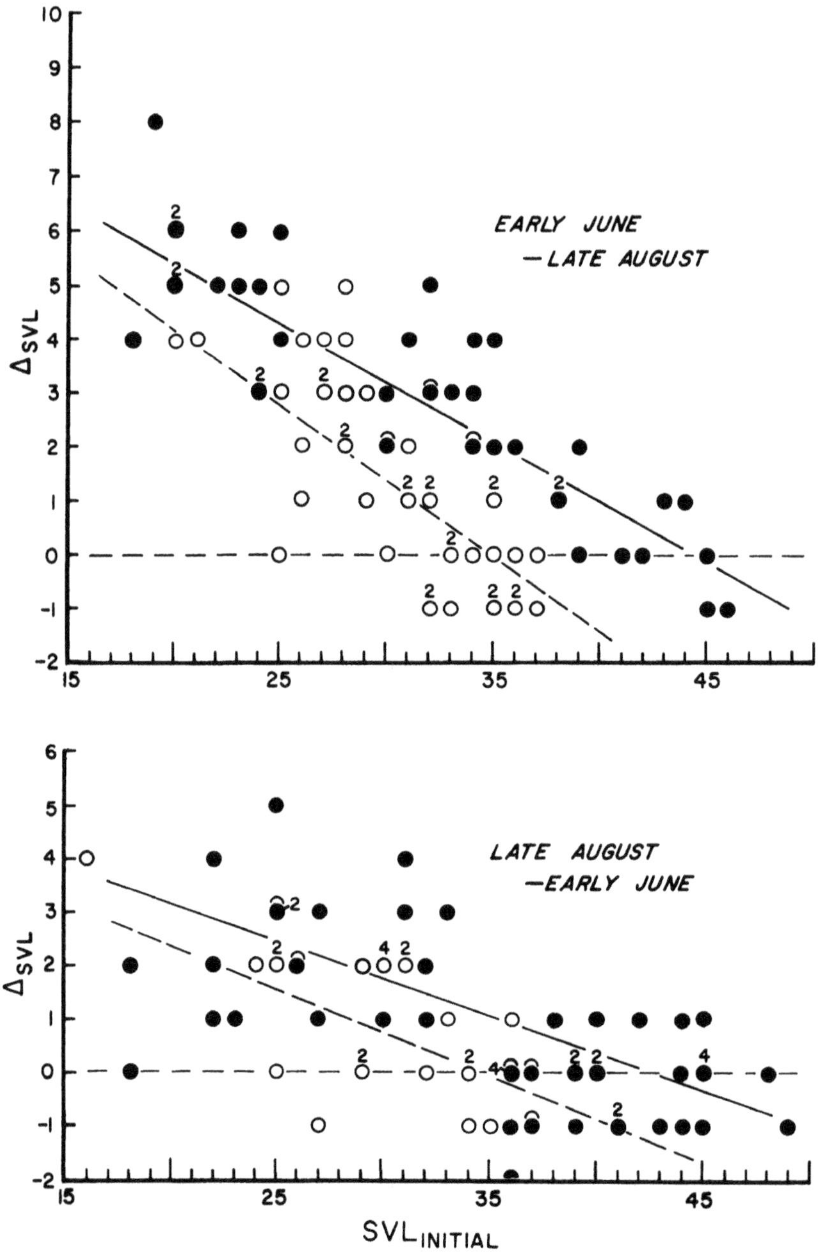

Fig. 1. Growth increments (ordinate) plotted against initial body lengths (abscissa) over two time intervals in the Whiteside population: early June through late August (top) and late August through early June (bottom). The data are based on specimens that were eventually sexed, captured at the beginning and end of a given interval in 1972, 1973, or 1974. Solid circles, males; hollow circles, females.

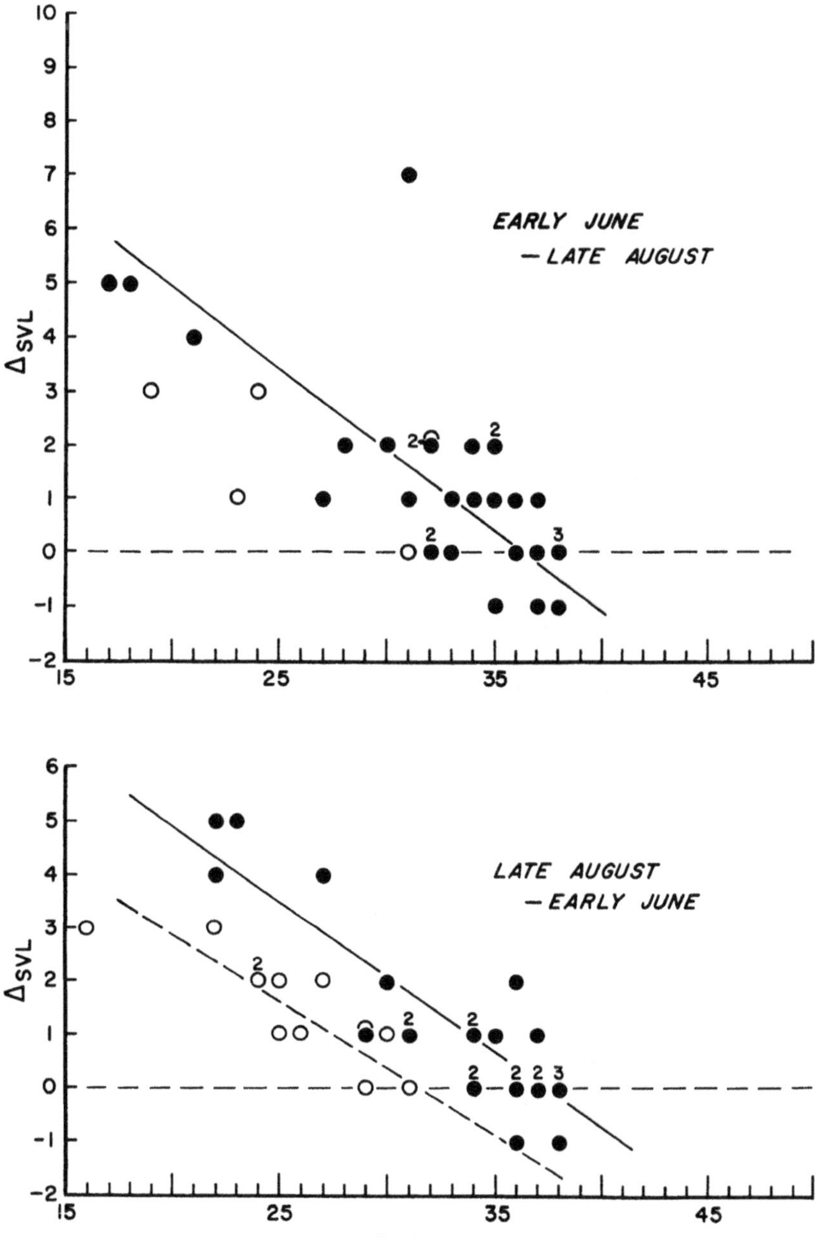

Fig. 2. Growth increments in the Dry Falls population. Symbols as in Fig. 1.

grow well after this stage in both populations, while females cease growth sooner.

Individuals of a given sex at Whiteside cease growth at body lengths 5-8 mm longer than individuals at Dry Falls. This variation in growth and adult body size has a profound effect on fecundities, via the relationship between female body size and clutch size. The relationships between body size and clutch size appear to be identical in the two populations, with body size variance accounting for about 40% of the variance in clutch size. Mean clutch sizes and their 95% confidence intervals, based on gravid females (follicles at least 2.0 mm in diameter) collected in June, 1975 or 1976, at each locality, are 11.272 ± 1.412 for Whiteside (N = 22) and 9.250 ± 0.624 for Dry Falls (N = 24). Confidence intervals for adult female body sizes, based on the same specimens, are 33.272 ± 1.222 and 28.500 ± 0.755. The variances are significantly greater among Whiteside females for both clutch size ($F = 4.651$, $P < 0.001$) and body size ($F = 2.375$, $0.025 > P > 0.01$), and 2-tailed Mann-Whitney U-tests were thus used to compare the clutch and body size distributions of the two populations. Significant U-statistics were obtained for both clutch sizes ($U_s = 364.5$, $0.05 > P > 0.02$) and body sizes ($U_s = 488$, $P < 0.001$). Differences in adult body sizes between the two populations account for about a 20% difference in average clutch sizes.

The regression lines fit to summer growth increments (Figs. 1-2) intersect the ordinates at similar Δ_{svl} values, and individuals 15-20 mm in snout-vent length appear to grow 5-6 mm from early to late summer in both populations. There are, however, considerable differences between the populations in growth increments over the late summer-early summer interval. In the Whiteside population, growth is depressed during the late August-early June interval, while at Dry Falls the regression lines for male growth over the two time intervals are virtually identical. Whiteside individuals, particularly juveniles 15-25 mm in body length, thus grow less during the fall, winter, and spring than Dry Falls individuals. Interpopulational variation in growth rates thus fails to explain the larger body sizes at which members of the Whiteside population mature and cease growth. All other things being equal, Whiteside individuals should attain smaller, not larger, body sizes through their lower growth rates during the colder months.

Ages at Maturity

Snout-vent length distributions of unsexed individuals at Whiteside and Dry Falls are shown in figure 3. Recapture records for individuals not sexable when first captured are shown in Tables 1-8. Some of these recapture records are plotted in figure 3 to

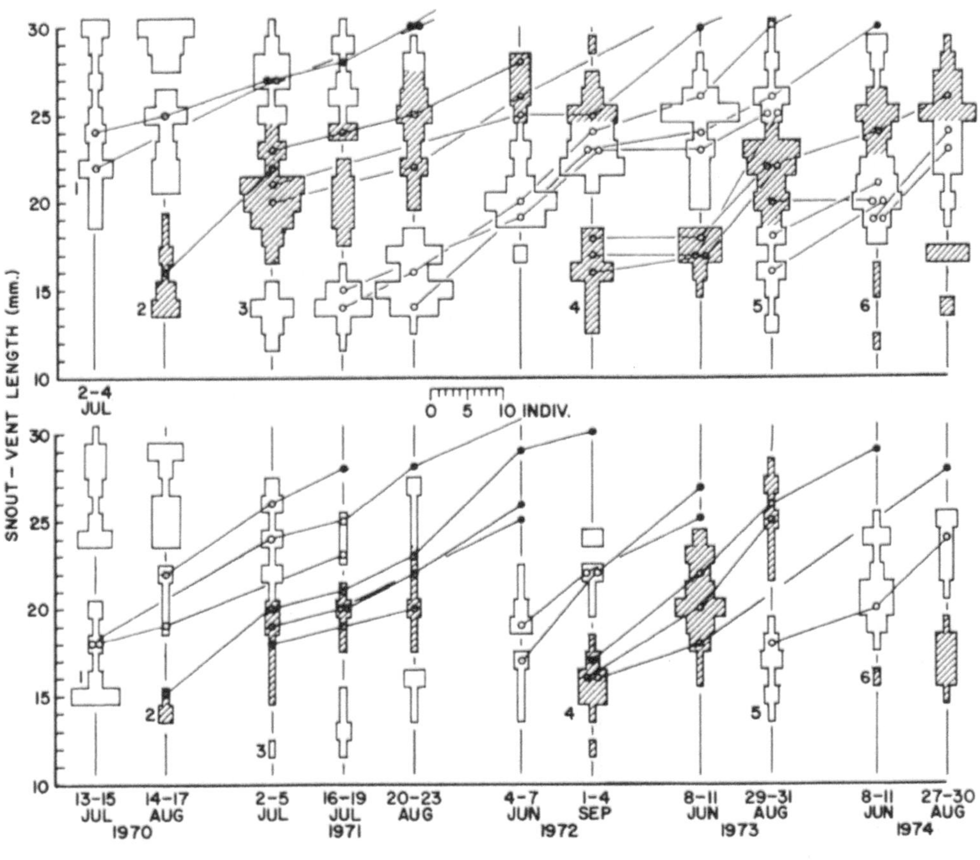

Fig. 3. Snout-vent length distributions with selected recapture records (see text) for all unsexed individuals taken during study at Whiteside (top) and Dry Falls (bottom). Hollow circles, unsexed individuals; solid circles, sexed individuals. Crosshatching delineates cohorts, numbered 1-6 in order of appearance.

Table 1. Snout-vent lengths and reproductive data for members of cohort 1 in the Whiteside population. Only individuals that were unsexable when first captured and subsequently recaptured at least once are included. Underlined snout-vent lengths followed by sex signs denote specimens that were externally sexable. Testis lobe counts are given for males that were preserved at last capture. Numbers in parentheses following female snout-vent lengths denote approximate diameters of ovarian follicles.

	1970		1971			1972		1973		1974	
	2-4 Jul	14-17 Aug	2-5 Jul	16-19 Jul	20-23 Aug	4-7 Jun	1-4 Sep	8-11 Jun	29-31 Aug	8-11 Jun	27-30 Aug
Est. age (mos.)[1]	22	23.5	33	33.5	36	45	48	57	60	70	72
1	21		25	25							
2		26	28	29	28						
3			26	26							
4			29	29							
5	22		27		30♂		38♂				
6		23	27	28♂			37♂				
7	22	22	27	28					36♂ (2.0 lobes)		
8	24	25	27♂	28♂	30♂	30♂	33♂		37♂ (1.5 lobes)		
9		25	27	26	27♀ (1.0)						
10		26	28	27♀(.5)	29♀ (1.0)			30♀ (2.0)			
11			26	26♀(.5)	26♀			30♀ (1.5)	32♀ (1.0)		
12			27		30♀ (.5)	32♀ (1.5)					
13				27				32♀ (1.5)			
14	22	22							34♀ (1.5)		
15			28		29♀ (.5)						
16		25	29♀(1.0)								

[1]Assuming hatching on September 1.

Table 2. Snout-vent length and reproductive data for members of cohort 2 in the Whiteside population. Methods and notation as in Table 1.

Est. age (mos.)[1]	1970		1971			1972		1973		1974	
	2-4 Jul	14-17 Aug	2-5 Jul	16-19 Jul	20-23 Aug	4-7 Jun	1-4 Sep	8-11 Jun	29-31 Aug	8-11 Jun	27-30 Aug
	10	11.5	21	21.5	24	33	36	45	48	58	60
1			21	22	24	26			31		
2			17		20	22					
3			19	21	23	25					
4			21				29				
5			19	15 (measurement error)							
6			18	20							
7			19	19							
8			21	20							
9		16	22								
10				19	22						
11			20			24♂↑	29♂↑			35♂↑ (1.5 lobes)	
12				18		23		33♂↑			
13			20		22		33♂↑				
14			19	19		24		31♂↑			
15			20		22	26		32♂↑	37♂↑ (1.5 lobes)		
16			21							33♀	
17			18			21	25		29♀ (1.0)	29♀ (3.0)	30♀ (0.5)
18			21			25	25	30♀ (2.0)	32♀ (2.0)		32♀ (0.5)
19			23		26		29♀ (1.5)	29♀ (2.0)			
20						28	32♀ (1.0)				

[1]Assuming hatching on September 1

Table 3. Snout-vent lengths and reproductive data for members of cohort 3 in the Whiteside population. Methods and notation as in Table 1.

	1970		1971			1972		1973		1974	
	2-4 Jul	14-17 Aug	2-5 Jul	16-19 Jul	20-23 Aug	4-7 Jun	1-4 Sep	8-11 Jun	29-31 Aug	8-11 Jun	27-30 Aug
Est. age (mos.)[1]			9	9.5	12	21	24	33	36	46	48
1							22	25			
2							21	21			
3							22	22			
4					15	20					
5					16	20					
6					18	20					
7								26	28		
8								21	26		
9										29	31
10				15		19	23	23	25		
11							25	24	29		
12					18	21	26	26			
13					17		21				
14						23		28			
15					15	20		27			
16			12					25			
17			15				24	26			
18			14				26				
19							23		28		
20				14			24		31♂ (0.5 lobes)		
21						20	25		31♂ (0.5 lobes)		
22								20	25	30♂ (1.0 lobe)	
23							22	23♂	28♂		
24					14		23	24	27♂	30♂ (1.0 lobe)	
25					14	17				30♂	32♂

Table 3 (cont.)

	1970		1971			1972		1973		1974	
	2-4 Jul	14-17 Aug	2-5 Jul	16-19 Jul	20-23 Aug	4-7 Jun	1-4 Sep	8-11 Jun	29-31 Aug	8-11 Jun	27-30 Aug
Est. age (mos.)[1]			9	9.5	12	21	24	33	36	46	48
26						19	27	28♂			36♂ (1.0 lobe)
27						20				34♂	
28								25	31⚥		
29								23	29♂		
30					15					30♀ (0.5)	
31							22		27♀ (1.0)		
32							22			29♀ (2.0)	
33								22			30♀ (1.5)
34								25	28♀ (1.0)		30♀ (<.5)
35								25	29♀ (1.0)	31♀ (1.5)	32♀ (2.0)
36				14	16	20	24	26	30	31♀ (0.5)	32♀ (1.5)
37										29	32♀ (1.0)
38										28	30♀(<0.5)
39								25	29♀ (0.5)		

[1]Assuming hatching on September 1

Table 4. Snout-vent lengths and reproductive data for members of cohort 4 in the Whiteside population. Methods and notation as in Table 1.

	1970		1971			1972		1973		1974	
	2-4 Jul	14-17 Aug	2-5 Jul	16-19 Jul	20-23 Aug	4-7 Jun	1-4 Sep	8-11 Jun	29-31 Aug	8-11 Jun	27-30 Aug
Est. age (mos.)[1]						9	12	21	24	34	36
1							16			25	
2							17	17	25		
3							16	17	22		
4								18	23	25	
5							18		26		
6								17	21		
7									21	24	
8									22	24	
9									23	25	
10									20	20	
11										25	29♂
12									22		27♂
13									20		29♂
14									22		29♂
15								18	23	26♂	
16							18	20	26	28♂	(1.0 lobe)
17							18	18	22	24	27♂
18									23	24	27♀ (<.5)
19									20		28♀ (0.5)
20								18		27	30♀ (0.5)
21										25	28♀ (<.5)

[1]Assuming hatching on September 1

Table 5. Snout-vent lengths and reproductive data for members of cohort 1 in the Dry Falls population. Methods and notation as in Table 1, except that testis lobe counts are unavailable.

	1970		1971			1972		1973		1974	
	13-15 Jul	14-17 Aug	2-5 Jul	16-19 Jul	20-23 Aug	4-7 Jun	1-4 Sep	8-11 Jun	29-31 Aug	8-11 Jun	27-30 Aug
Est. age (mos.)[1]	22	23.5	33	33.5	36	45	48	57	60	70	72
1	15	22									
2	20		25♂	26♂							
3			24	24♂	26♂						
4			25	26♂							
5		22	26	28♂							
6			26		26♂			30♂			
7	18	19		23			32♂				
8				24	25	27♀ (1.5)	28♀ (1.0)				
9			22	25	24	26♀ (2.0)					
10	18		24	25	28♀ (0.5)						
11			24			28♀ (>2.0)		29♀ (2.0)			

[1]Assuming hatching on September 1

Table 6. Snout-vent lengths and reproductive data for members of cohort 2 in the Dry Falls population. Methods and notation as in Table 1, except that testis lobe counts are unavailable.

	1970		1971			1972		1973		1974	
	13-15 Jul	14-17 Aug	2-5 Jul	16-19 Jul	20-23 Aug	4-7 Jun	1-4 Sep	8-11 Jun	29-31 Aug	8-11 Jun	27-30 Aug
Est. age (mos.)[1]	10.5	11.5	21	21.5	24	33	36	45	48	58	60
1					16	19					
2					19	20					
3			19	20		25♂					
4				20	22	26♂					
5			21			25♂		30♂			
6				18				28♂			
7					20				31♂	32♂	
8		14	20	21	23	28♂	30♂				
9			18	19	20						

[1]Assuming hatching on September 1

Table 7. Snout-vent lengths and reproductive data for members of cohort 3 in the Dry Falls population. Methods and notation as in Table 1, except that testis lobe counts are unavailable.

	1970		1971			1972		1973		1974	
	13-15 Jul	14-17 Aug	2-5 Jul	16-19 Jul	20-23 Aug	4-7 Jun	1-4 Sep	8-11 Jun	29-31 Aug	8-11 Jun	27-30 Aug
Est. age (mos.)[1]			9	9.5	12	21	24	33	36	46	48
1						21	24				
2				15		22					
3						17	22	27			
4				13	15						
5							24	26♀ (1.5)		27♀ (2.0)	
6						19	22	25♀ (1.0)			
7						20		25♀ (1.5)			

[1]Assuming hatching on September 1

Table 8. Snout-vent lengths and reproductive data for members of cohort 4 in the Dry Falls population. Methods and notation as in Table 1, except that testis lobe counts are unavailable.

	1970		1971			1972		1973		1974	
	13-15 Jul	14-17 Aug	2-5 Jul	16-19 Jul	20-23 Aug	4-7 Jun	1-4 Sep	8-11 Jun	29-31 Aug	8-11 Jun	27-30 Aug
Est. age (mos.)[1]						9	12	21	24	34	36
1								22	25		
2								16	24		
3							16	20	25		
4							15	20			
5							18	21			
6								21	23		
7								21	25♂		
8								18	23♂		
9								19			31♂
10							17	22	26	29♀ (2.0)	
11							16	19			28♀ (0.5)
12								18		26♀ (0.5)	
13								23	24♀ (1.0)		

[1]Assuming hatching on September 1

aid in visualizing cohorts of equal-aged individuals in successive samples. Plots displaying all the recapture records were too complex to be reproduced here, and did not clarify the cohorts further. Figure 3 reveals six different cohorts in each population, numbered 1-6 in their order of appearance in the population. These cohorts correspond to fairly well-defined size classes in many of the samples, but translating size classes to age classes would have been difficult without recapture records, particularly for the Dry Falls population.

The size-frequency data indicate that growth increments of cohorts vary both between populations for the same time interval, and within populations among different time intervals. For instance, in early July, 1970, members of cohort 1 at Whiteside were larger than were members of that same cohort at Dry Falls by mid-July of that same year. They were also considerably larger than members of Whiteside cohort 2 in early July, 1971.

Recently-hatched juveniles occur in both populations in both spring and late summer. The proportion of females that brood during the winter and produce spring hatchlings is not known, but summer oviposition and late summer hatching appear to be the prevailing patterns (see Table 11). Cohorts were thus assigned ages (Tables 1-10) by assuming that all individuals hatch on September 1 of a given year. In fact, a member of a cohort that hatched in late summer might be as many as 8 or 9 months older than a member that hatched the following spring. Whenever an individual hatches, the following summer represents its first full growing season. It is thus convenient to distinguish unsexed individuals according to whether they are first, second, or third summer juveniles.

Figure 3 reveals a clear tendency for Whiteside individuals to remain unsexable through their third summers of life, that is through about 36 months of age assuming hatching on September 1. The samples from Dry Falls are smaller, but it is still apparent that relatively few individuals are still unsexable by their third summers. Early summer samples, especially those for 1972-74, tend to contain a single size class, representing a cohort of individuals about 21 months old. Most individuals one year older, entering their third summer of life, have evidently been recruited into the sexable component of the population by early summer of each year.

A more detailed comparison of ages at maturity can be made by assigning each individual that was unsexed when first captured and that was recaptured at least once to a cohort, according to its snout-vent length on given dates and its recapture record. Ages can then be assigned to cohorts and individuals by assuming September 1 hatching. In Tables 1-8 individuals are grouped and aged according to cohort membership. Individuals whose cohort assignments were ambiguous are omitted. In table 9 the data in tables 1-8 are classified according to an individual's population, its age, and whether or not it was sexable at that age. A given individual appears in table 9 as many times as it was recaptured, and the number of observations in the table thus exceeds the number of different individuals seen. Two-way log likelihood, or G-, tests of independence (Sokal and Rohlf 1969) were employed to compare relative frequencies of sexable and unsexable individuals of a given age class in the two populations. Few individuals younger than 33 months were sexable and few older than 36 months were unsexable in either population. Three G-tests were thus performed (Table 9) to test independence of population and sexability in individuals aged 33 months, 33-34 months, and 33-36 months. The first two tests yielded the most significant deviations from independence, indicating that the populations differ with respect to probabilities of juveniles becoming externally sexable by early in their third summers of life. When individuals 36 months old are included, the difference is less pronounced but still marginally significant. For 33-month juveniles in the Whiteside population, the probability of being externally sexable and its 95% confidence

Table 9. Relative numbers of externally sexable and unsexable individuals of various ages at Whiteside and Dry Falls. See text for description of G-tests of independence.

	Cohort Age (mos.)[1]						
	9-12	21-24	33	33.5	34	36	45-48
Whiteside							
Sexable:	0	0	5	4	2	29	29
Unsexable:	24	100	41	7	11	11	5
	24	100	46	11	13	40	34
Dry Falls							
Sexable:	0	3	8	4	2	6	9
Unsexable:	9	39	10	4	1	2	1
	9	42	18	8	3	8	10

$G_{adj}=6.472$, $0.025 > P > 0.01$

$G_{adj}=9.284$, $P < 0.005$

$G_{adj}=2.858$, $0.1 > P > 0.05$

[1]Assuming hatching on Sept. 1

interval is 0.109 (5/46) ± 0.090. Combining 33-, 33.5-, and 34-month juveniles, it becomes 0.157 (11/70) ± 0.085. Corresponding probabilities for Dry Falls are 0.444 ± 0.230 and 0.359 ± 0.150.

Snout-vent length distributions of 33-34-month juveniles in each population that were recaptured at least once are compared in table 10. The variances of these distributions are artificially low, because individuals whose snout-vent lengths lay near their cohorts' boundaries were omitted. The two distributions do not differ significantly by a 2-tailed Mann-Whitney U-test (U_s = 1179.5, d.f. = 70 and 29, P ∿ 0.2). This would appear to contradict the observation made earlier that Whiteside individuals grow more slowly during the cold months than those at Dry Falls. This discrepancy is explained by the fact that Dry Falls individuals 33-34 months of age are approaching sexual maturity, and the corresponding diversion of energy from growth to reproductive activities evidently compensates for the greater annual growth increments attained earlier in life.

Table 10 also indicates that among 33-34-month individuals the sexable animals tend to be the larger ones, both at Whiteside (U_s = 459.5, d.f. = 11 and 59, 0.05 > P > 0.02) and at Dry Falls (U_s = 168.5, d.f. = 14 and 15, 0.01 > P > 0.001). This could be due to either a direct relationship between body size and maturation, or to a tendency for larger individuals to be older, perhaps having hatched in the fall as opposed to the following winter or spring.

Table 10. Relative numbers of externally sexable and unsexable individuals of various snout-vent lengths, aged 33-34 mos. assuming hatching on Sept. 1, at Whiteside and Dry Falls.

	Snout-vent Length (mm)											
	19	20	21	22	23	24	25	26	27	28	29	Total
Whiteside												
Sexable:					1	1		2	2	4	1	11
Unsexable:		2	3	3	3	7	14	10	8	6	3	59
Total:		2	3	3	4	8	14	12	10	10	4	70
Dry Falls												
Sexable:						1	5	5		2	1	14
Unsexable:	1	1		1	1	5	3	2	1			15
Total:	1	1		1	1	6	8	7	1	2	1	29

Whereas Whiteside males are often externally sexable by late in their third summers of life, they evidently do not reproduce until their fourth summers (see below under "testis lobation and age"). Females at Whiteside appear to typically spend their fourth summers undergoing oogenesis, with the initial oviposition occurring in the fifth summer. For instance, individual 17 in Table 2 was still unsexable late in her third summer, had 1.0 mm ovarian follicles by late in her fourth, and oviposited in her fifth. Individual 18 appears to have retained enlarged follicles through her fourth summer and to have first oviposited in her fifth. Individual 19 had enlarged follicles by early in her fourth summer, but it is not known whether or not she oviposited them that summer or the next. Individual 34 in Table 3 evidently did oviposit in her fourth summer while two other members of her cohort (nos. 35 and 36) were still undergoing their initial oogenesis.

At Dry Falls, age at first reproduction in males is less certain than at Whiteside since no males from there have been dissected. Several females appear to have been ready to oviposit by early in their fourth summers: nos. 9 and 11 in Table 5 and no. 5 in table 7. Other females had distinctly enlarged follicles by early in their third summers, and should have been able to oviposit in their fourth (nos. 6 and 7, Table 7 and no. 10, Table 8).

Oogenic Cycles

Table 11 shows the relative numbers of females in the two populations with enlarged, intermediate, and small ovarian follicles in early and late summer 1973 or 1974. The prevailing pattern in both populations is for females to enter the summer with enlarged follicles, oviposit, and be spent in late summer samples. A few females, however, are spent in early summer and a few are gravid in late summer. These, and the presence of hatchlings on the rockfaces in early summer, indicate that at least a few females oviposit during the colder months. The relative proportions of such individuals are apparently lower than in D. ochrophaeus populations in southwestern Virginia (Organ 1961, Tilley and Tinkle 1968), Mt. Mitchell, North Carolina (Tilley 1973b), and Ohio (Fitzpatrick 1973). The data in table 11, and those of Huheey and Brandon (1973), indicate that in rockface populations of D. ochrophaeus near Highlands, N. C., smaller proportions of females oviposit outside the summer months than elsewhere in the species' range.

Tables 12 and 13 show follicle diameters on different dates for females captured at least four times at Whiteside (Table 12) and Dry Falls (Table 13). Females which were judged, on the basis of their body sizes and recapture records, to be yolking their first clutches during the period of study are excluded to clarify oogenic patterns in females that have reproduced at least once.

Table 11. Reproductive conditions of females in June and August, 1973-74, at Whiteside and Dry Falls. Small females, which might have been approaching their first oviposition, are not included. Reproductive conditions assigned according to estimated diameters of ovarian follicles visible through ventral body wall, as follows: "gravid," ≥ 2.0 mm; "intermediate," ca. 1.5 mm; "spent," ≤ 1.0 mm.

	8-11 June 1973 or 1974	27-31 Aug. 1973 or 1974
Whiteside[1]		
Gravid	28	4
Intermediate	7	8
Spent	13	57
Dry Falls[2]		
Gravid	15	1
Intermediate	2	1
Spent	1	8

[1]Females ≥ 30 mm snout-vent length.

[2]Females ≥ 27 mm snout-vent length.

Females are evidently capable of enough oogenesis between late and early summers of successive years to permit annual oviposition. It is very difficult to prove that an individual has fulfilled this potential, since this requires capturing her in both gravid and spent conditions in successive years. The necessary recapture records are available for only one female, individual 2 in table 12. There is strong evidence for annual oviposition in several others: numbers 3, 9, 10, 12, and 17 in table 12 and numbers 1-3 in table 13.

Several patterns other than oviposition in successive summers are evident. A number of individuals developed or carried enlarged follicles rather than ovipositing them during at least one summer. At Whiteside (Table 12) these include females 1 (in 1971), 2 (in 1970), 5 (in 1970 and 1972), 6 (in 1970), 7 (in 1971), 9 (in 1970), 10 (in 1971), 11 (in 1971), 15 (in 1973), and 18 (in 1971). At Dry Falls (Table 13) they include female 5 (in 1973). Two other females at Dry Falls carried small follicles through the summer with little or no detectable oogenesis: numbers 1 (in 1970) and 5 (in 1973) in table 13. Female 4 in table 12 had enlarged follicles in mid-August, 1970, early July, 1971, and mid-July, 1971. She either carried the same batch of enlarged follicles from August of one year through mid-July of the next, or oviposited and developed another clutch of enlarged follicles between mid-August and early July of successive years. Two females evidently carried small follicles from late

Table 12. Follicle diameters on different dates for females captured at least four times, excluding individuals yolking their first clutches, in the Whiteside Mtn. population. Follicle diameters estimated by viewing follicles through ventral body wall, and are thus subject to ca. ± 0.5 mm measurement error.

Spec.	1970		1971			1972		1973		1974	
No.	2-4 Jul	14-17 Aug	2-5 Jul	16-19 Jul	20-23 Aug	4-7 Jun	1-4 Sep	8-11 Jun	29-31 Aug	8-11 Jun	27-30 Aug
1		<1.0	1.5	2.0	2.0		1.5		1.0		<0.5
2	2.0	2.0				2.5	1.0	2.0	0.5		1.0
3		<1.0			<1.0	2.0	0.5	1.5		2.0	0.5
4		3.0	2.5	2.5		1.0	2.0		0.5		
5	<1.0	2.0				1.0	2.0		<1.0	1.0	
6	1.5	1.5			1.0		1.0		1.0		1.0
7		<1.0	2.0	1.5	1.5		0.5		0.5		
8		1.0		<1.0	<1.0		1.0		1.0		
9	2.0	2.0			1.0	3.0		2.0			
10			1.0		1.5	2.0	<1.0	2.0			
11			1.0		2.0		1.0	1.0			1.0
12			3.0				1.0	>2.0	0.5		1.0
13			<1.0	<1.0		2.0			1.0		1.0
14		<1.0	>2.0			2.0			0.5		1.0
15						2.0	0.5	1.0	1.5		
16						2.0	1.0		2.0		1.0
17							1.0	2.0	0.5	2.0	
18		1.0	1.5	1.5	2.0						
19		<1.0			<1.0		1.0		0.5		

Table 13. Follicle diameters on different dates for females in the Dry Falls population. Females chosen and diameters estimated as in Table 12.

Spec.	1970		1971			1972		1973		1974	
	13-15	14-17	2-5	16-19	20-23	4-7	1-4	8-11	29-31	8-11	27-30
No.	Jul	Aug	Jul	Jul	Aug.	June	Sep	Jun	Aug	Jun	Aug
1	<1.0	<1.0	2.0		<1.0	2.0		2.0		2.0	
2		1.0			0.5	2.0	0.5	2.5			1.0
3	<1.0		1.8	1.5	0.5	2.0					
4			1.0	0.5	0.5				1.0		
5			3.0			1.5		1.0	1.5		

summer of one year through early summer of the next: numbers 5 (in 1973-74) and 11 (in 1972-73) in table 12.

The data thus indicate that, while oviposition in successive years may be the prevalent pattern, there is considerable individual variation in oogenic cycles. The capacity for annual oviposition in D. ochrophaeus verifies the speculations of Tilley and Tinkle (1968), Fitzpatrick (1973), and Tilley (1973b), and the findings of Forester (1977). To characterize the females of a population as ovipositing annually, however, obscures much interesting variation. The number of females failing to oviposit in a given summer indicates that relatively few females adhere to strictly annual oviposition schedules throughout their lives.

The available data do not, unfortunately, clarify the phenomenon of oviposition during the fall, winter, and spring. Gravid individuals appear in August samples and spent individuals occur in early summer, but none of the 24 females in tables 12-13 can be conclusively shown to have oviposited during the fall, winter, or spring.

Testis Lobation and Age

Humphrey (1922) believed that male Desmognathus required three years to develop two complete testis lobes, and that successive lobes matured at two-year intervals. Organ (1961) used this hypothesis to assign males to biennial age classes in a population study. He considered males with a single lobe to be either immature or in their first reproductive year, two-lobed individuals to be in their second or third reproductive years, and so on. This method was followed by Spight (1967) and Danstedt (1975). Tilley (1973a, 1974) and Huheey and Brandon (1973) attempted to classify male D. ochrophaeus into annual rather than biennial age groups by distinguishing between "mature" lobes with caudal appendages indicating sperm evacuation and "immature" lobes lacking caudal appendages. Tilley speculated that the resultant lobe categories approximated year classes: individuals with a single mature lobe were in their first reproductive years, individuals with an immature lobe posterior to a single mature one were in their second reproductive years, individuals with two mature lobes were in their third reproductive years, and so forth. This coincided with Humphrey's statement that two years were required for the development of a mature lobe. Huheey and Brandon (1973) used a similar scheme to estimate the number of previous reproductive seasons for a given male. The chief difficulty in estimating ages from using lobe categories was thought to be the finite space available for lobe development on the spermatogenic cord, so that very old males might have fewer lobes than expected (Collins 1964).

It is possible to estimate from recapture data, with varying degrees of accuracy, the ages of 55 of the males preserved in the course of the removal of adult males from the Whiteside population. Each of these males was dissected and assigned to one of the lobe categories developed earlier (Tilley 1973a). These categories are numbered as follows: 1.0 = one mature lobe, 1.5 = one mature and one immature lobe, 2.0 = two mature lobes, 2.5 = two mature and one immature lobes, 3.0 = three mature lobes, 3.5 = three mature and one immature lobes. When lobe counts differed between spermatogenic cords of the same individual, the greater count was recorded.

The estimated ages of these males are shown in figure 4, together with their body lengths (abscissa) and testis lobe classes (ordinate). The numbers on the graph give the estimated number of reproductive seasons for each individual, based on recapture records. The letters following these numbers denote whether an individual was captured and preserved early (E) or late (L) in the summer. Thus, the designation "2-3E" denotes a male preserved early in his second or third summer of reproductive activity. Some individuals were large males the first time they were captured in 1970 or 1971. When such males were recaptured and preserved in 1973 or 1974 they were in at least their fourth or fifth reproductive summers and probably considerably older than that. They are designated "4+" or "5+" on the graph. The zeros near the graph's origin correspond to males with immature testis lobes when preserved. These, by definition, had not yet reproduced even though pigmented vasa deferentia were visible through the ventral body wall. All four of these individuals were taken in late summer.

Whiteside males become externally sexable at body lengths of about 30 mm. Of the eight males 27-31 mm dissected, half had mature and half had immature lobes, as indicated on the graph. The four specimens with immature lobes, however, were all preserved in late summer, and two of them (nos. 20-21, Table 3) were known to be in their third summers of life. The four individuals in the same size range with mature lobes were preserved in early summer, and two of them (nos. 22 and 24, Table 3) were known to be in their fourth summers. This indicates that young Whiteside males that are sexable in late summer undergo sperm evacuation and reproduce the following summer, in their fourth year of life. One male (no. 16, Table 4) had a mature lobe on each testis early in its third summer. No males have been preserved and dissected from the Dry Falls population, but males there are typically first sexable by early in their third summers. Huheey and Brandon (1973) estimated that males first reproduced during their third summers in other rockface populations near Highlands. If Dry Falls males do likewise, they would be reproducing almost a year earlier than most males of the same age at Whiteside.

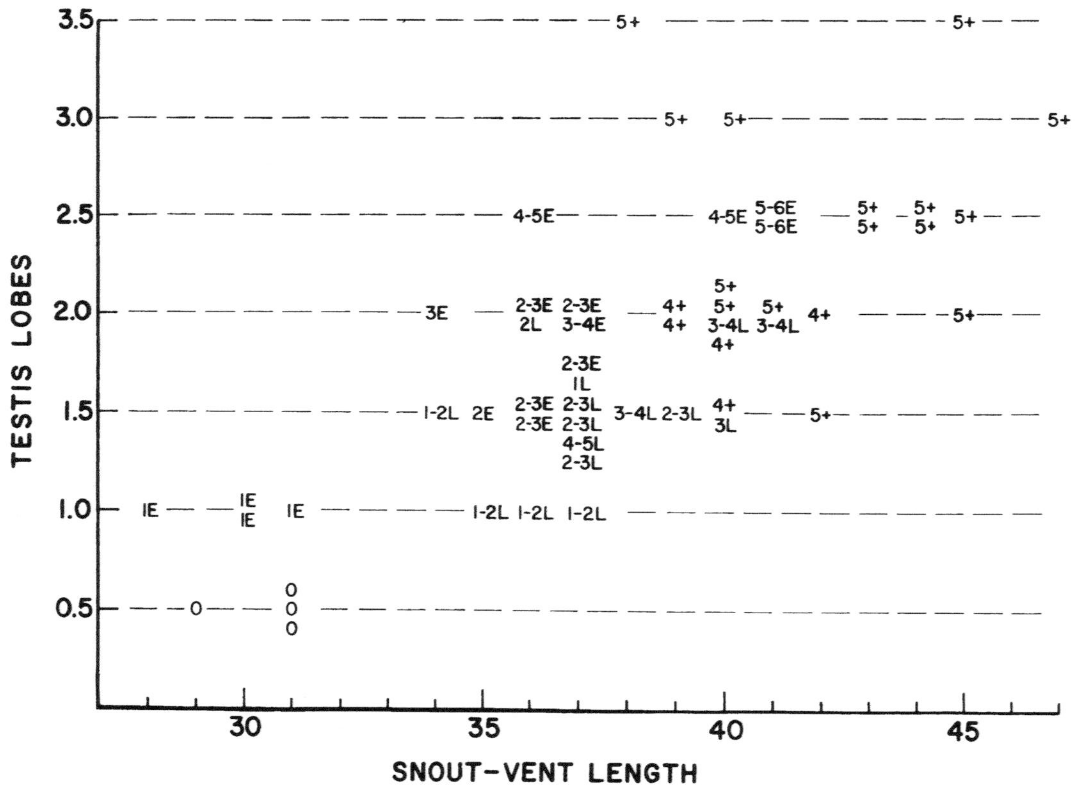

Fig. 4. Numbers of testis lobes plotted against body sizes for preserved males from the Whiteside population, whose ages could be estimated from recapture data. Numbers denote the estimated number of reproductive years completed by an individual. Letters following the numbers denote whether an individual was preserved early (E) or late (L) in the summer. See text for further explanation.

Three individuals ranging from 35 to 37 mm in body length also fell into the 1.0-lobe class. All were preserved in late summer, and it could not be determined whether these individuals were completing their first or second reproductive years. If they were still in their first years of maturity, they had evidently grown quite rapidly. The growth data (Fig. 1) indicate that while one male did, in fact, grow from 30 to 35 mm between early June and late August, 3 mm of summer growth is more typical for males 30 mm in body length in early summer. The other possibility is that these individuals were in their second reproductive summers, but had not yet added additional lobes to their testes.

Each of the other lobe categories contains males of varying ages. Even allowing for uncertainty of aging specimens, these varieties of ages within testis lobe classes do not support the simple relationship between lobation and age proposed by Humphrey. His speculation that three years are required to produce two complete lobes is only partially verified. Several members of the 2.0-lobe class in figure 4 could indeed have been in their third reproductive summers, but that class also contains a group of larger males in at least their fourth. Three years may thus approximate the minimum number of reproductive years for members of the 2.0-lobe class, but some members are considerably older. Likewise, the 2.5-lobe class contains four males that were in their fourth, fifth, or sixth reproductive years, and five larger, older males. The data also indicate that the numbers of reproductive years previously assigned to other lobe classes, 1 year for 1.0 lobes, 2 years for 1.5 lobes, etc., are almost certainly minimum estimates, and each of these classes contains males that are older than predicted. It is possible to make two potentially useful generalizations: males in the 3.0-lobe or higher classes are in at least their fifth reproductive years, and probably considerably older, and individuals with a single lobe are in their first or conceivably their second reproductive years. Testis lobation thus provides means of distinguishing young from old males, and possibly of assigning males minimum ages. Beyond this, testis lobation and age do not appear to be correlated in a simple way.

SUMMARY AND CONCLUSIONS

Capture-recapture studies of the Dry Falls and Whiteside populations corroborate earlier conclusions (Tilley 1973a) that high elevation populations of D. ochrophaeus delay their maturity relative to low elevation populations. This results in the same phenomena, larger adult body sizes and larger clutches, that were observed in high elevation populations in the Mt. Mitchell area. One would predict from the life history data presented here and the hypothesis of Williams (1966) that Dry Falls females are subjected to higher mortality, and this results in selection for earlier reproduction (Tilley 1973a). The fact that the Dry Falls population shares its habitat with a dense population of D. monticola lends support to this interpretation. High elevation D. ochrophaeus populations in the Mt. Mitchell area likewise face fewer salamander predators than low elevation populations (Tilley 1973a). It is possible, however, that environmental differences affect gonadal maturation directly, and transplant studies will be necessary to evaluate these evolutionary and environmental hypotheses.

The results presented here indicate considerable intra-populational variability in sexual cycles. It no longer seems adequate to characterize female D. ochrophaeus as annual or biennial breeders, and the male cycle is apparently so variable that testis lobe counts provide only minimum age estimates. One of the weaknesses of past studies is that they have done little to illuminate variation among individuals of the same population. It seems easy to forget, even while speculating on the evolutionary significance of life history variation among populations, that variation within populations is a requirement of evolution.

ACKNOWLEDGEMENTS

The field work on rockface populations of Desmognathus ochrophaeus has been supported by grants from the Highlands Biological Station, Inc.; the Committee on Aid to Faculty Scholarship and Science Advisory Committee, Smith College; the Penrose Fund of the American Philosophical Society; and N. S. F. grants GB-2496 to the Highlands Biological Station, Inc., and GU-3444 to Smith College. The manuscript reflects, I hope, the helpful comments of Richard C. Bruce and an anonymous reviewer. Perhaps my heaviest debt is to the Smith College students with whom I've shared the inspirations and discomforts of field work in the southern Appalachians: Liza Andrew, Carol Collier, Linda Fidnick, B. Reed Goossen, Suvia T. Judd, Kate Markin, Ellen Rivoir, Jacqueline Suitor, Mary G. Tyler, and Barbara Wu.

REFERENCES

Anderson, J. D. and P. J. Martino. 1966. The life history of Eurycea l. longicauda associated with ponds. Am. Midl. Natur. 75:257-279.

Angle, J. P. 1969. The reproductive cycle of the northern ravine salamander, Plethodon richmondi richmondi in the valley and ridge province of Pennsylvania and Maryland. J. Washington Acad. Sci. 57:192-202.

Ayala, F. J. 1976. Molecular evolution. Sinauer Associates, Sunderland, Massachusetts.

Brandon, R. A. 1971. North American troglobitic salamanders: some aspects of modifications in cave habitats, with special reference to Gyrinophilus palleucus. Nat. Speleological Soc. Bull. 33:1-21.

Bruce, R. C. 1969. Fecundity in primitive plethodontid salamanders. Evolution 23:50-54.

Bruce, R. C. 1971. Life cycle and population structure of the salamander Stereochilus marginatus. Copeia 1971:234-246.

Bruce, R. C. 1972. Variation in the life cycle of the salamander Gyrinophilus porphyriticus. Herpetologica 28:230-245.

Bruce, R. C. 1974. Larval development of the salamanders Pseudotriton montanus and P. ruber. Am. Midl. Natur. 92:173-190.

Bruce, R. C. 1975. Reproductive biology of the mud salamander, Pseudotriton montanus, in western South Carolina. Copeia 1975:129-137.

Bruce, R. C. 1976. Population structure, life history, and evolution of paedogenesis in the salamander Eurycea neotenes. Copeia 1976:242-249.

Burton, T. M. and G. E. Likens. 1975a. Energy flow and nutrient cycling in salamander populations in the Hubbard Brook Experimental Forest. Ecology 56:1068-1080.

Burton, T. M. and G. E. Likens. 1975b. Salamander populations and biomass in the Hubbard Brook Experimental Forest, New Hampshire. Copeia 1975:541-546.

Cole, L. C. 1954. The population consequences of life history phenomena. Quart. Rev. Biol. 29:103-137.

Collins, D. 1964. An investigation of the correlation of testicular lobing in Desmognathus with age. J. Ohio Herp. Soc. 4:107 (abstract).

Danstedt, R. T., Jr. 1975. Local geographic variation in demographic parameters and body size of Desmognathus fuscus (Amphibia: Plethodontidae). Ecology 56:1054-1067.

Fitzpatrick, L. C. 1973. Energy allocation in the Alleghany Mountain salamander, Desmognathus ochrophaeus. Ecol. Monogr. 43:43-58.

Forester, D. C. 1977. Comments on the female reproductive cycle and philopatry by Desmognathus ochrophaeus (Urodela: Plethodontidae). J. Herp.(in press).

Giesel, J. T. 1976. Reproductive strategies as adaptations to life in temporally heterogeneous environments. Ann. Rev. Ecol. and Syst. 7:57-79.

Hall, R. J., and D. P. Stafford. 1972. Studies in the life history of Wehrle's salamander, Plethodon wehrlei. Herpetologica 28: 300-309.

Harrison, J. R. 1967. Observations on the life history, ecology, and distribution of Desmognathus aeneus aeneus Brown and Bishop. Am. Midl. Natur. 77:356-370.

Harrison, J. R. 1973. Observations on the life history and ecology of Eurycea quadridigitata (Holbrook). H. I. S. S. News-Journal 1:57-58 (abstract).

Highton, R. 1956. The life history of the slimy salamander, Plethodon glutinosus, in Florida. Copeia 1956:75-93.

Highton, R. 1962. Geographic variation in the life history of the slimy salamander. Copeia 1962:597-613.

Huheey, J. A. and R. H. Brandon. 1973. Rockface populations of the mountain salamander, Desmognathus ochrophaeus, in North Carolina. Ecol. Monogr. 43:59-77.

Humphrey, R. R. 1922. The multiple testis in urodeles. Biol. Bull. 43:45-67.

Ireland, P. H. 1971. Systematics, reproduction and demography of the salamander, Eurycea multiplicata (Cope). Unpub. Dissertation, Univ. Arkansas.

Ireland, P. H. 1976. Reproduction and larval development of the gray-bellied salamander Eurycea multiplicata griseogaster. Herpetologica 32:233-238.

Lewontin, R. C. 1965. Selection for colonizing ability. IN: H. G. Baker and G. L. Stebbins (eds.), The genetics of colonizing species. Academic Press, New York, New York, pp. 77-94.

Lewontin, R. C. 1974. The genetic basis of evolutionary change. Columbia Univ. Press, New York, New York.

MacArthur, R. and E. O. Wilson. 1967. The theory of island biogeography. Princeton Univ. Press, Princeton, New Jersey.

Martof, B. S. 1962. Some aspects of life history and ecology of the salamander Leurognathus. Am. Midl. Natur. 67:1-35.

Martof, B. S. and F. L. Rose. 1963. Geographic variation in southern populations of Desmognathus ochrophaeus. Am. Midl. Natur. 69:376-425.

McKenzie, D. S. 1970. Aspects of the autecology of the plethodontid salamander, Aneides ferreus. Unpubl. Dissertation, Oregon State Univ.

Mertz, D. B. 1971. Life history phenomena in increasing and decreasing populations. IN: G. P. Patil, E. C. Pielou and W. E. Waters (eds.). Statistical Ecology, Vol. II. Penn. State Univ. Press, University Park, Pennsylvania, pp. 361-369.

Nei, M. 1975. Molecular population genetics and evolution. North-Holland, Amsterdam.

Noble, G. K. 1927. The plethodontid salamanders; some aspects of their evolution. Am. Mus. Nov. 249:1-26.

Organ, J. A. 1961. Studies of the local distribution, life history, and population dynamics of the salamander genus Desmognathus in Virginia. Ecol. Monogr. 31:189-220.

Pianka, E. 1970. On "r" and "K" selection. Am. Natur. 104:595-597.

Salthe, S. N. 1969. Reproductive modes and the numbers and sizes of ova in the urodeles. Am. Midl. Natur. 81:467-490.

Salthe, S. N. and J. S. Mecham. 1974. Reproductive and courtship patterns. IN: B. Lofts (ed.). Physiology of the Amphibia. Vol. II. Academic Press, New York, New York, pp. 309-521.

Sayler, A. 1966. The reproductive ecology of the red-backed salamander, Plethodon cinereus, in Maryland. Copeia 1966:183-193.

Sever, D. M. 1974. The occurrence of multiple testes in the genus Eurycea (Amphibia: Plethodontidae). Herpetologica 30:187-193.

Sokal, R. R. and F. J. Rohlf. 1969. Biometry. W. H. Freeman and Co., San Francisco, California.

Spight, T. M. 1967. Population structure and biomass production by a stream salamander. Am. Midl. Natur. 78:437-447.

Stearns, S. C. 1976. Life-history tactics: a review of the ideas. Quart. Rev. Biol. 51:3-47.

Tilley, S. G. 1968. Size-fecundity relationships and their evolutionary implications in five desmognathine salamanders. Evolution 22:806-816.

Tilley, S. G. 1969. Variation in the dorsal pattern of Desmognathus ochrophaeus at Mt. Mitchell, North Carolina, and elsewhere in the southern Appalachian Mountains. Copeia 1969:161-175.

Tilley, S. G. 1972. Aspects of parental care and embryonic development in Desmognathus ochrophaeus. Copeia 1972:532-540.

Tilley, S. G. 1973a. Life histories and natural selection in populations of the salamander Desmognathus ochrophaeus. Ecology 54:3-17.

Tilley, S. G. 1973b. Observations on the larval period and female reproductive ecology of Desmognathus ochrophaeus (Amphibia: Plethodontidae) in western North Carolina. Am. Midl. Natur. 89: 394-407.

Tilley, S. G. 1973c. Desmognathus ochrophaeus Cope. Mountain dusky salamander. Cat. Am. Amphib. and Rept. 129.1-129.4.

Tilley, S. G. 1974. Structures and dynamics of populations of the salamander Desmognathus ochrophaeus Cope in different habitats. Ecology 55:808-817.

Tilley, S. G. and J. S. Hausman. 1976. Allozymic variation and occurrence of multiple inseminations in populations of the salamander Desmognathus ochrophaeus. Copeia 1976:734-741.

Tilley, S. G. and D. W. Tinkle. 1968. A reinterpretation of the reproductive cycle and demography of the salamander Desmognathus ochrophaeus. Copeia 1968:299-303.

Trivers, R. L. 1972. Parental investment and sexual selection. IN: B. Campbell (ed.), Sexual selection and the descent of man --1871-1971. Aldine-Atherton, Chicago, Illinois, pp. 137-179.

Tupa, D. D. and W. K. Davis. 1976. Population dynamics of the San Marcos salamander Eurycea nana Bishop. Texas J. Sci. 27: 180-195.

Werner, J. K. 1971. Notes on the reproductive cycle of Plethodon cinereus in Michigan. Copeia 1971:161-162.

Williams, G. C. 1966. Natural selection, the costs of reproduction, and a refinement of Lack's principle. Am. Nat. 100:687-690.

Williams, G. C. 1974. Sex and evolution. Princeton Univ. Press, Princeton, New Jersey.

LIFE HISTORY PATTERNS AND REPRODUCTIVE BIOLOGY OF NEOTROPICAL SALAMANDERS

Lynne D. Houck[1]

Museum of Vertebrate Zoology and Department of Zoology
University of California, Berkeley, California 94720

The life cycles and reproductive biology of salamanders have been studied for over 75 years (e.g. Ritter and Miller 1899, Ritter 1903). Initial investigations have emphasized descriptive natural history (e.g. Blanchard 1928, Burger 1935, Lynn and Dent 1941, Pope and Pope 1951, Gordon 1952, Taub 1961), but more recent workers have attempted to understand and interpret life history phenomena in terms of the selective pressures bearing on salamander populations (e.g. Organ 1961, Tilley 1968, Bruce 1969, Salthe 1969, Highton 1970, Maiorana 1976). Implicit in this recent approach is the view that a life history pattern is a complex set of interdependent features representing an adaptive and dynamic compromise to pertinent selective pressures.

The goal of many current studies is to understand the relationship between life history features and the selective regimes experienced by different populations. This goal is not easily realized. Frequently, very different selective pressures may account for the same life history features, and it may not always be possible to directly evaluate the significance of each selective force (see Stearns 1976). As new and more detailed life history information becomes available, however, we can revise hypotheses relating observed life history phenomena to selective regimes, and tractable aspects can be investigated (e.g. see below, Intraspecific Variation

---

[1]Current address: Department of Biology
University of Chicago
Chicago, Illinois 60637

in Clutch Size). Although we can not document precisely all selective pressures responsible for the evolution of life history phenomena, the level of our understanding can be greatly increased.

Recent studies (Valdivieso and Tamsitt 1965; Vial 1968; Houck 1977a, 1977b) have added a new dimension to the investigation of salamander life histories. These studies concern salamanders which occur in the Neotropics, and are the first life history studies to focus on salamanders in an area other than the temperate zone. Emerging information pertaining to neotropical salamanders not only is beginning to reveal life history patterns associated with tropical climatic and biotic regimes, but also provides a basis for latitudinal comparisons. Studies of neotropical salamanders provide additional information about the association of life history features with different climatic conditions. Assuming that life history differences reflect differences in selection, observations concerning local and latitudinal variation can lead to new ideas of ways in which tropical and temperate selective regimes are different.

Neotropical salamanders are closely related taxonomically, belonging to one supergenus (Bolitoglossa) of the Plethodontidae (Wake 1966). The primary life history feature of most species in this family, including all neotropical species, is a life cycle which is completely terrestrial. Young hatch from eggs laid on land and there is no aquatic larval stage (Mode III, Salthe 1969). Almost all other salamander species are either permanently aquatic or have land-dwelling adults which periodically return to ponds or streams to lay eggs. Completely terrestrial plethodontids do not return to water, but are restricted to areas in which surface moisture is available (e.g. Rosenthal 1954). Subterranean or surface refugia are utilized when moisture or temperature conditions are unfavorable. Like other salamander species, plethodontids are carnivorous and opportunistically feed on a variety of small invertebrates. Although some species actively move about the surface during relatively warm, humid nights, most terrestrial salamanders remain close to burrows, logs, leaf litter, or other cover objects. Studies of density show that terrestrial plethodontids can be relatively abundant and may rival or exceed birds and small mammals in biomass density (Merchant 1972, Burton and Likens 1975).

In addition to having species characterized by completely terrestrial development, the Plethodontidae is the only salamander family which is widespread in the Neotropics. Plethodontids are believed to have evolved in the Appalachian Mountains of eastern North America (Wilder and Dunn 1920, Dunn 1926) with subsequent dispersal across North America, and into the New World tropics (Wake 1966). Some species of salamandrids and ambystomatids occur at the northernmost tropical latitudes (e.g. in Mexico and SE Asia), but salamanders other than plethodontids essentially are restricted to the Holarctic

(Wake and Lynch 1976). While other salamander families presumably had physical access to neotropical dispersal routes during the course of evolutionary time, the Plethodontidae may have been the only family with ecological access: the neotropical distribution of plethodontids doubtless is related to their completely terrestrial life cycle and resulting independence of free-standing water (Wake 1966, 1970).

Neotropical plethodontids are extremely abundant and diverse. The seven neotropical genera currently recognized include approximately 150 species (Wake and Lynch 1976). Close to half of all living salamander species are plethodontids found in the Neotropics (Wake 1970). The patterns of distribution and diversity of these salamanders have been documented extensively by Wake and Lynch (1976).

As a group, then, neotropical salamanders are closely related taxonomically and all share the life history feature of completely terrestrial development. Their occurrence in a wide variety of habitats and climatic regimes provides a basis for life history comparisons between different neotropical populations as well as between neotropical and temperate zone populations.

Two early studies of neotropical salamanders revealed intriguing latitudinal differences in certain life history features. Vial's (1968) study of *Bolitoglossa subpalmata* in Costa Rica showed that these salamanders were active during all months of the year and that courtship and oviposition also were aseasonal. Valdivieso and Tamsitt (1965) presented evidence of a similar aseasonal pattern for *B. adspersa* in Colombia. These aseasonal reproductive cycles differ sharply from the strongly seasonal patterns observed for extratropical plethodontids (e.g. see Peacock and Nussbaum 1973).

One might expect that aseasonal patterns are common among neotropical salamanders, but *B. subpalmata* and *B. adspersa* were studied at high elevations, primarily above 2500 m, in areas where moisture conditions permit surface activity throughout the year. Rainfall patterns vary significantly in most neotropical areas (Rumney 1968) and salamander activity can be severely restricted by long periods without rain. In addition, neotropical species occur at a variety of elevations from sea level to over 3800 m (Wake and Lynch 1976). Average temperatures experienced by salamanders at elevational extremes can differ by over 20° C (Feder 1976), although average temperature fluctuations at a given locality may be relatively small ($\leq$ 5° C; McBryde 1942). Variation in temperature and rainfall alone require that species from a more representative sample of habitats and climatic conditions be considered before generalizations emerge from life history studies of neotropical salamanders. Work recently completed (Houck 1977a, 1977b) provides life history descriptions for 11 plethodontid species from western Guatemala. These species occur in different vegetation types and microhabitats over an elevational range from 1000 m to 3800 m. These life history descriptions, together with previous studies (Vial 1968, Valdivieso and Tamsitt 1965), provide a more substantial data base for identifying and comparing life history patterns of neotropical plethodontids.

The goals of this paper are to present life history patterns of a group of neotropical species and, at the same time, to compare these patterns with those of North American plethodontids. Four groups of life history parameters are considered: (1) generalized patterns of surface activity, courtship and spermatogenesis, and oviposition; (2) growth rates, size and age at sexual maturity; (3) clutch size, egg size, and clutch volume; (4) intraspecific variation in clutch size. The following comparisons reveal latitudinal variation in certain life history features, and raise new hypotheses concerning the nature of selective pressures contributing to this variation. Although the underlying causes of many life history differences remain open to speculation, some hypotheses are considered which seek to explain variation in particular life history features.

GENERAL ACTIVITY PATTERNS

General patterns of surface activity, spermatogenesis and courtship, and timing of oviposition are compared for plethodontid salamanders in North America and in Guatemala. Because climatic regimes differ greatly within North America, salamander species from the eastern and western United States are considered separately. Different plethodontid species living under similar climatic conditions often show similar patterns of overall activity. The generalized patterns described below are a reflection of these similarities. Seasonal activity patterns of some salamanders can differ, however, from those of most sympatric species; these individual cases are discussed.

Salamander species from each area included in the following comparisons are listed in Appendix I along with primary literature references. Information concerning these species usually is based on the collection of chronological samples from field populations and represents the "vertical" approach discussed by Tilley (this volume). Tilley examines problems associated with this approach, but appreciates the fact that more desirable capture-recapture studies are feasible only in certain cases.

Elsewhere in this volume, Tilley considers life history parameters of North American plethodontids, and emphasizes those of a species having aquatic larvae. Parameters can vary greatly with reproductive mode, however. Species with aquatic sites of oviposition, for example, generally have larger clutch sizes than strictly terrestrial species (Salthe 1969). Since all neotropical species have completely terrestrial development, only North American plethodontids having terrestrial development are included in the following comparisons when reproductive mode influences life history variations.

Eastern United States

Most available life history information concerns eastern plethodontids (species of Plethodon and Aneides) from areas climatically characterized as follows. Snow and rain are common during the winter but most rainfall occurs during late spring and summer months (roughly late March through October). Temperatures are relatively high during these months, are lower during fall (October-November), and are close to or below freezing during winter and early spring (December through March).

General activity patterns of salamanders which live in this seasonally temperate climate are depicted in Figure 1A. The significance of surface activity for eastern and other plethodontids is that most feeding occurs while salamanders are at or near the surface (Maiorana 1976; see also Growth Rates, below). Eastern salamanders are active from spring until fall when surface moisture is available and temperatures are not excessively hot or cold. As temperatures become colder in the fall, salamanders are increasingly rare at the ground surface and seek deep rock crevices or subterranean retreats in which to overwinter. Salamanders may become active in the winter if temperatures rise above freezing (Caldwell 1975), but commonly are restricted from surface activity until the onset of warmer temperatures in the spring. Within this general pattern of surface activity, large and small plethodontid species (sensu Grobman 1944) may have different peaks of activity (Highton 1972). Large Plethodon species (e.g. P. glutinosus and P. jordani apparently are relatively more active during warm summer months while small species (e.g. P. cinereus and P. hoffmani are more likely to be found at the surface when temperatures are above freezing during the winter (Highton 1962b, 1972; Taub 1961).

Courtship activity, leading to sperm transfer, occurs during the late summer and fall months in the eastern United States; this activity is interrupted by winter but can occur again in the spring (Fig. 1A). Sperm production is accompanied by the development in males of a submandibular (mental) gland and enlarged cloacal papillae (Howell and Hawkins 1954). These secondary sexual characteristics are seasonally developed (Hall and Stafford 1972) and reflect periods when mature spermatozoa are produced. The spermatogenic cycle for eastern species is similar to that described for P.

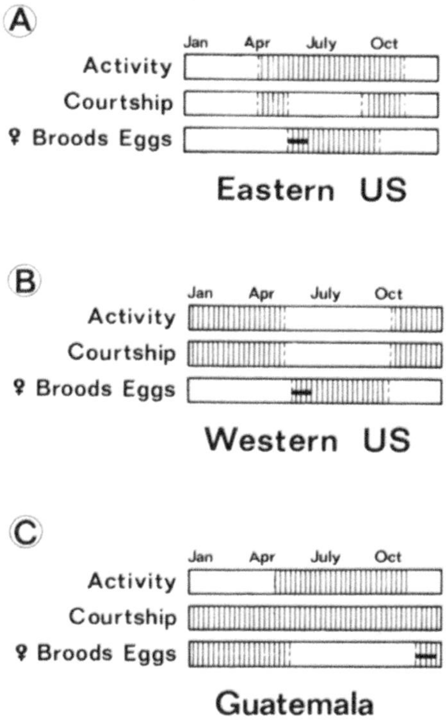

Fig. 1. General patterns of surface activity, courtship), and brooding for groups of plethodontid salamanders in North America (A and B) and the Neotropics (C). The horizontal bar indicates when oviposition usually occurs. The vertical lines indicate the general range of time over which each activity occurs. Surface activity for Guatemalan salamanders applies only to terrestrial species. See text for discussion.

glutinosus (Highton 1962a): spermatogenesis may begin in April, but mature sperm are present in the vasa deferentia only from August or September until the following April. Females possess a specialized cloacal structure, the spermatheca, in which sperm apparently can be stored for many months so courtship need not directly precede oviposition (Sayler 1966, Marynick 1971).

The timing of oviposition is highly seasonal and occurs in late spring or early summer. In any given area, oviposition takes place over a relatively short period, perhaps two weeks, although the precise timing varies annually and geographically depending on climatic conditions (Highton 1962a). Most egg clutches are laid underground and eggs of many species have not been observed in the field. Although oviposition occurs annually within a population, not every female oviposits each year. Reports of individual ovipositional patterns for Plethodon hoffmani (under the name richmondi, Angle 1969) and P. glutinosus (Highton 1962a) are typical of information available for most other eastern species. For these species, two classes of adult females usually can be distinguished: those which have enlarged ova or are brooding eggs, and those which have comparatively small, immature ova (possibly in the initial stages of development). On this basis, individual females are thought to oviposit every other year. This biennial pattern may be a consequence of the cost of brooding eggs. Females remain with their clutches for 2-3 months until embryonic development is complete and hatchlings emerge from the egg capsules in late summer or early fall. Brooding females essentially are deprived of food (Organ 1961, Forester 1974) relative to conspecifics which can be actively feeding at the surface. The short activity period which remains after brooding probably is not long enough for spent females to acquire sufficient food to develop a clutch of eggs for the following spring (Organ 1961, Highton 1962a, Salthe and Mecham 1974). Females which will not oviposit the following spring or early summer apparently are not receptive to courtship activity (Sayler 1966, Arnold 1976).

These overall patterns of surface activity, courtship and oviposition are derived from available life history data (see Appendix I) although local climatic variation may account for shifts in timing from place to place. Species which are widespread may be exposed to a variety of climatic conditions, however, and life history patterns of populations in some areas may be markedly different from those of other populations. For example, Sayler (1966) studied a population of Plethodon cinereus (from approximately 300 m above sea level in Maryland) in which surface activity and courtship occurred throughout the winter, at least for some individuals. In other populations of P. cinereus (collected by Highton from elevations near sea level in Maryland and Pennsylvania), females I examined appeared to have an annual instead of biennial pattern of oviposition: 42 of 43 adult females collected in the late spring had enlarged ovarian eggs. These two populations probably differ from the general pattern of eastern United States plethodontids because they experience winter conditions which permit surface activity and therefore feeding and courtship. Where it occurs, annual oviposition may be a reflection of a female's ability to feed at the surface during more months of the year.

Variation in life history patterns of P. cinereus in different parts of its range suggest that climatic rather than physiological restrictions govern levels of activity, courtship, and individual timing of oviposition. This conclusion is supported by life history data available for P. glutinosus (Highton 1956, 1962a): populations in Florida experience a climatic regime which permits more months of surface activity relative to populations in Maryland or Pennsylvania. Among other life history parameters which vary, juvenile growth rates of Florida P. glutinosus are relatively larger and individual females have an annual pattern of oviposition.

Where microclimatic conditions permit greater yearly activity, then, inter- or intraspecific differences may be observed for growth rates, individual patterns of oviposition, and other life history parameters. The general activity patterns outlined in figure 1A primarily are characteristic of species in temperate areas where climatic conditions strongly restrict surface activity during the winter and where most activity occurs during the rainy period between spring and early fall.

Western United States

Species considered here include those in four genera (Aneides, Batrachoseps, Ensatina, and Plethodon) which are found primarily within the states of California, Oregon, and Washington. In this area, summer months (June, July, August) are characterized by warmer temperatures and by drought conditions resulting in surface habitats which essentially are completely dry. Rains usually begin in the fall (September or October) and continue intermittently through winter until spring (April or May). Cooler temperatures prevail in winter months although snow and freezing temperatures are infrequent or at least not continuous throughout the winter.

The activity patterns of salamanders from this area are depicted in figure 1B. Individuals are active at the surface from the first heavy fall rains until spring when rains cease. Rosenthal (1954), for example, found Aneides lugubris active from November through early May. Temperature and moisture conditions normally permit activity throughout the winter, but unusually low temperatures or lack of rainfall preclude surface activity. Moisture is believed to be a more critical factor in limiting activity than is temperature (Rosenthal 1954). During periods of summer drought, salamanders retreat to subterranean refugia.

The spermatogenic cycle of species in the western United States is similar to that of plethodontids in the eastern United States. Information reported for Ensatina eschscholtzii (Stebbins 1949) is typical of that known for most other western species: males begin the process of spermatogenesis in the early summer but are capable of producing mature sperm only from early October until March or April. Courtship activity reflects sperm production (Fig. 1B), and usually occurs throughout the rainy period when salamanders are active. Sperm masses were found in the cloacae of female E. eschscholtzii from November through March (Stebbins 1949). Peacock and Nussbaum (1973) report that male Plethodon vehiculum also are in breeding condition from fall (September) to early spring (March), although the period from November to January may be when most mating activity occurs. Information concerning the frequency of individual mating is rare, but laboratory and field observations for two female E. eschscholtzii indicate that an individual female may be inseminated more than once during the courtship season (Stebbins 1954). Females of all western terrestrial species are thought to store sperm for many months so courtship need not directly precede oviposition.

Females in the western United States oviposit in late spring or early summer and thus are similar to eastern plethodontids. The timing of oviposition may vary from year to year, but in any given year females are relatively synchronous. Embryonic development takes from two months (P. vehiculum, Peacock and Nussbaum 1973) to four months (E. eschscholtzii, Stebbins 1954), so some females may remain with their clutches for a longer period in the west than in the east. Although brooding time can be longer, western females brood eggs during the dry summer months when conspecifics also are constrained in surface feeding activity. Thus females of western species may spend more time brooding eggs than females of eastern species, but this increased brooding time may not reflect a disadvantage in terms of relative access to food.

Two species, Batrachoseps attenuatus and B. major, have patterns of activity similar to those of other western salamanders, but their courtship and ovipositional patterns are markedly different (Maiorana 1976). Females of these species oviposit in the fall when the first heavy rains occur (Maslin 1939, Davis 1952), rather than early in the summer. Parental care (brooding) has been found for all other plethodontids with direct development which have been studied, but females of B. attenuatus do not rigorously brood their eggs and tend to abandon them in communal nests (Maiorana 1974; Arnold, pers. comm.). Secondary sexual characteristics (specialized premaxillary teeth, enlarged neck musculature) are developed only in males collected during the summer or very early in the fall (Maiorana, pers. comm.), and male B. attenuatus produce mature sperm only during this period (Yanev, pers. comm.). Courtship is inferred to take place when individuals are underground during the summer or early fall. Unlike the example of P. cinereus populations in the eastern United States, these two Batrachoseps species occur in the same climatic area (Maiorana 1976) as the western plethodontid species upon which general activity patterns are based. Within a given climatic regime, then, sympatric species may exhibit very different patterns of activity. Nevertheless, all western patterns strongly reflect the seasonality of temperature and rainfall.

Neotropics (Guatemala)

The neotropical plethodontids considered here include six terrestrial and five arboreal species collected between 1000 m and 3800 m from an area in western Guatemala (Fig. 2; Houck 1977a, 1977b; Wake and Lynch 1976). A distinction is made between arboreal and terrestrial species because of seasonal differences in moisture availability in each microhabitat (see below). Three of the arboreal species (Bolitoglossa englehardti, B. franklini, Chiropterotriton bromelicia) were collected primarily from bromeliads, but a fourth species (B. occidentalis) was taken from the moist vertical surfaces within the leaf axils of banana plants. The fifth species (B. flavimembris) included in the arboreal group was not collected in truly arboreal microhabitats but was found off of the ground surface under the bark of logs and stumps (Wake and Lynch 1976). Of the six terrestrial species, five (B. morio, B. rostrata, Pseudoeurycea brunnata, P. goebeli, P. rex) usually are found directly on the ground surface under rocks or logs or in the leaf litter. The sixth species (B. resplendens) actually occurs both in arboreal and terrestrial microhabitats (Wake and Lynch 1976), but there is some indication that eggs are laid in a terrestrial site and that terrestrial microclimatic regimes are most important (Houck 1977b).

The Guatemalan area can be characterized as having a dry season from November to May followed by a period of almost daily rains from June to October (McBryde 1942). Above approximately 1600 m, a low cloud cover is brought in by convectional air currents. This cloud cover brings almost daily rains in the wet season; during the dry season, moisture condenses on taller vegetation but precipitation rarely reaches the ground. Water-trapping arboreal plants, such as bromeliads, remain wet during the dry season as well as during the rainy season, while the ground surface becomes very dry once rains cease. Temperature varies more with elevation than with season (McBryde 1942); salamanders at 1050 m may experience temperatures over 12° C warmer than salamanders at altitudes near 3500 m (Feder 1976).

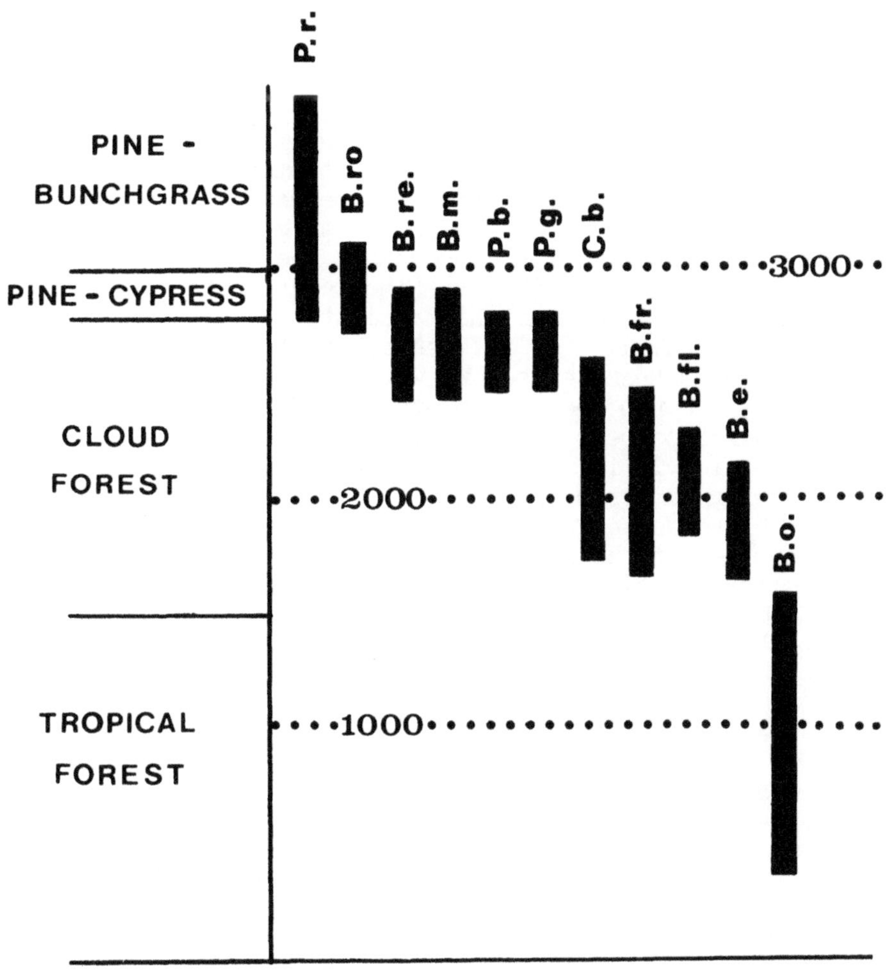

Fig. 2. Altitudinal and habitat distribution of 11 plethodontid species in Guatemala (Houck 1977a, 1977b). Initials refer to species names listed in Appendix I. Elevations are given in meters. Figure adapted from Wake and Lynch 1976.

Temporal variation in activity patterns has not been determined for the arboreal species. These species were collected from the same microhabitats (e.g. bromeliads) throughout the year, and the lack of alternate microhabitats precludes a direct evaluation of seasonal patterns of activity. Differential microhabitat utilization observed for the terrestrial species reveals that surface activity is strongly correlated with seasonal rainfall. Individuals generally are active from May until November (Fig. 1C) and can be collected under surface cover objects during these months. When the ground surface becomes dry, salamanders retreat to subterranean refugia or the moist central areas of rotten logs. Terrestrial salamanders are at the surface for approximately six months of the year and their activity apparently is limited primarily by rainfall rather than temperature extremes.

Spermatogenesis occurs continually for all Guatemalan species studied, terrestrial or arboreal: mature spermatozoa are produced by all adult males during all months of the year. Secondary sexual characteristics (presence of a submandibular gland and enlarged cloacal papillae) are developed throughout the year in males which are producing sperm (Houck 1977b). [The relationship of secondary sexual characteristics to courtship activity is discussed by Arnold (1976, and this volume).] By inference, courtship activity is aseasonal and sperm transfer occurs without respect to wet or dry season. Like the North American plethodontids discussed above, Guatemalan females possess sperm storage organs and presumably can store and maintain viable spermatozoa for many months before oviposition. The combination of aseasonal spermatogenesis and courtship with highly seasonal surface activity and oviposition (see below) has not been found for extratropical plethodontids. Two other neotropical species (B. subpalmata in Costa Rica and B. adspersa in Colombia) also have aseasonal spermatogenesis and courtship, but both have aseasonal patterns of oviposition and surface activity as well (Vial 1968; Valdivieso and Tamsitt 1965).

The pattern of oviposition and brooding presented in figure 1C applies to at least six (and probably nine) of the 11 Guatemalan species studied. The six species for which data are most complete include terrestrial species (B. resplendens, B. rostrata, P. brunnata, P. goebeli) as well as arboreal species (B. englehardti, B. franklini). Young of these species hatch in May or June. Females of B. rostrata oviposit in November or December and brood eggs through the dry season until embryonic development is complete (Houck 1977a). The ovipositional pattern of B. rostrata probably describes that of all species which have hatchlings emerging in May or June. This 5-6 month period of brooding is longer than that known for any North American plethodontid. It is not known why embryonic developmental time is relatively longer for these species, especially since egg size (discussed below) is distinctly smaller than for extratropical plethodontids of similar size. A seventh species, Pseudoeurycea rex, also shows a seasonal pattern of oviposition. This pattern is temporally shifted, however, such that young first emerge in November or December, a difference of six months relative to the other species. This difference may be related to the fact that P. rex occurs at high elevations (2850-3800 m) where microclimatic conditions may be sufficiently different from those experienced by other species. There are at least seven species then, both terrestrial and arboreal, having highly seasonal patterns of oviposition. Data concerning three additional species (B. flavimembris, B. occidentalis, B. morio) are not complete although the first two species probably oviposit seasonally as illustrated in figure 1C. In summarizing the scanty ovipositional information for other neotropical salamander species, McDiarmid and Worthington (1970) concluded that seasonal oviposition may be timed such that young emerge near the beginning of the rainy period. Data available for the Guatemalan species support this conclusion. Juveniles which hatch when conditions are wet presumably will have a greater chance of survival.

Data available for only one of the 11 Guatemalan species (Chiropterotriton bromeliacia) suggest that oviposition occurs aseasonally. Although no egg clutches were found, size-frequency histograms show that juveniles near hatchling size were collected at different times of the year; adult females with yolky ovarian eggs were also collected during many months of the year (Houck 1977b). If aseasonal oviposition is the case for C. bromeliacia, then species with broadly overlapping microhabitat utilization, and which frequently are microsympatric, respond differently under the same moisture and temperature conditions.

Summary of General Activity Patterns

General patterns of activity have been presented and compared for plethodontid species groups from three climatically different areas: eastern United States, western United States, and Guatemala. Many species conform to a characteristic pattern for each area (Figure 1), but marked differences have been observed for two species in the western United States and one species in Guatemala. Widespread species from any area may include populations which experience different climatic conditions and therefore have different patterns of activity.

GROWTH RATES, SIZE AND AGE AT SEXUAL MATURITY

Growth Rates

Estimates of average growth rates for neotropical and temperate zone terrestrial plethodontids are based primarily on differences in the average size of salamanders in different age groups. Size classes revealed from size-frequency histograms are taken to represent age groups, but size classes are distinguishable only for the first few years. Age groups of older animals blend together as a result of individual variation in yearly growth and the reduced growth rates associated with reproduction. The most accurate estimates, therefore, are of early growth rates; adult growth rates usually must be inferred. Size-frequency histograms for species which breed seasonally (e.g. Bolitoglossa subpalmata, Vial 1968) do not show distinct size classes and growth estimates reflect mark-recapture data.

First year growth rates reported for 11 plethodontid species range from 5 to 21 mm per year (Table 1). Growth rates for successive years commonly are 10-12 mm per year until sexual maturity is reached (Highton 1962a, Peacock and Nussbaum 1973, Lynch 1974, Houck 1977a). Since adult size ranges are not large relative to the annual growth of immature individuals, it is assumed that adult growth rates are relatively low. Mark-recapture data for Ensatina eschscholtzii (Stebbins 1954) indicate that adults may grow 0-3 mm per year. Thus growth rates appear to be highest during the first year, moderately high until sexual maturity is reached, and relatively low for adults. Growth rates are higher during months when surface activity occurs, and annual growth rates are greater for species which spend a greater proportion of the year active at the surface (Plethodon glutinosus in Florida, Highton 1956). Some growth can occur even during periods when animals retreat from the ground surface.

The growth rate for one species, Bolitoglossa subpalmata, is exceptionally low in view of rates for other plethodontids: the annual growth rate for juveniles and adults combined is only 3 mm per year (range = 0.0 - 6.0 mm per year; Vial 1968). It is tempting to dismiss this anomalous rate as inaccurate but Vial's data are substantial: his growth estimate is based on 75 marked individuals recaptured after six months or more. Also, B. subpalmata occurs in relatively high densities at high elevations and cool temperatures (Vial 1968), and these conditions could be reflected in a low growth rate. On the other hand, individuals are active at the surface, presumably feeding, throughout the year so one would expect the overall growth rate to be higher than 3 mm per year. Moreover, this growth rate estimate translates size at sexual maturity into age estimates for males and females which are exceptionally high compared to those for other plethodontids (see below). The suggestion that reproduction is greatly delayed in B. subpalmata relative to other plethodontids implies profound differences in age-specific mortality schedules (Charnov and Schaffer 1973; Giesel 1976; Stearns 1976). Although an attempt to substantiate or disprove the 3 mm per year growth rate estimate will necessarily involve an extensive mark-recapture study, the implications of strong differences in selective pressures experienced by B. subpalmata demand a reevaluation of this growth estimate.

Size at Sexual Maturity

Plethodontid females frequently are larger than males both in average adult size (Houck 1977b) and in size at sexual maturity (Table 1). Size dimorphism is most extreme for certain large neotropical species, genus Bolitoglossa, and is least noticeable or absent in relatively small species. Since there is evidence to suggest that growth rates are similar for juvenile females and males (Highton 1962a; Houck, unpublished), then females achieve a

Table 1. Initial growth rates, and estimates of size and age at sexual maturity for neotropical and temperate zone plethodontid salamanders having direct development. See text and Appendix I for literature references for each species. To standardize data, snout-vent length (SL) values from literature sources were increased by 5% if measurements were made to the anterior (rather than posterior) margin of the vent, and were decreased by 5% if measurements were made before animals were preserved. Where ranges were given, the midpoint was taken as age at sexual maturity.

	Growth Rate (mm/1st yr):	Size at Sexual Maturity (mm SL):		Age at Sexual Maturity (yrs):	
		Males	Females	Males	Females
NEOTROPICAL SPECIES					
Bolitoglossa rostrata	15	45	58	2-1/2	3-3/4
B. subpalmata	5	31	45	6	12
Pseudoeurycea brunnata	18	43	49	2	2-1/2
TEMPERATE SPECIES					
Aneides flavipunctatus	12	50	65	3-1/4	4-1/4
Batrachoseps attenuatus	12	26	40	2-1/2	3
Ensatina eschscholtzii	12	51	61	3	3-1/2
Plethodon cinereus	16	38	38	1-3/4	1-3/4
P. glutinosus (Maryland)	16	62	65	4-1/2	5
P. glutinosus (Florida)	21	48	51	2-1/2	3
P. hoffmani	15	41	43	2-1/4	3-1/4
P. vehiculum	10	44	46	2-3/4	3-1/4

relatively larger adult size by reaching maturity a year or more later than males. Females continue growing at relatively high juvenile rates while males of the same age become sexually mature and grow more slowly.

For both males and females, size at maturity presumably represents a compromise between the advantages of delaying reproduction and attaining a larger size and the advantages of reproducing earlier but at a smaller size. Possible selective factors influencing the timing of reproduction have been reviewed elsewhere (Stearns 1976), and the following discussion is restricted to some considerations which may be most critical for salamanders. The relationship between body size and clutch size, for example, may help explain why some plethodontid females grow to a larger size than males before reproducing. Clutch size is positively correlated with body size so that larger females commonly have larger clutch sizes (Salthe 1969). Since growth rates apparently are low once sexual maturity is reached, a female which first reproduces at a larger size can retain the advantage of larger clutch sizes throughout her lifetime. Independent of clutch size, large size may also be favored to the extent that predation or other causes of mortality are inversely related to body size. The advantage of delaying reproduction in favor of growth at some point is outweighed by the advantage of producing offspring without further delay. At the extreme, females which wait too long before attaining maturity may not survive to reproduce at all.

Large size apparently is less critical for males than for females. During courtship, for instance, relatively small males can successfully inseminate females which are much larger (Arnold, pers. comm.). In fact, the body size of some adult Bolitoglossa resplendens females exceeds that of the largest males by over 20 mm (Houck 1977b). Although small males may be capable of inseminating females, relative size differences among males could confer advantages to those which are larger. Large males might be able to produce more spermatophores per unit time, or more sperm per spermatophore. In addition, there is evidence that males in breeding condition interact aggressively (Arnold 1976, and this volume; Thurow 1976) and larger males may be more successful at excluding rivals from courtship activities. Although large size may increase the mating success of males in some respects, the advantages of delaying reproduction and growing larger evidently are outweighed by the probability of leaving more offspring if sexual maturity is not deferred.

Age at Sexual Maturity

Direct evidence concerning age at sexual maturity is largely unavailable for terrestrial plethodontids. Most studies present estimates which are based on average growth rates and average size at sexual maturity. Because of the difficulties in tracking marked animals (Tilley, this volume), individual variation in age at sexual maturity has not been documented.

Available estimates of age at sexual maturity are reported in table 1. In most cases, males are producing mature sperm at least six months before oögenesis begins in females.

Despite observed differences in first year growth rates (Table 1), age at sexual maturity is similar for species of similar size. If the relatively high age estimates for Bolitoglossa subpalmata males and females are excluded (see Growth Rates, above), there appears to be a direct linear relationship between average size and age at sexual maturity ($y = -1.97 + 0.1x$; $r = 0.93$; $p \leq 0.01$) for both males and females. The high correlation between size and age implies that overall differences among species in juvenile growth rates are not large, despite differences in seasonal activity patterns.

CLUTCH SIZE, EGG SIZE, AND CLUTCH VOLUME

Average clutch sizes of neotropical plethodontids commonly are larger than those of temperate zone species of similar body size (Fig. 3). In particular, average clutch sizes for four Bolitoglossa species in Guatemala range from 34 to 46 eggs (Houck 1977b) and are the largest averages recorded for terrestrial salamanders. These average clutch sizes represent a two- to three-fold increase over clutch sizes of similarly sized temperate species. The differences in clutch size between neotropical and temperate salamanders are unexpected in view of the opposite latitudinal trend found for mammals (Spencer and Steinhoff 1968), birds (Cody 1966), lizards (Tinkle et al. 1970), and frogs (Inger and Bacon 1968).

Although clutch sizes usually are larger, the neotropical species which have been examined have ova which are small relative to temperate species of comparable size. Maximum diameters of ova typical of the largest neotropical species studied range from 3.5 - 3.75 mm while those known for comparable temperate species range from 4.0 - 5.5 mm (Table 2).

Estimates of clutch volume were calculated from maximum diameters of ova and average clutch sizes for seven neotropical and eight North American plethodontid species. There is a significant positive correlation ($r = 0.77$; $p \leq 0.01$) between standard length (snout to posterior margin of vent; SL) and the log of clutch volume for all species considered (Fig. 4). While it is not surprising that larger salamanders have larger clutch volumes, figure 4 reveals that, holding body size constant, neotropical species uniformly

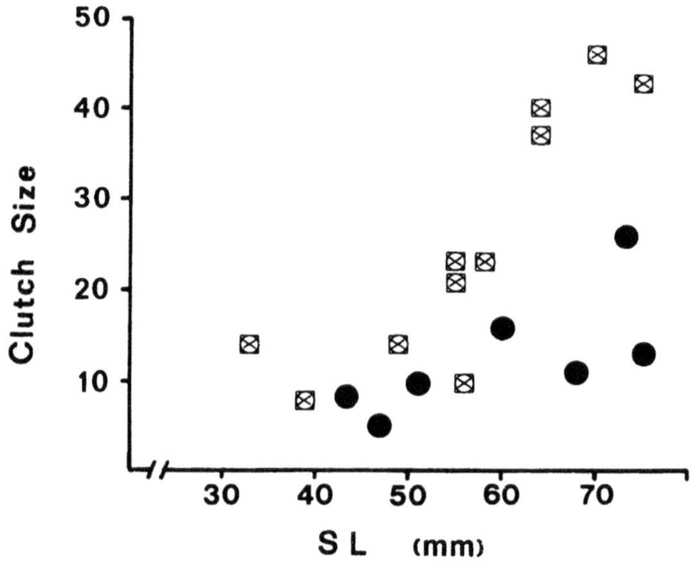

Fig. 3. Average adult female size (SL = snout-vent length) compared with average ovarian clutch size for plethodontid salamander species occurring in the Neotropics (boxes with cross) and in the temperate zone (solid circles).

have a larger clutch volume than extratropical plethodontids. An analysis of covariance shows that linear regressions for temperate and neotropical species have similar slopes ($F = 1.33$; $df = 1, 11$) but that the y-intercept for the neotropical regression line is significantly higher ($F = 16.33$; $df = 1, 12$). To the extent that clutch volume reflects reproductive effort, neotropical species not only have a greater reproductive effort than comparably sized temperate species, but they are "packaging" this output in a very different way.

The observed trends of decreased egg size, increased clutch size, and increased clutch volume found for neotropical species are largely unexplained. Assuming that latitudinal differences in clutch parameters reflect differential selective pressures experienced by temperate and neotropical salamanders, a variety of hypotheses can be invoked which seek to explain these differences. Latitudinal differences in egg size, for example, are consistent with the hypothesis that juvenile mortality for neotropical species is not significantly increased by having smaller eggs (hence smaller hatchlings). If this is true, it follows that females of neotropical species which have decreased egg size will have increased the overall number of surviving offspring since more young are produced.

Table 2. Female size, clutch size, and related data for some neotropical and temperate zone species of plethodontid salamanders having completely terrestrial development. See text and Appendix I for literature references for each species. An asterisk indicates values which were estimated from other data given for the species.

	Average Size Adult Females (mm SL):	Mean Clutch Size	N	Range	Max. Dia. Eggs (mm)
NEOTROPICAL SPECIES					
Bolitoglossa rostrata (San Marcos)	64	37	30	22-54	3.5
B. rostrata (Cuchumatanes)	49	24	7	20-28	3.5
B. subpalmata	55*	23	31	13-38	5.0
Chiropterotriton bromeliacia	33	13	92	6-20	3.0
Pseudoeurycea brunnata	55	21	52	14-29	3.75
P. goebeli	55	23	64	10-30	3.75
P. rex	49	13	31	9-20	3.5
TEMPERATE ZONE SPECIES					
Aneides flavipunctatus	70*	11	--	---	5.5
Batrachoseps attenuatus	43	7	--	2-12	3.0
Ensatina eschscholtzii	68	11	26	8-17	5.0
Plethodon cinereus	43	7	42	3-11	3.0
P. glutinosus (Maryland)	70*	26	10	16-34	4.0
P. glutinosus (Florida)	58*	16	7	10-22	4.0
P. hoffmani	47	5	19	3- 8	3.5
P. vehiculum	52	10	65	6-19	4.0

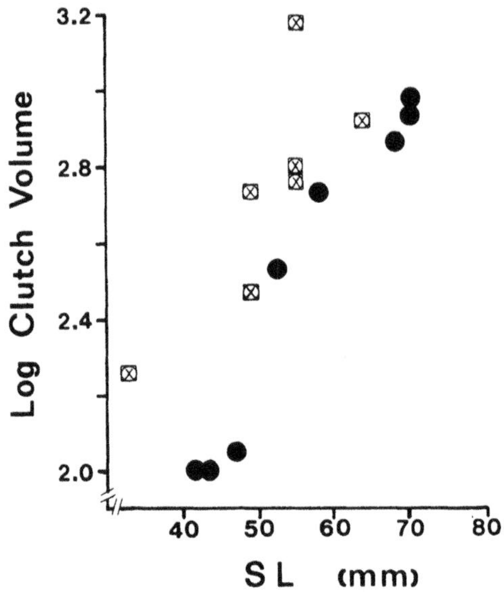

Fig. 4. Average adult female size (SL = snout-vent length) compared with the log of clutch volume for plethodontid salamander species occurring in the Neotropics (boxes with cross) and in the temperate zone (solid circles).

Other hypotheses outline selective pressures which could lead to the evolution of increased clutch volume. It is possible, for example, that clutch volume differences are related to absolute or proportional differences in energy available for reproduction. Alternatively, increased clutch volume may be a function of relatively high adult mortality rates if neotropical females which produce as many offspring as possible each reproductive period leave more offspring than females which produce fewer offspring each time but which might survive to reproduce during another period. Stearns (1976) reviewed these and other hypotheses in detail, and concluded that for any given situation more than one plausible hypothesis exists and, in fact, a combination of hypotheses may explain observed trends. Until data are gathered to test these hypotheses, the precise nature of differential selective pressures leading to latitudinal differences in clutch parameters remains unknown.

INTRASPECIFIC VARIABILITY IN CLUTCH SIZE

Clutch size variability within species is extremely high for all terrestrial plethodontids which have been studied. Differences in female body size (snout-vent length) rarely account for more than half of the variance in clutch size (Houck 1977b). Bruce

(1969) reported a similar phenomenon for Gyrinophilus porphyriticus, Pseudotriton ruber and P. montanus, plethodontids with aquatic larvae: only 36-59% of the variance in clutch size is related to differences in body size. Variation in clutch size is great despite the positive and usually significant correlation between body size and clutch size.

Clutch size variation observed for Pseudoeurycea goebeli (Fig. 5) is typical of other species for which data are available. The product-moment correlation coefficient between snout-vent length and clutch size is positive and highly significant ($r = 0.64$; $p \leq 0.01$) for P. goebeli, so that larger females tend to have larger clutch sizes. Nevertheless, much of the clutch size variation is unexplained by differences in body length ($S_{y/x}^2 = 21.7$); females with the same snout-vent length may have clutches which differ by more than 10 eggs (Fig. 5).

One possible source of variation in clutch size may be differences in the amount of food acquired by individual females and thus in the number of ova which can be yolked. Food abundance may vary spatially as well as seasonally, and some salamanders may encounter areas of high food abundance while other individuals do not. Preliminary evidence indicates that clutch size may be correlated with food intake. Maiorana (1974) has demonstrated for a California plethodontid, Batrachoseps attenuatus, that tail length, an indirect measure of stored nutrients, is significantly correlated with clutch size when body size is held constant. Observations by Maiorana

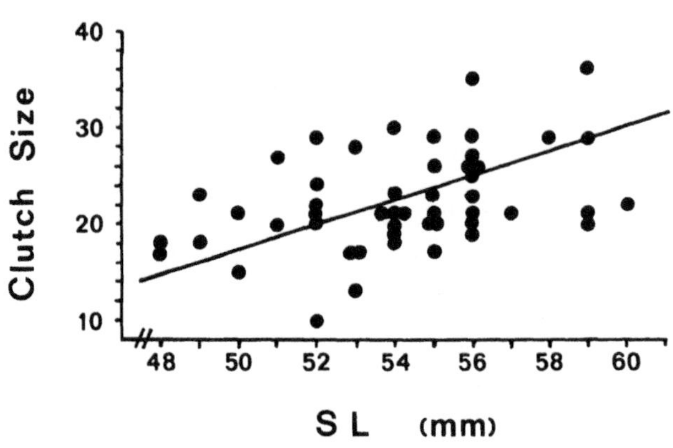

Fig. 5. Adult female size (SL = snout-vent length) compared with ovarian clutch size for Pseudoeurycea goebeli, a neotropical plethodontid. Linear regression equation: $y = -45.63 + 1.24x$.

(1974) also indicate that food abundance may be patchy: B. attenuatus in wetter microhabitats have more food in their stomachs than individuals in drier microhabitats. Additional evidence suggests that females do not forage widely within a given habitat and thus may not encounter areas of high food abundance to the extent that these areas naturally occur (a "coarse-grained" environment, sensu Levins 1968). Studies of marked individuals show that plethodontid females usually can be recaptured within a relatively small area (e.g. 30 ft^2, Plethodon jordani; 70 ft^2 P. glutinosus; Merchant 1972), and movements involving long horizontal distances apparently are rare (Hendrickson 1954, Taub 1961, Madison 1969, Merchant 1972). Assuming that food availability is patchy and that females are relatively sedentary, then intraspecific variation in food intake could be sufficiently high to account for much of the observed variation in clutch size.

Other sources of variation in clutch size may also be important. Females of the same body size might have very different clutch sizes as a reflection of genetic differences. Alternatively, the proportion of energy devoted to reproduction may vary with age such that clutch size differences occur among differently-aged females which have the same body size. The significance of these and other factors is more difficult to determine than that of variable food intake. Laboratory experiments are currently being conducted to determine if variation in clutch size is reduced among females which are fed ad lib as compared with females in a field situation. The importance of alternate explanations of clutch size variability can be more accurately estimated after the effects of food differences have been evaluated.

SUMMARY AND CONCLUSIONS

A comparison of the life histories of neotropical and temperate zone terrestrial plethodontid salamanders reveals patterns of reproduction which strongly reflect seasonal variation in climatic conditions. Seasonal patterns of activity and oviposition characterize most neotropical and all temperate zone plethodontids studied, including several neotropical species found in arboreal microhabitats essentially buffered from moisture and temperature extremes throughout the year.

Most species which occur within a given climatic regime have similar life history patterns, but a few species have reproductive and activity cycles which differ greatly from those of other sympatric plethodontids. Widespread species may include populations which differ in timing of surface activity and other life history parameters if temperature and rainfall conditions vary for these populations.

Although the timing of surface activity and oviposition largely reflects climatic variability, other activities may vary without respect to seasonal differences in climatic regimes. Spermatogenesis and courtship occur throughout the year in neotropical populations, for example, although these activities occur seasonally in temperate zone populations.

Growth rates and age at sexual maturity are broadly similar for species of comparable size, but clutch parameters show striking latitudinal variation: females of neotropical species generally have larger clutches, smaller eggs, and larger clutch volumes than females of comparable extratropical salamanders. Variation observed for life history parameters, whether genetic or phenetic, can be viewed as a reflection of adaptive flexibility in individual life histories, and presumably increases the level of adaptation to local and variable environmental conditions. Latitudinal differences, especially those concerning clutch parameters, suggest significant differences in selective pressures bearing on temperate and neotropical salamander populations.

The overall patterns of reproductive biology and life history considered in this paper essentially are populational patterns. These patterns are important in characterizing and comparing different species, but they are based primarily upon average tendencies (e.g. average growth rate) for a group of individuals. Information pertaining to the range of variation associated with particular life history parameters is largely unavailable. Future investigations concerning life histories can increase our understanding of the relationship between life history parameters and selective pressures by focusing on the underlying individual patterns and variation within populations. Studies are needed which investigate specific life history parameters (e.g. age at sexual maturity) and which attempt to determine the extent and nature of individual variation.

In this paper the life history parameters of neotropical and temperate plethodontids have been compared. At this point, it is worth stepping back from this comparison to emphasize that the species considered here share a unifying life history feature, completely terrestrial development. In one sense, terrestrial development is the foundation of most plethodontid life histories and is embellished by variation in other life history parameters. The evolutionary significance of completely terrestrial development has been repeatedly stressed by Wake (1966, 1970); reproduction which is not dependent upon free-standing water probably is <u>the</u> distinguishing feature which permitted the widespread dispersal and subsequent high diversity of the Plethodontidae.

ACKNOWLEDGEMENTS

I greatly appreciate the constructive criticism of manuscript drafts offered by S. J. Arnold, S. Emerson, R. Highton, R. Wassersug, D. B. Wake, and an anonymous reviewer. R. Nussbaum generously shared results from his current study concerning western *Plethodon* species. I especially thank D. B. Wake for his enthusiastic and untiring support during the course of my research. Portions of this work were completed in partial fulfillment of the requirements for the Ph. D. Degree, University of California, Berkeley. Research was supported by NSF Grants GB 17112 and BMS74-20922 to D. B. Wake, and by grants from the Society of Sigma Xi and the Center for Latin American Studies.

Appendix I. Generalized patterns of activity discussed in the text are based primarily on life history information concerning the terrestrial plethodontid species listed below. References are restricted to those providing the most detailed information for each species.

EASTERN UNITED STATES

Aneides aeneus	Gordon 1952
Plethodon cinereus	Blanchard 1928, Lynn and Dent 1941, Sayler 1966, Highton 1972
P. glutinosus (Maryland)	Highton 1962a, 1972
P. hoffmani	Angle 1969, Highton 1962b
P. jordani	Howell and Hawkins 1954, Schwartz 1957
P. ouachitae	Pope and Pope 1951
P. richmondi	Duellman 1954
P. wehrlei	Hall and Stafford 1972
P. welleri	Organ 1960, Thurow 1963
P. yonahlossee	Pope 1950

WESTERN UNITED STATES

Aneides flavipunctatus	Lynch 1974
A. lugubris	Ritter and Miller 1899, Ritter 1903, Miller 1944, Rosenthal 1954
Batrachoseps attenuatus	Hendrickson 1954, Maiorana 1974, 1976
Ensatina eschscholtzii	Stebbins 1949, 1954
Plethodon dunni	R. Nussbaum, pers. comm.
P. elongatus	"
P. larselli	"
P. stormi	"

WESTERN UNITED STATES (cont.)

P. vandykei	R. Nussbaum, pers. comm.
P. vehiculum	Peacock and Nussbaum 1973

GUATEMALA

Bolitoglossa englehardti	Houck 1977b
B. flavimembris	"
B. franklini	"
B. morio	"
B. occidentalis	"
B. resplendens	"
B. rostrata	Houck 1977a
Chiropterotriton bromeliacia	Houck 1977b
Pseudoeurycea brunnata	"
P. goebeli	"
P. rex	"

REFERENCES

Angle, J. P. 1969. The reproductive cycle of the northern ravine salamander, Plethodon richmondi richmondi, in the Valley and Ridge Province of Pennsylvania and Maryland. J. Wash. Acad. Sci. 59:192-202.

Arnold, S. J. 1976. Sexual behavior, sexual interference and sexual defense in the salamanders Ambystoma maculatum, Ambystoma tigrinum and Plethodon jordani. Z. Tierpsychol. 42:247-300.

Blanchard, F. N. 1928. Topics from the life history and habits of the red-backed salamander in southern Michigan. Am. Nat. 62:156-164.

Bruce, R. C. 1969. Fecundity in primitive plethodontid salamanders. Evolution 23:50-54.

Burger, J. W. 1935. Plethodon cinereus (Green) in eastern Pennsylvania and New Jersey. Am. Nat. 69:578-586.

Burton, T. M. and G. E. Likens. 1975. Salamander populations and biomass in the Hubbard Brook Experimental Forest, New Hampshire. Copeia 1975:541-546.

Caldwell, R. S. 1975. Observations on the winter activity of the red-backed salamander, Plethodon cinereus, in Indiana. Herpetologica 31:21-22.

Charnov, E. L. and W. M. Schaffer. 1973. Life history consequences of natural selection: Cole's result revisited. Am. Nat. 107:791-793.

Cody, M. L. 1966. A general theory of clutch size. Evolution 20: 174-184.

Davis, J. 1952. Observations of the eggs and larvae of the salamander, Batrachoseps pacificus major. Copeia 1952:272-274.

Duellman, W. E. 1954. The salamander Plethodon richmondi in southwestern Ohio. Copeia 1954:40-45.

Dunn, E. R. 1926. The salamanders of the family Plethodontidae. Smith College Fiftieth Anniv. Publ., Northampton, Massachusetts.

Feder, M. 1976. Oxygen consumption and body temperature in neotropical and temperate zone lungless salamanders (Amphibia: Plethodontidae). J. Comp. Physiol. 110:197-208.

Forester, D. C. 1974. Parental care in Desmognathus ochrophaeus Cope (Urodela: Plethodontidae): A behavioral study. Unpubl. Ph. D. Diss., N. Carolina State Univ., Raleigh.

Giesel, J. T. 1976. Reproductive strategies as adaptations to life in temporally heterogeneous environments. Ann. Rev. Ecol. Syst. 7:57-79.

Gordon, R. E. 1952. A contribution to the life history and ecology of the plethodontid salamander Aneides aeneus (Cope and Packard). Am. Midl. Nat. 47:666-701.

Grobman, A. B. 1944. The distribution of the salamanders of the genus Plethodon in eastern United States and Canada. Ann. N. Y. Acad. Sci. 45:261-316.

Hall, R. J. and D. P. Stafford. 1972. Studies in the life history of Wehrle's salamander, Plethodon wehrlei. Herpetologica 28:300-309.

Hendrickson, J. R. 1954. Ecology and systematics of salamanders of the genus Batrachoseps. Univ. Calif. Publ. Zool. 54:1-46.

Highton, R. 1956. The life history of the slimy salamander, Plethodon glutinosus, in Florida. Copeia 1956:75-93.

Highton, R. 1962a. Geographic variation in the life history of the slimy salamander. Copeia 1962:597-613.

Highton, R. 1962b. Revision of North American salamanders of the genus Plethodon. Bull. Fla. State Mus. 6:235-267.

Highton, R. 1970. Evolutionary interactions between species of North American salamanders of the genus Plethodon. Part 1. Genetic and ecological relationships of Plethodon jordani and P. glutinosus in the southern Appalachian Mountains. In: Evol. Biol., Vol 4, Appleton-Century Crofts, New York, New York, pp. 211-241.

Highton, R. 1972. Distributional interactions among eastern North American salamanders of the genus Plethodon. In: The distributional history of the biota of the southern Appalachians. Part III: Vertebrates. Va. Poly. Inst., Res. Div. Monogr. 4: 139-188.

Houck, L. D. 1977a. Reproductive biology of a neotropical salamander, Bolitoglossa rostrata (Caudata: Plethodontidae). Copeia 1977:70-83.

Houck, L. D. 1977b. Reproductive patterns in neotropical salamanders. Unpubl. Ph. D. Diss., Univ. Calif., Berkeley.

Howell, T. and A. Hawkins. 1954. Variation in topotypes of the salamander Plethodon jordani melaventris. Copeia 1954:32-36.

Inger, R. F. and J. P. Bacon, Jr. 1968. Annual reproduction and clutch size in rain forest frogs from Sarawak. Copeia 1968: 602-606.

Levins, R. 1968. Evolution in changing environments. Princeton Univ. Press, Princeton, New Jersey.

Lynch, J. F. 1974. Ontogenetic and geographic variation in the morphology and ecology of the black salamander (Aneides flavipunctatus). Unpubl. Ph. D. Diss., Univ. Calif., Berkeley.

Lynn, W. G. and J. N. Dent. 1941. Notes on Plethodon cinereus and Hemidactylium scutatum on Cape Cod. Copeia 1941:113-114.

Madison, D. M. 1969. Homing behavior of the red-cheeked salamander, Plethodon jordani. Anim. Behav. 17:25-39.

Maiorana, V. C. 1974. Studies in the behavioral ecology of the plethodontid salamander Batrachoseps attenuatus. Unpubl. Ph. D. Diss., Univ. Calif., Berkeley.

Maiorana, V. C. 1976. Size and environmental predictability for salamanders. Evolution 30:599-613.

Marynick, S. P. 1971. Long term storage of sperm in Desmognathus fuscus from Louisiana. Copeia 1971:345-347.

Maslin, T. P., Jr. 1939. Egg-laying of the slender salamander (Batrachoseps attenuatus). Copeia 1939:209-212.

McBryde, F. W. 1942. Studies in Guatemalan meteorology. 1. The climates of southwest Guatemala. Bull. Amer. Meteor. Soc. 23: 254-263.

McDiarmid, R. W. and R. D. Worthington. 1970. Concerning the reproductive habits of tropical plethodontid salamanders. Herpetologica 26:57-70.

Merchant, H. 1972. Estimated population size and home range of the salamanders Plethodon jordani and Plethodon glutinosus. J. Wash. Acad. Sci. 62:248-257.

Miller, L. H. 1944. Notes on the eggs and larvae of Aneides lugubris. Copeia 1944:224-230.

Organ, J. A. 1960. Studies on the life history of the salamander, Plethodon welleri. Copeia 1960:287-297.

Organ, J. A. 1961. Studies of the local distribution, life history, and population dynamics of the salamander genus Desmognathus in Virginia. Ecol. Monogr. 31:189-220.

Peacock, R. L. and R. A. Nussbaum. 1973. Reproductive biology and population structure of the western red-backed salamander, Plethodon vehiculum (Cooper). J. Herp. 7:215-224.

Pope, C. H. 1950. A statistical and ecological study of the salamander Plethodon yonahlossee. Bull. Chicago Acad. Sci. 9:79-106.

Pope, C. H. and S. H. Pope. 1951. A study of the salamander Plethodon ouachitae and the description of an allied form. Bull. Chicago Acad. Sci. 9:129-152.

Ritter, W. E. 1903. Further notes on the habits of Autodax lugubris. Am. Nat. 37:883-886.

Ritter, W. E. and L. Miller. 1899. A contribution to the life-history of Autodax lugubris Hallow, a California salamander. Am. Nat. 33:691-704.

Rosenthal, G. M. 1954. The role of moisture and temperature in the local distribution of the plethodontid salamander Aneides lugubris. Univ. Calif. Publ. Zool. 54:371-420.

Rumney, G. R. 1968. Climatology and the world's climates. MacMillan Co., New York, New York.

Salthe, S. N. 1969. Reproductive modes and the numbers and sizes of ova in the Urodeles. Am. Midl. Nat. 81:467-490.

Salthe, S. N. and J. S. Mecham. 1974. Reproductive and courtship patterns. In: Physiology of Amphibia, Vol. 2, Academic Press, Inc., New York, New York, pp. 309-521.

Sayler, A. 1966. The reproductive ecology of the red-backed salamander, Plethodon cinereus, in Maryland. Copeia 1966:183-193.

Schwartz, A. 1957. Variation and natural history of Plethodon jordani clemsonae Brimley. Copeia 1957:94-107.

Spencer, A. W. and H. W. Steinhoff. 1968. An explanation of geographic variation in litter size. J. Mam. 49:281-286.

Stearns, S. C. 1976. Life-history tactics: A review of the ideas. Quart. Rev. Biol. 51:3-47.

Stebbins, R. C. 1949. Courtship of the plethodontid salamander Ensatina eschscholtzii. Copeia 1949:274-281.

Stebbins, R. C. 1954. Natural history of the salamanders of the plethodontid genus Ensatina. Univ. Calif. Publ. Zool. 54:47-123.

Taub, F. B. 1961. The distribution of the red-backed salamander, Plethodon c. cinereus, within the soil. Ecology 42:681-698.

Thurow, G. R. 1963. Taxonomic and ecological notes on the salamander, Plethodon welleri. Univ. Kansas Sci. Bull. 44:87-108.

Thurow, G. R. 1976. Aggression and competition in eastern Plethodon (Amphibia, Urodela, Plethodontidae). J. Herp. 10:277-291.

Tilley, S. G. 1968. Size-fecundity relationships and their evolutionary implications in five desmognathine salamanders. Evolution 22:806-816.

Tinkle, D. W., H. M. Wilbur and S. G. Tilley. 1970. Evolutionary strategies in lizard reproduction. Evolution 24:55-74.

Valdivieso, D., and J. R. Tamsitt. 1965. Reproduction in a neotropical salamander, Bolitoglossa adspersa (Peters). Herpetologica 21:228-236.

Vial, J. L. 1968. The ecology of the tropical salamander, Bolitoglossa subpalmata, in Costa Rica. Rev. Biol. Trop. 15:13-115.

Wake, D. B. 1966. Comparative osteology and evolution of the lungless salamanders, family Plethodontidae. Mem. So. Calif. Acad. Sci. 4:1-111.

Wake, D. B. 1970. The abundance and diversity of tropical salamanders. Am. Nat. 104:211-213.

Wake, D. B. and J. F. Lynch. 1976. The distribution, ecology, and evolutionary history of plethodontid salamanders in tropical America. Nat. Hist. Mus., Los Angeles Co., Sci. Bull. 25: 1-65.

Wilder, I. L. W. and E. R. Dunn. 1920. The correlation of lunglessness in salamanders with a mountain brook habitat. Copeia 1920:63-68.

THE REPRODUCTIVE BIOLOGY OF CAECILIANS: AN EVOLUTIONARY PERSPECTIVE

Marvalee H. Wake

Department of Zoology and Museum of Vertebrate Zoology
University of California, Berkeley, California 94720

Caecilians (Amphibia: Gymnophiona) are elongate, limbless, burrowing or swimming animals. They are pan-tropical in distribution and, until recently, infrequently seen by collectors. The relationships of caecilians to other amphibians are enigmatic (Carroll and Currie 1975), and relationships among caecilian genera and species are poorly understood. The acquisition of the attenuate habitus has been accompanied by changes of suites of morphological features, including those of the reproductive system. Concomitantly, a number of aspects of the physiology and ecology of the group have adapted, so that features of the reproductive biology and life histories of caecilians provide considerable information of use in analyzing patterns of evolution.

Certain trends in reproductive biology appear repeatedly in lineages of amphibians. These include: 1) external to internal fertilization; 2) large to small clutches; 3) aquatic to terrestrial breeding; 4) simple to elaborate mate recognition and courtship patterns; 5) abandonment of clutch to forms of parental care; 6) oviparity to viviparity; 7) free-living aquatic larval stage to forms of terrestrial direct development. Some aspects of caecilian reproductive biology are unique among vertebrates with regard to these trends, and several are unique among amphibians. This paper emphasizes the pattern of evolution of caecilian reproductive biology, especially as it relates to fecundation in terrestrial environments, reproductive cycles, and development and maintenance of young. The position of caecilians relative to the above trends is contrasted with that of other amphibians. Comparative morphology provides the data base for much of this evaluation, and the caecilian pattern is compared with that of anurans and urodeles.

COPULATION AND FERTILIZATION

Several features of the morphology of the adult reproductive systems of caecilians provide means of achieving internal fertilization by facilitating sperm transport.

Intromittent organs

All species of caecilians have an intromittent organ. It is the posterior part of the cloaca, termed the phallodeum. Other vertebrates, including such birds as ducks, evert part of the male cloaca during copulation, but not to the extent and complexity seen in caecilians. The organ is everted by a combination of contractions of the body wall and of contractions of the musculature of the cloaca itself. Cloacae of various species have particular muscular and connective tissue arrangements that facilitate extrusion of the phallodeum (Tonutti 1932, 1933, Wake 1972). In some species vascular sinuses fill to promote extrusion of the organ (Tonutti 1931, 1933, Taylor 1965, Wake 1972). The phallodeum in all species is retracted by a stout *m. retractor cloacae*, which originates on the posterior body wall and inserts on the phallodeum. Its presence may be used to sex juveniles and late fetuses since it is evident before gonads mature and internal cloacal morphology is apparent.

The phallodeum has a species-specific morphology (Tonutti 1932, 1933, Taylor 1960, 1968, Wake 1972). Musculature and fibrous connective tissue form a number of cloacal structural features, including antero-posterior tracts, possibly for conduction of sperm and waste products, sac-like structures probably used in phallodeal eversion, flaps that shield the mesonephric duct openings into the cloaca from intestinal waste, and various knobs, transverse furrows, and other ornamentation. The cloacae of females of the same species, although less ornate in morphology, have an internal pattern that corresponds to that of the male (Fig. 1). Because all species have intromittent organs, fertilization is presumed to be internal in all species, although copulation has rarely been observed (Barrio 1969). Internal fertilization in all species of the order is a considerable contrast to the very few species of frogs which appear to practice internal fertilization, and the urodele situation in which primitive species practice external fertilization. Only the frog *Ascaphus truei* practices copulation by means of an intromittent organ, and the few other frogs, such as *Nectophrynoides*, that have internal fertilization apparently transfer sperm by cloacal apposition. The intromittent organ of *Ascaphus* is a structure comprised of muscles and cartilage retained from the tadpole stage. It is inserted into the vent of the female during amplexus. Internal fertilization in advanced salamander species is effected by male deposition of a spermatophore, which is picked up from the substrate by the female

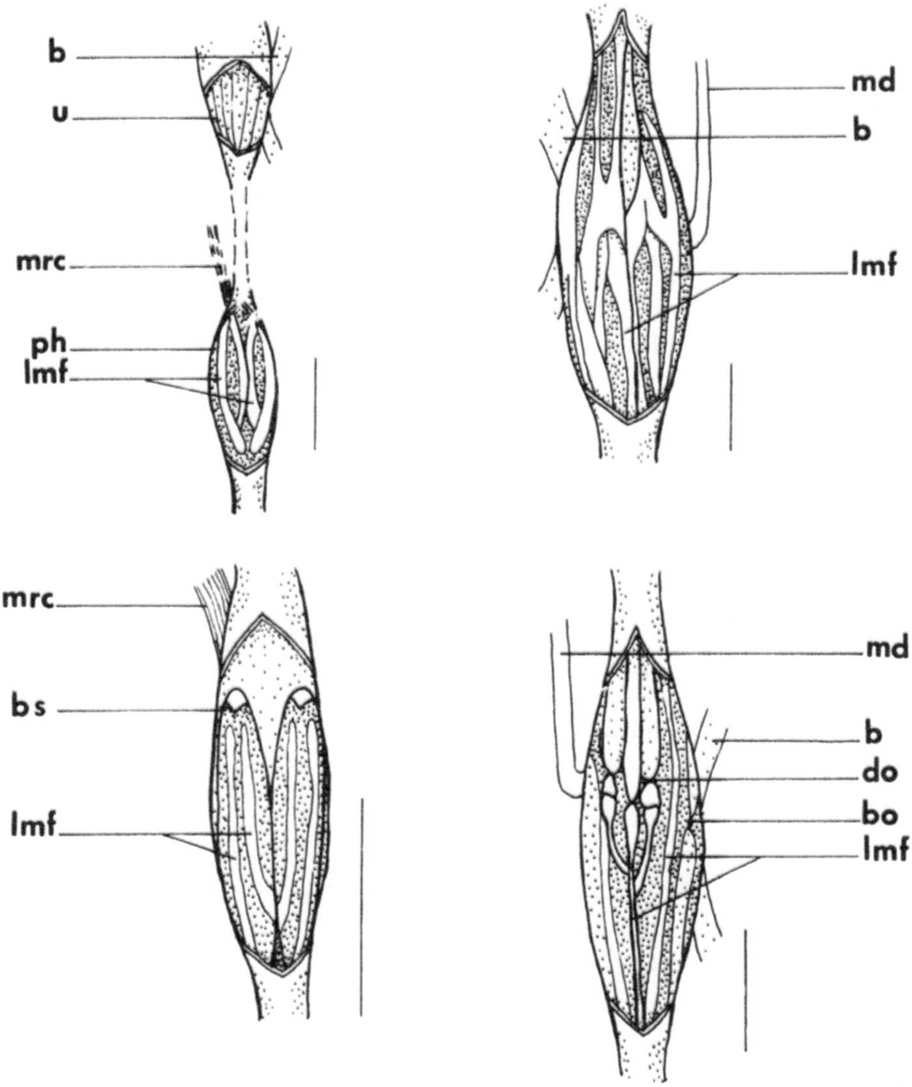

Fig. 1. Cloacae of Scolecomorphus kirkii (male = A, female = B) and Scolecomorphus uluguruensis (male = C, female = D). Cloacae are opened ventrally and the internal configuration is illustrated. Abbreviations: b = bladder, bo = bladder orifice, bs = blind sac, do = duct orifice, lmf = longitudinal muscular folds, md = Mullerian duct, mrc = musculus retractor cloacae, ph = phallodeum, u = urodeum. Bar = 5 mm.

with her vent. Therefore, the position of caecilians with regard to the trend toward internal fertilization is unique among amphibians in that all species practice internal fertilization, and all use an intromittent organ. Little is known of mate recognition or courtship in caecilians. Barrio (1969) noted a particular coiling and touching behavior in the aquatic Chthonerpeton indistinctum that precedes copulation as used in these apparently non-vocal, non-visual animals. Such mechanisms are in contrast to the well known courtship and recognition behaviors of other amphibians.

Mullerian glands

Another reproductive structure that is unique to caecilians is the male Mullerian gland. The paired Mullerian ducts, which appear transitorily during the development of all vertebrates of both sexes, are characteristically absent or represented only by strips of connective tissue in adult males. In females the Mullerian ducts develop into the oviducts. Male caecilians are characterized by having functional Mullerian ducts. The anterior two-thirds of each duct is reduced to a connective tissue strand, but the posterior several millimeters are large glandular structures the lumens of which empty into the cloaca. During the period of active spermatogenesis the structure is comprised internally of transverse simple tubular glands which empty into the lumen of the Mullerian structure (Fig. 2). The central lumen is lined with ciliated columnar epithelium. Secretion is active during the entire period of spermatogenesis and cells are apocrine, expelling secretory granules with some breakdown of cellular structure. The glands regress during the period of testicular quiescence. Cells are much reduced in size, no secretory granules are produced, and cilia are not present on the epithelial cells lining the lumen during the quiescent period.

I suggest that the functional significance of the retained Mullerian duct and of its elaboration as a gland is to provide a means of fluid transport, and perhaps nutrition, for sperm, and that this structure evolved with the acquisition of internal fertilization. Many organisms, aquatic and terrestrial, have fluid-secreting structures associated with sperm transport (such as shark seminal vesicles, turtle and mammalian prostates). These structures, however, usually develop as outgrowths of the urethra or the mesonephric ducts, in conjunction with surrounding mesenchyme (Nelsen 1953). The caecilian structure is unique in that it appears to be entirely derived from the Mullerian duct. It is of interest, though, that components of the Mullerian duct are known to be incorporated into the mammalian prostate gland. It was long thought that the inclusion was vestigial and non-functional, but it is now suspected that the Mullerian structure is an active component of

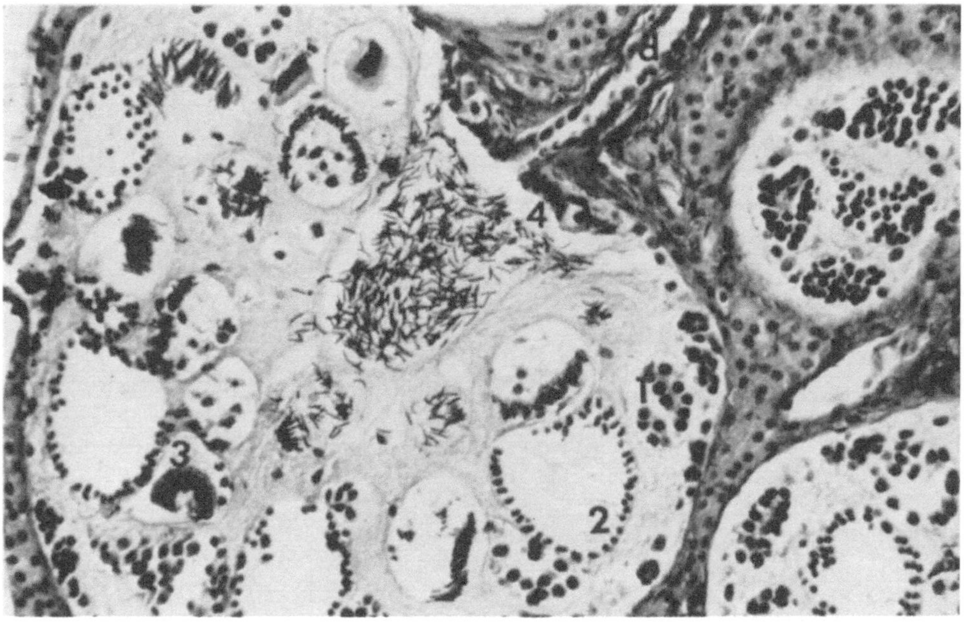

Fig. 2. Testis of Schistometopum gregorii 42.5 X. Reprinted by permission of Journal of Morphology. One lobule, containing several cell nests in different stages of development, is featured. Abbreviations: d = interlobular sperm duct, i = interstitial tissue, 1 = primary spermatocytes, 2 = secondary spermatocytes, 3 = spermatids, 4 = nearly mature spermatozoa ready to evacuate the testis.

the mammalian prostate gland (Bloom and Fawcett 1968). A direct comparison of the secretion of the caecilian Mullerian gland with that of the mammalian prostate would be of interest in determining the evolution of sperm-transport structures.

Following copulation, fertilization apparently takes place in the anterior part of the oviduct (Wake 1970). Development then ensues, and eggs are either laid soon after fertilization or are retained through metamorphosis.

Spermatogenesis

The testes of caecilians are paired, elongate, and multi-lobed. Species considered primitive on other grounds are characterized by testes comprised of numerous lobes of nearly equal size; more derived species have testes comprised of long anterior lobes and a few small posterior lobes. In such viviparous species as Gymnopis multiplicata and Geotrypetes seraphini, anterior testis lobe fusion and addition of some posterior lobes occurs during the first year after birth. Testes appear to mature in the second year, and the number of lobes remains fixed from that time, as determined from examination of testes of specimens representing apparently different age classes. In contrast, a number of salamander species are known to add testis lobes with age. The variation in number of testis lobes in caecilian species therefore reflects an evolutionary trend toward reduction of lobe number through fusion in derived species rather than an increase in numbers of lobes with age as in salamanders.

The lobes of the testis are connected by a mesentery and by transport ducts. Each lobe has one to several ducts that lead directly to the anterior part of the kidney. Sperm are transported from the testis through such ducts, through tubules of the anterior part of the kidney, thence to the mesonephric duct and to the cloaca. The sperm-bearing kidney tubules apparently remain fully functional osmoregulatory units, in contrast to the situation in many frogs and salamanders in which the anterior part of the kidney loses urinary function and functions exclusively for sperm transport.

Spermatogenesis has been described for several species (Seshachar 1963, 1937, 1942a, 1943, 1945, Wake 1968). Active spermatogenesis is characterized by the presence of a number of dilated lobules forming each testis lobe. Each lobule contains a number of locules. These locules are the cell nests of Lofts (1974) and others. In contrast to the situation in most temperate and many tropical anuran species and all urodele species, locules in each lobule contain sperm at all stages of maturation. In a single lobule clusters of cells may be found ranging from primary spermatogonia to maturing spermatids (Fig. 3). In fact, mature spermatids are present in the testis of G. multiplicata for several months of the year, and ducts from the locules that connect to the interlobular sperm transport ducts are well developed during that period. During the period of testis regression, interstitial tissue increases, lobule stromata disintegrate, only a few spermatocytes are present peripherally in the lobule, and the intra-lobular ducts degenerate. Most of the testes examined of some 50 species from all families of caecilians showed active spermatogenesis. Most species were represented by single specimens, however, so information about cyclicity is not available.

Oogenesis

The ovaries are paired elongate sac-like structures. Adult females may have a large number (50-100) of pre-vitellogenic and

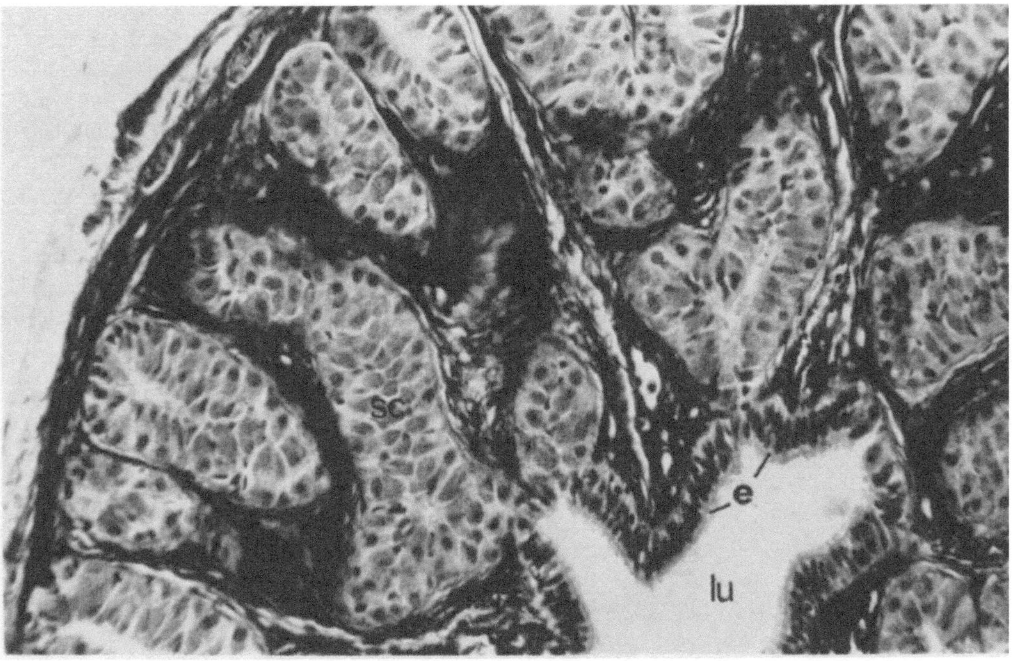

Fig. 3. Active male Mullerian duct of Gymnopis multiplicata. 42.5 X. Note the deep tubular glands. Abbreviations: e = epithelium, lu = lumen, sc = secretory cells.

early vitellogenic ova, but not all of these will become fully yolked. Data from samples of G. multiplicata indicate that ovum development cycles are also confined to part of the year. Only pre-vitellogenic ova are present during other parts of the year. In samples of other caecilian species (Dermophis mexicanus, Typhlonectes compressicauda) in which several females were collected on a particular date, ovum development is at the same stage in all females. I therefore infer synchrony of ovum maturation in these populations, which suggests seasonality of the cycle. Other species, such as certain Ichthyophis, are thought to be acyclic (Taylor 1960; see section on population biology herein for discussion).

Development of ova in caecilians is typical of that of amphibians generally (Lofts 1974). Developing ova lie in follicles that are associated with the thin wall of the ovary. As vitellogenesis proceeds, the layers of the follicle distend, and a mature ovum has a discrete egg membrane, a narrow zona pellucida, a surrounding follicular layer one cell deep, and a follicular stalk adherent to the ovary wall (Fig. 4).

The mature ova of most caecilians are large compared with those of other amphibians. Several of the egg-laying species have ova 10 mm long and up to 6-8 mm deep. These include Ichthyophis glutinosus, 9 x 3 mm (Sarasin and Sarasin 1887-90) or 9 x 6 mm (my data); Hypogeophis rostratus, 10 mm (Tonutti 1931) or 9 x 6 mm (my data); and Siphonops annulatus, 10 x 8.5 mm (Goeldi 1899). Some egg layers have small eggs, however, such as Siphonops paulensis (4.3-4.5 mm in diameter; Gans 1961). Live-bearing species have smaller eggs (Dermophis mexicanus, 2 x 1 mm; Typhlonectes compressicauda, 3 x 2.5 mm; Geotrypetes seraphini, 3-5 x 2-3.7 mm; Schistometopum thomensis, 3 x 2 mm; Gymnopis multiplicata, 2 mm in diameter).

Clutch size. Ovarian clutches are largest in the egg-laying species. Sarasin and Sarasin (1887-90) reported 54 mature ova in Ichthyophis glutinosus and I counted 41 in a 340-mm specimen with fully yolked ova. Clutches laid range from 25-38 ova (W. C.

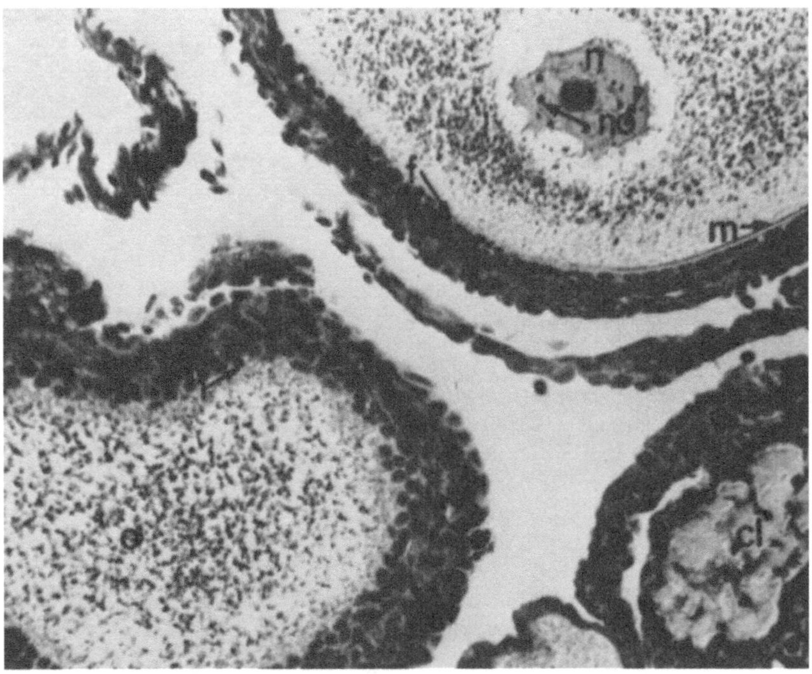

Fig. 4. Ovary of Scolecomorphus vittatus. 42.5 X. Reproduced by permission of Journal of Morphology. Note the corpus luteum. Abbreviations: cl = corpus luteum, f = follicle, m = membrane, n = nucleus, no = nucleolus, o = ovum.

Breckenridge, pers. comm.). Laid clutches of six are reported for the diminutive (90-110 mm) Idiocranium russeli (Sanderson 1937) and for the larger (to 400 mm) Siphonops annulatus (Goeldi 1899) and S. paulensis (Gans 1961). Hypogeophis rostratus has ovarian clutches of 15 (my data on specimens averaging 200 mm total length) to 20 (Tonutti 1931) and lays 15-18. The ovarian clutches of live-bearing species are of similar size. Geotrypetes seraphini has clutches of 18-30 ova, Gymnopis multiplicata 25-30, Dermophis mexicanus 28-32. Oviducal young or young born in these species, however, are much reduced in number. My data indicate that G. seraphini bears 1-4 young, G. multiplicata 2-9 and D. mexicanus 2-13. Oviducal development imposes constraints on the number of eggs that can develop.

Ovarian clutch data exist for individuals of three species of unknown reproductive mode. Gegeneophis ramaswamii has 13 ova that are 4 x 2 mm and Caecilia occidentalis has 7 ova that are 4 x 3 mm (pers. obs.); Caecilia corpulenta has 18 ova that are 4 x 3 mm (Taylor 1968). These few fully yolked ova of relatively small size lead one to postulate tentatively that they are live-bearers, but their status remains unknown.

I do not have samples that allow me to assess whether or not there is an increase in egg size and/or egg number with an increase in body size, either within a species due to growth, or among species of different size ranges. The information at hand suggests that both large and small species may lay small clutches, and the reduced ovum size in live-bearers, without reduction of ovum number, indicates that females may be investing energy for maintenance of the young by other means of maternal nutrition (see below). In any case, generalizations about clutch size-clutch volume-body size relationships such as have been made for frogs and salamanders cannot be made for caecilians.

The clutch of egg-laying species is laid as a strand of equidistant ova coated with jelly and constricted between ova (Sarasin and Sarasin 1887-90, W. C. Breckenridge, pers. comm., Ichthyophis glutinosus; Seshachar 1942a, Ramaswami 1954, Gegeneophis carnosus; Brauer 1897, and pers. obs., Hypogeophis rostratus; Sanderson 1937, Idiocranium russeli). The clutch is laid in a burrow (Sarasin and Sarasin 1887-90, I. glutinosus) or under the upper grass and root stratum (Sanderson 1937, I. russeli). I. glutinosus lays its clutch in a heap on the bottom of the burrow and then coils around it to brood; I. russeli builds a small pyramid of earth on the floor of its hole, lays the clutch atop it, and coils about it (Sanderson 1937). Sanderson reported that the females defended their clutches by spitting water at him as he collected them. Brauer (1897) noted that the female was at the side of the clutch in Hypogeophis rostratus, and Gans (1961) commented on the maternal presence in

Siphonops paulensis. In many of the known egg-layers, females appear to brood the developing clutch.

Corpora lutea. Ovulation occurs through the wall of the ovary into the coelom, and ova are picked up by the infundibula of the oviducts. Corpora lutea are present in 11 of 12 species examined histologically. Most of the species known to be viviparous have been demonstrated to have corpora lutea (G. multiplicata, D. mexicanus, G. seraphini, T. compressicauda, S. uluguruensis, S. gregorii). Several species with unknown reproductive modes also have corpora lutea (Boulengerula boulengeri, Oscaecilia ochrocephala). The egg-layers Siphonops annulatus, Hypogeophis rostratus, and Idiocranium russeli have them as well (Fig. 4). Only the oviparous Epicrionops bicolor lacked corpora lutea among species surveyed. This could be an artifact of the time of collection or a demonstration of total absence. Corpus luteum development in caecilians is similar to that in other vertebrates. The follicle involutes, the cavity fills with stroma, lipids aggregate, and a cellular structure characterized by large nuclei is established. The function of the corpora lutea in the viviparous frog Nectophrynoides is well known (Vilter and Lugano 1959, Lamotte et al. 1964, Xavier et al. 1970) as it is for the viviparous Salamandra atra (Vilter and Vilter 1960, 1964, Joly 1961, Joly and Picheral 1972). Lodge and Smith (1960) presented evidence for hormonal control of oviducal secretion in amphibians, and one can postulate that the corpora lutea function to maintain oviducal structure and secretion, and have other effects that help maintain the pregnancy in the live-bearing caecilians. Their function in the egg-layers is more questionable. Since fertilization is internal in all of these species, however, and some post-fertilization development occurs in the oviduct before the clutch is laid, the corpora lutea may have some transitory effect on maintenance of the fertilized ova.

Development and maintenance of young

Viviparity and maternal brooding. The evolution of the reproductive biology of caecilians is characterized by a trend toward protection of developing ova, either through brooding the clutch or retaining the developing young in the oviducts during their developmental period, and by suppression of the larval state. Since all species practice internal fertilization, retention of ova is facilitated. In no other amphibian lineage is live-bearing so exploited. At least three genera of frogs include some live-bearing species (Nectophrynoides, Eleutherodactylus, and Rheobatrachus); the former two have oviducal development, the latter gastric (Corben et al. 1974). These include five of the approximately 4500 species of frogs. Among salamanders, only Salamandra atra of the several hundred urodele species is an obligate live-bearer. Among caecilians, however, 50% of the species having a known reproductive mode are live-bearers (Table 1). This assessment is conservative, because direct evidence for the reproductive mode was required for it to be considered known. Species of questionable adult or larval identification and species for which only indirect evidence of reproductive mode is available are not included for this evaluation. I suspect that further data will demonstrate that the majority of caecilian species are live-bearers. Further, the live-bearing caecilians appear to practice intra-oviducal nutrition of the developing young (see section on fetal development). Viviparity in caecilians includes maternal nutrition of oviducal young following resorption of the yolk, but not placentation. The species of Nectophrynoides and Salamandra atra practice intra-oviducal nutrition as well, and the mechanism of nutrient supply is similar in members of all three amphibian orders. Certain components of the oviducal epithelium become secretory under hormonal control (Xavier et al. 1970) and a material comprised of lipids and mucoproteins expressed from the epithelium is ingested by the fetuses. This form of viviparity in the frog and the salamander species is thought to have arisen as an evolutionary response to severe physical habitat stresses (Gallien 1959). The viviparous species dwell in high mountains where there is a long season of extreme cold and dryness. The period of gestation can be quite long, nine months in Nectophrynoides occidentalis (Lamotte and Xavier 1972) and three to five years in Salamandra atra (Vilter and Vilter 1960).

A number of caecilian species practice viviparity, however, and the majority of them are lowland tropical species. Many of these species are confined to the lowlands, but some species have wider ranges and occur at elevations up to 1000 m; at least two viviparous species occur from 1000 to 2500 m. One need not postulate that viviparity arose in highland situations. Several of the genera and one family (Typhlonectidae) which are viviparous are exclusively lowland. None of the caecilian species occur in habitats as rigorous as those in which Nectophrynoides and Salamandra occur. Viviparity is found in species of three of the four caecilian families (fide Taylor 1968, 1969), so it is not restricted taxonomically as it is in other amphibians. In fact, one of the features defining the most primitive family, the Ichthyophiidae, is the absence of live-bearing and the presence of a free-living larval stage. Several species of the Caeciliidae and probably all of the Typhlonectidae and the Scolecomorphidae are viviparous. Internal fertilization doubtless was involved in the acquisition of viviparity, but the caecilian condition stands in contrast to the salamanders, many species of which in several families have internal fertilization via spermatophores, but only one of which acquired viviparity. Direct development is the theme in more derived salamander lineages. Viviparity in caecilians need not arise as a direct response to climatic severity. The reproductive modes involving viviparity in a number of species and maternal care and

Table 1. Reproductive Modes of Caecilian Species

	Mode	Evidence
Ichthyophiidae		
Caudacaecilia		
asplenia	O	L ?
larutensis	O	L ?
weberi	O	L
Epicrionops		
bicolor	O	C
peruvianus	O	C
petersi	O	C
Ichthyophis		
acuminatus	O	L ?
glutinosus	O	C,L
hypocyaneus	O	L
mindanaoensis	O	L
monochrous	O	L
peninsularis	O	L
supachii	O	L
youngorum	O	L
Caeciliidae		
Caecilia		
subnigricans	V	FT
tentaculata	V	FT
Dermophis		
mexicanus	V	OF
oaxacae	V	OF
parviceps	V	OF
Gegeneophis		
carnosus	O	C
Geotrypetes		
angeli	V	OF
grandisonae	O?	L
seraphini	V	OF
Grandisonia		
alternans	O	C,L
brevis	O	C,DD
diminutiva	O	C,DD
larvata	O	L
sechellensis	O	C,L
Gymnopis		
multiplicata	V	OF
Hypogeophis		
rostratus	O	C,DD
Idiocranium		
russeli	O	C,DD
Schistometopum		
thomense	V	OF
Uraeotyphlus		
oxyurus	O	C
Scolecomorphidae		
Scolecomorphus		
uluguruensis	V	OF
Typhlonectidae		
Typhlonectes		
compressicauda	V	OF
natans	V	OF
obesus	V	OF
Chthonerpeton		
indistinctum	V	OF
viviparum	V	OF
Nectocaecilia		
petersi	V	OF
ladigesi	V	F

Abbreviations: C = clutch; DD = direct development; F = fetus; FT = fetal teeth; L = larvae; O = oviparous; OF = oviducal fetus; V = viviparous; ? indicates some question regarding identification of species or of reproductive mode.

suppression of the larval stage in others may be adaptations correlated with success in terrestrial environments for the low vagility caecilian lineages.

Larval development. Ontogeny is best known among certain egg-laying species of caecilians because clutches at different stages of development have been collected in the field, and clutches have been reared in the laboratory. The sequence of development of *Ichthyophis glutinosus* is known through the magnificent work of Sarasin and Sarasin (1887-90), as well as that of Brauer (1897, 1899), Semon (1892), Spengel (1876), and Wiedersheim (1879). The development of *Hypogeophis rostratus* was described in some 30 papers by Marcus and his colleagues from 1900 to 1945 (certain pertinent papers are cited elsewhere in this review in the appropriate context), and more recently by Lawson (1963, 1965a, 1965b, 1966a, 1966b). Other early embryological work includes that of Peters (1874, 1875) on *Typhlonectes* and Goeldi (1899) on *Siphonops*.

Ichthyophis burrows are near a stream, and larvae hatch from the eggs, emerge from the burrow, and wriggle into the stream (Sarasin and Sarasin 1887-90). Larvae are characterized by the presence of long, triramous gills, a lateral line system, and an adult-type dentition. Lawson (1965a, 1965b) and Marcus (1908, 1909, 1920, 1922) described the pre-adult teeth of the egg-laying *Hypogeophis rostratus*. Sanderson (1937) reported collecting clutches and attendant females of *Idiocranium russeli* in a marsh in Mamfe, Cameroons. He noted that various clutches contained developing young at a number of stages, from large yolky eggs with a tiny, poorly-formed embryo atop the yolk to gilled 'larvae,' to 'larvae' that had resorbed their gills and appeared ready to emerge from the egg fully metamorphosed. Brauer (1897, 1899, 1902) reported that *Hypogeophis rostratus* passes its development through metamorphosis in the egg membrane, growing and resorbing gills and laterial line organs before hatching. Parker (1958) noted that some immature hatched specimens of *H. rostratus* had open gill clefts and pits believed to be lateral line organs. Direct development, circumventing a free-living larval stage, may occur in several species.

Fetal development. Development in the oviduct is characterized by several features unique to caecilians, although the general pattern is similar to that of live-bearing fish, amphibians, and reptiles. Clutch size and ovum size are reduced compared to those of egg-layers. The young are retained through their metamorphosis in the live-bearing species; birth of less than fully metamorphosed young has not been observed, although Largen *et al*. (1972) suggested that it may occur in *Geotrypetes grandisonae*. Yolk is resorbed relatively early in development, whereupon embryos hatch from the egg membrane in which they had been coiled and assume a more linear position in the oviducts. Almost invariably, developing young are found in both ducts. Following resorption of yolk considerable growth ensues and the young obtain nutrition from the secretion of the oviducal epithelium (Wake 1976, 1977).

The demands upon the female for nutrition are apparently great. Yolk is resorbed in Geotrypetes seraphini at 25-40 mm total length of embryo (Parker 1956); in Geotrypetes angeli at 38 mm; in Schistometopum thomense at 32-36 mm (Parker and Dunn 1964); in Typhlonectes compressicauda at 29-40 mm; and in Gymnopis multiplicata at 24-38 mm (the last two species from my data). Sizes at birth indicate substantial growth following yolk resorption. Geotrypetes young of 77 mm may be born of a 235 mm female; each fetus is 32% of the total length of the mother (Parker and Dunn 1964). Young of Schistometopum at 112 mm are 57% of the length of the mother; 125 mm young may be born of a 250 mm Gymnopis (50% of maternal length), and 200 mm Typhlonectes of 330 mm females (60% of maternal length). These measurements are from specimens in my collection. The percentages may reflect situations in which young are born of small females so that the proportion of fetal to maternal size is great. Such percentages are reduced for larger females; for example, a 200 mm young born of a 500 mm Typhlonectes is 40% of the maternal length; but these ratios are nonetheless very significant. A female Typhlonectes may bear as many as nine young. The embryos resorb their yolk at approximately 30 mm; they are born at approximately 200 mm, having increased their lengths 6.6 times during the period of maternal nutrition. With nine fetuses, each increasing 6.6 times in length, and each 40% of the female's total length at birth, the energetic demands on her for nutrition are great. My data do not indicate that females carrying large numbers of young have smaller young than females of the same size with fewer young, nor that females with large numbers of young give birth any earlier than females with fewer young. Smaller females may, however, have fewer young. In any case, considerable maternal nutrition is supplied.

Oviducal embryos have a morphology that generally resembles that of larvae. They have gills through much of their development, a thin skin, and external dimensions in which the length to breadth ratio is considerably less than in adults. The eyes develop relatively normally and then are roofed by skin, or skin and bone, and certain of the muscles and nerves of the eyes degenerate. The tentacle develops late in the metamorphic period. Some of the extrinsic eye muscles and their motor nerves serve as tentacular retractors and gland constrictors (Norris and Hughes 1918). The pattern of muscle-nerve organization varies from species to species, although the retractor bulbi, innervated by the abducens, becomes the retractor muscle of the tentacle in each species examined by Norris and Hughes (1918). In some forms, the tentacle perforates after birth. Oviducal young lack the lateral line system found in aquatic larvae (Sarasin and Sarasin 1887-90, Taylor 1960, 1970) and in some direct developers such as Hypogeophis rostratus (Parker 1958).

The gills of scolecomorphid and caeciliid fetuses are tri-ramous and fimbriated (Fig. 5), similar to those of free-living larvae and direct developers (Marcus 1939, Seshachar 1942b, Rama-swami 1954, Wake 1967, 1969). The gills of all typhlonectid species examined, however, are large sacciform structures (Fig. 6; Peters 1874, 1875, Wake 1977). These hollow sacs are vascularized by three aortic arches as are the triramous gills of other species. The gills are often observed in dissections to be appressed to the oviduct wall, extending both anterior to the fetus and around the body of the fetus and against the duct wall. Typhlonectids are totally aquatic. It is possible that their oxygen requirements cannot be met with the "terrestrial" pattern of respiratory organs. Perhaps the typhlonectid fetal gill structure evolved to provide a better surface for intra-oviducal gaseous exchange in aquatic inhabitants, although it is difficult to imagine why the result of stress on the triramous gill, presumed to be the primitive condition by comparison with larval morphology and gill structure in other groups, would not have resulted in a lengthening of rami and fimbriae as it does in salamander larvae (Bond 1960). The adult typhlonectid system also differs from the condition in terrestrial species. Adult typhlonectids are the only caecilians in which both lungs are elongate and functional; in terrestrial forms, only the right lung is developed (Wake 1974).

Metamorphosis is characterized by resorption of gills, thickening of the skin with some keratinization of the outer layers of cells, tentacular development and the resorption of some or all of the fetal teeth. The adult dentition erupts shortly after birth.

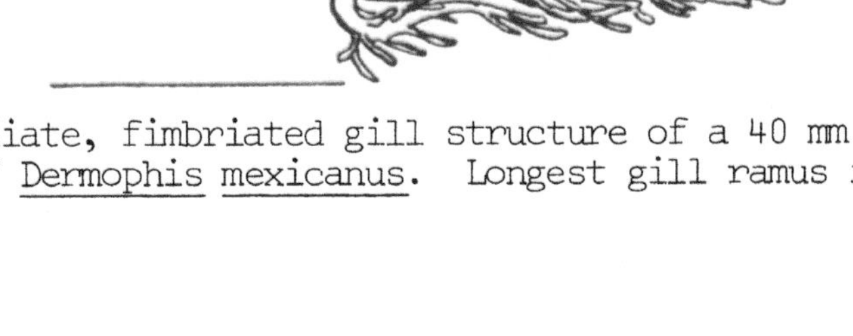

Fig. 5. Triradiate, fimbriated gill structure of a 40 mm total length fetus of Dermophis mexicanus. Longest gill ramus is 8 mm. Bar = 5 mm.

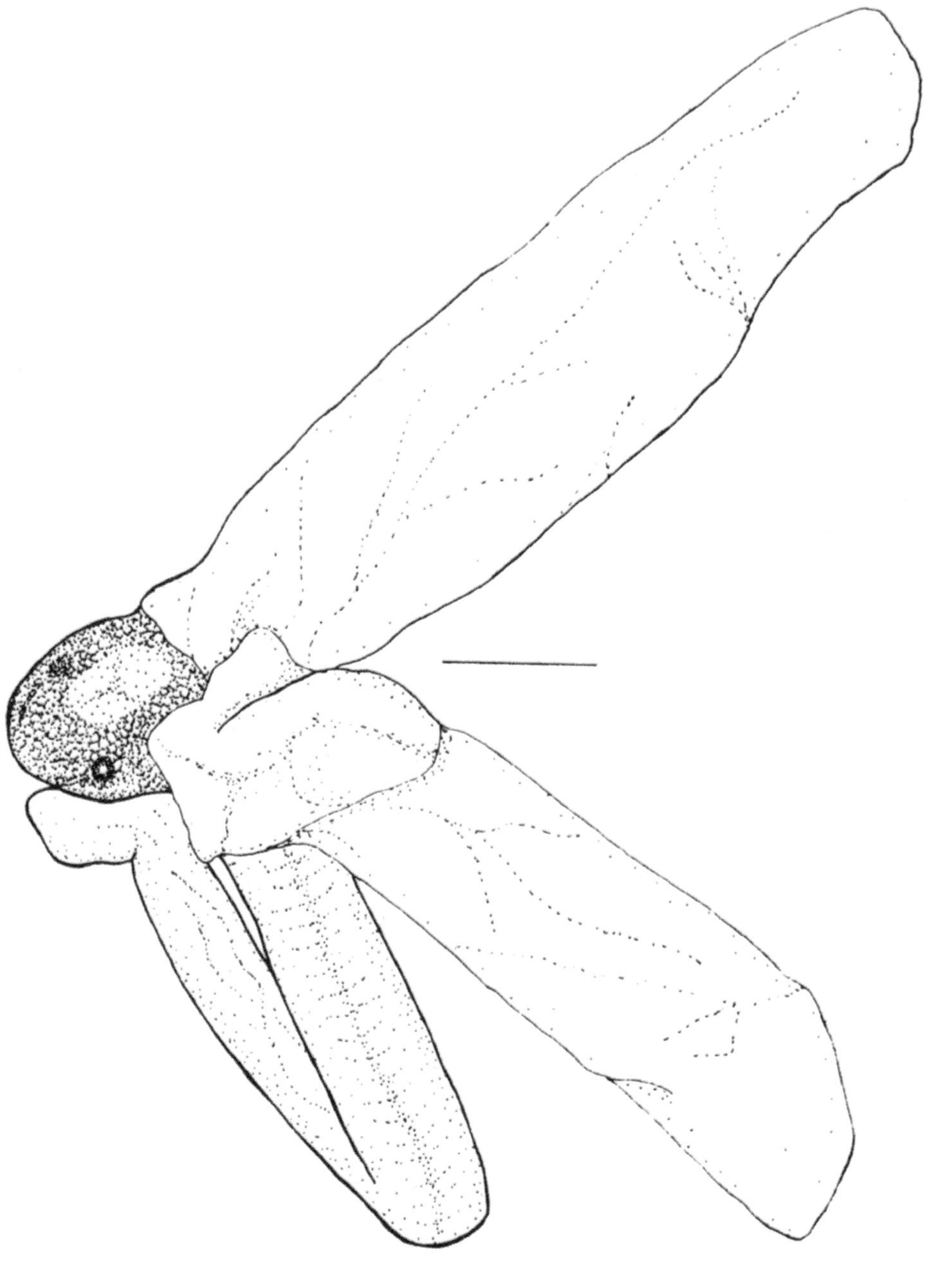

Fig. 6. Large saccular gills of a 36 mm total length fetus of *Typhlonectes compressicauda*. Compare structure and proportions with triradiate gills of *G. multiplicata* in figure 5. Gills are 24 mm long. Bar = 5 mm.

Fetal nutrition. The means by which fetuses obtain the nutrients supplied in the oviduct is of particular interest. The oviducal epithelium proliferates and, before embryos hatch from the egg membrane, is several cell layers deep. As the fetus grows, the duct dilates and pockets of connective tissue form which are lined with secretory epithelium (Wake 1970, 1977). The non-pocketed epithelium is also secretory, but is progressively denuded by the action of the fetuses (Wake 1970). Parker (1956) characterized the secretion as fatty and amorphous.

At the time that fetuses hatch from the egg, a peculiar dentition erupts. The fetal dentition is species-specific, and unlike that of the adult in crown morphology, number, and arrangement. These teeth aggregate at the outer margin of the jaw and are used to rasp the oviducal epithelium to ingest the secretion and, more significantly, to stimulate secretion and cellular proliferation by irritating the epithelial cells (suggested by Salthe and Mecham 1974, Wake 1977). The teeth are very small and numerous. Parker (1956), Parker and Dunn (1964), and Wake (1976) illustrated crown morphology in various species, and Wake (1976) analyzed tooth development and replacement in fetuses of the viviparous species. The fetal teeth of Gymnopis multiplicata and Typhlonectes compressicauda demonstrate two types of crown morphology and tooth arrangement. Gymnopis teeth have crowns topped with multiple spikes of enamel (Fig. 7); Typhlonectes crowns are flattened and topped with a single row of rounded knobs (Figs. 7 and 8). Gymnopis teeth are arranged in three or four rows on both upper and lower jaw (Fig. 9); in Typhlonectes they occur as paired plates (Fig. 10) on the lower jaw formed of the fused pedicels of some nine rows of teeth (Wake 1976).

Although these teeth have species-specific morphologies, their function appears to be the same in all species. I suggest that the variation in morphology of the fetal dentition is a response to selection operating on a species-level substrate. Species reflect variation through slight modification of a common developmental field, particularly by elaborating the paired cusps of the crown, and by retaining teeth on the jaws to form multiple rows and patches.

Population biology

Seasonality. Gallien (1959) reviewed the evidence for seasonality of reproduction in anurans and urodeles and concluded that reproduction is nearly always correlated with "spring" and regimes of increased moisture and usually of increased, or optimum, temperature in both temperate and tropical species.

Considerable evidence exists that breeding is seasonal in many caecilian species, and correlated with inception of the tropical

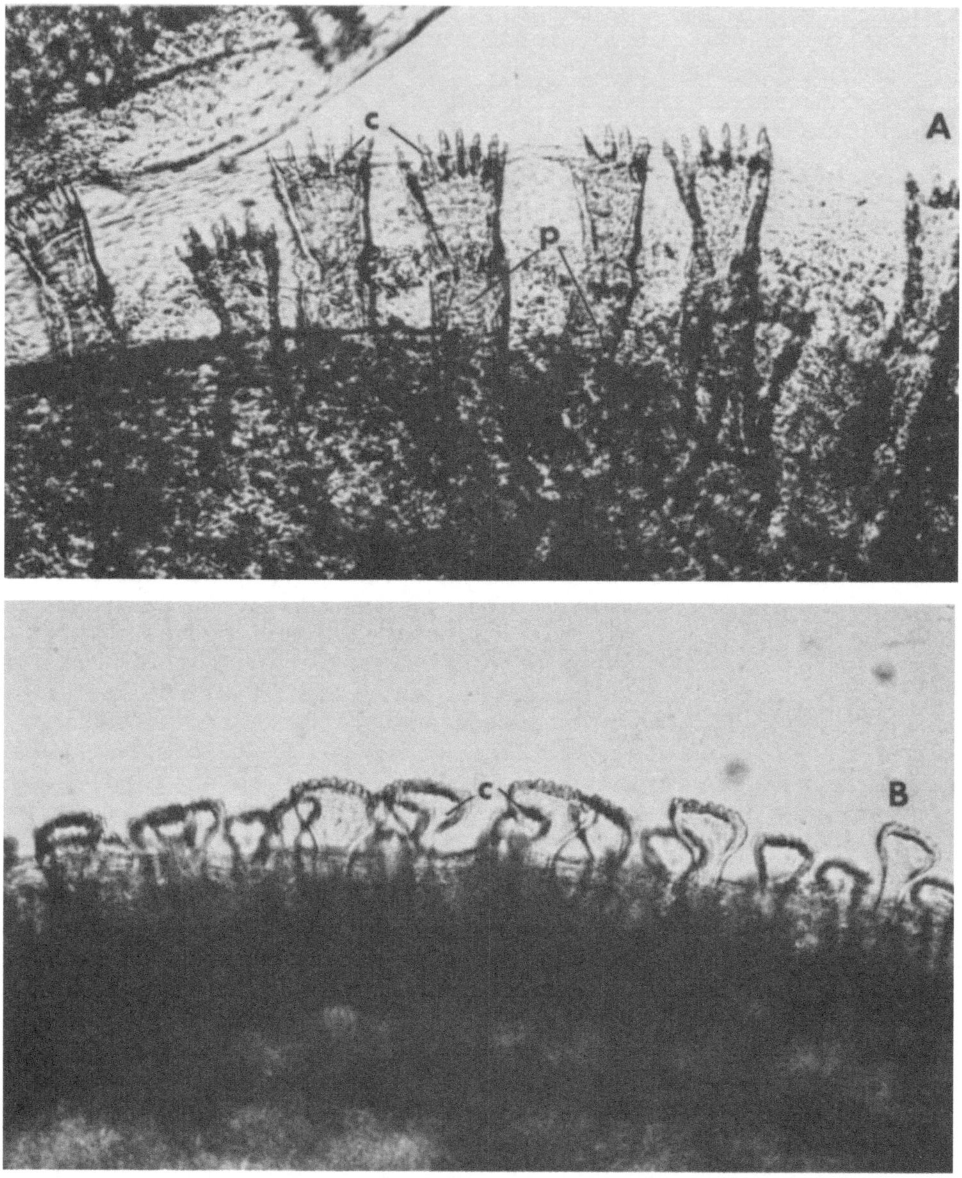

Fig. 7. A. Teeth of a 57 mm fetus of Gymnopis multiplicata. 156 X. Printed with permission of Journal of Morphology. Abbreviations: c = crown, p = pedicel. Cleared and stained specimen. B. Teeth of a 75 mm fetus of Typhlonectes compressicauda. 156 X. Printed with permission of Journal of Morphology. Compare crown structure of A and B. Abbreviation: c = crown. Cleared and stained specimen.

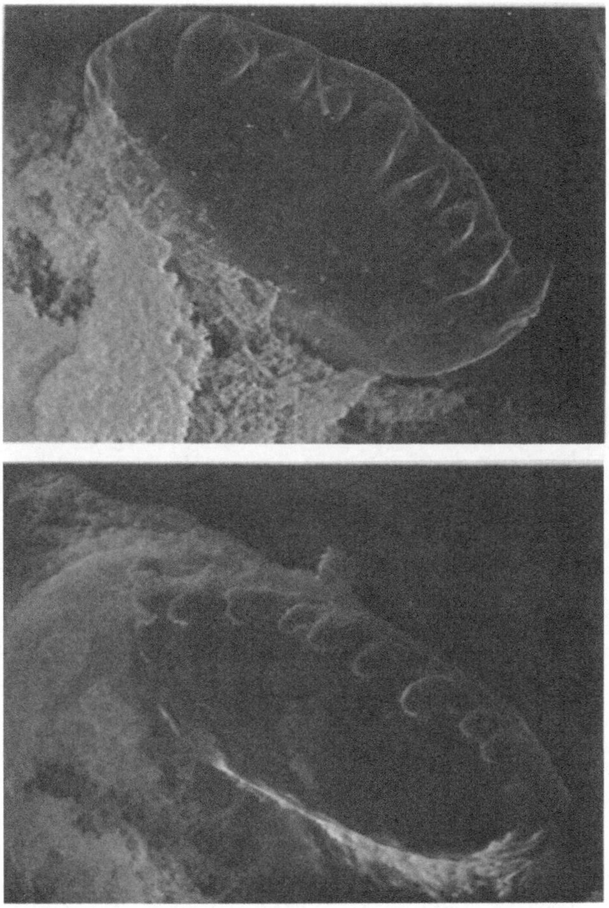

Fig. 8. Scanning electron photomicrograph of labial (top) and lingual (bottom) views of tooth crown of 75 mm total length fetus of Typhlonectes compressicauda. 980 X. Compare details of structure with that shown in figure 7B of tooth crowns from an oviducal sibling (cleared and stained specimen).

rainy season. Sarasin and Sarasin (1887-90) reported that egg laying in Ichthyophis glutinosus in Ceylon occurs from July through September, although Seshachar (1936) believed it to be from December to March in time for the monsoons. Two possibilities may contribute to this discrepancy: 1) the workers were dealing with different species, since five species of Ichthyophis are now recognized from Ceylon (Taylor 1968); 2) the brooding period is actually much longer, and both workers sampled only a part of the period. Taylor

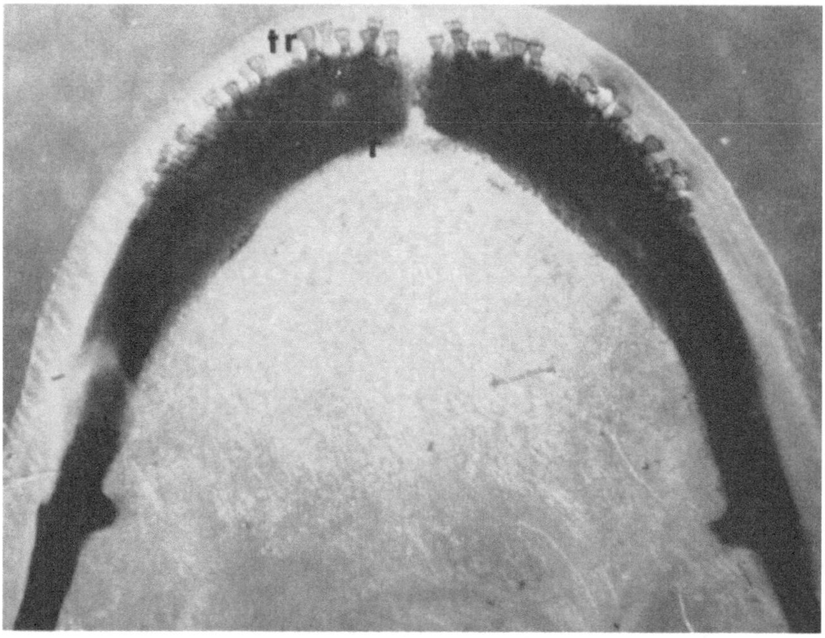

Fig. 9. Jaws and teeth of an 84 mm total length fetus of Gymnopis multiplicata. 16 X. Cleared and stained specimen. Note that teeth are in four rows on both jaw rami. Abbreviations: r = ramus, tr = tooth rows.

(1960) examined a series of Ichthyophis larvae from the Philippines and found that they represented a great range of total lengths and growth states. He found that they separated roughly into three age classes with substantive size ranges for each, and he suggested that the breeding period might last for much of the year. It is therefore possible that breeding might last for several months, and could be seasonal and acyclic in some species and among populations within species. Brauer (1897) reported that he collected egg clutches of Hypogeophis rostratus primarily from May through July in the Seychelles, but that he also found clutches containing older embryos occasionally in August and September.

Wake (1968) characterized six phases of the spermatogenic cycle in G. multiplicata. These include: 1) January to March: lobules contain large, peripheral spermatocytes; testes have considerable inter-lobular interstitial tissue. 2) April to June: spermatocytes divide mitotically to form cell nests or locules; locular ducts re-form. 3) June through August: maturing locules migrate centrally in the lobules, intra-locular stroma increases, new spermatocytes develop in the region of the necks of the transport

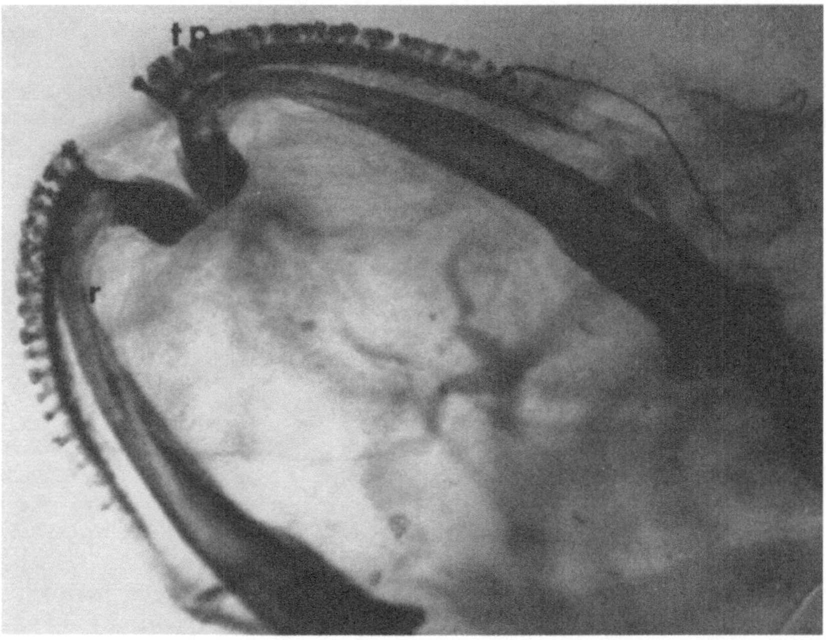

Fig. 10. Jaws and teeth of a 75 mm fetus of Typhlonectes compressicauda. 16 X. Compare structure of jaw rami and dentigerous ridges with those in figure 9. Note especially that Typhlonectes teeth are disposed as a plate formed of fused pedicels of the 8-9 rows of teeth. Abbreviations: r = ramus, tp = tooth plate.

ducts of each lobule. 4) July through October: spermatocytes continue to develop, mature spermatids exit the lobule, interstitial tissue begins to develop. 5) October through November: lobules contain primarily nests of maturing and exiting spermatids and extensive transport ducts. 6) December and January: remaining mature sperm leave the testis, new spermatocytes form peripherally, ducts degenerate, interstitial tissue is extensive. Further, I observed (unpubl.) that several females of a species collected at a given point in time have ova at the same stage of maturation, implying synchrony of vitellogenesis in the population. In several species, especially Gymnopis multiplicata and Dermophis mexicanus (Wake 1968), Hypogeophis rostratus (Brauer 1897), Ichthyophis glutinosus (Seshachar 1936), hatching or birth appears to be correlated with the inception of the rainy season and moisture and perhaps temperature and food optima. It would be of great interest to have sufficient material to examine timing of ovum maturation and of birth in typhlonectids, which are aquatic and presumably are not subjected to seasonal moisture regimes as are terrestrial caecilians.

Population profile. Little is known of the population biology of caecilians. Few large samples have been reported upon, and these have not been analyzed in terms of population characteristics. Largen et al. (1972) presented data on external morphological characters of a sample of *Geotrypetes grandisonae*, and compared the relationship of total length with other characters in order to analyze ontogenetic variation in the population. A sample of 118 *Geotrypetes seraphini* from Tafo, Ghana, collected in July-August, 1972, 1973, and 1975 presents the population profile shown in figure 11. Samples were collected so that virtually all animals observed through extensive digging were taken in order to collect as unbiased a population series as possible. When analyzed separately, the three collections show the same characteristics, so data are aggregated on the assumption that the samples represent the characteristics of the population at that time of year. Seventy-three of 118 animals (62%) are immature juveniles; juveniles are characterized by immature gonads and relatively less ossification in skull, vertebrae, and ribs than in adults, as well as their smaller size. I infer that these juveniles represent one age class. Birth in this species occurs when fetuses are 70-81 mm total length (Parker 1936, 1956, Parker and Dunn 1964), so this segment of the population may represent the first year class. The next segment (180-240 mm) may represent the second year class. Gonads of these specimens are mature, although males have slightly fewer testis lobes than do larger animals. The smallest female known to carry oviducal young occurs in this class (215 mm), so age at first reproduction may be as early as two years after birth. It is difficult to ascertain age classes for larger animals in the sample since there are so few, and I do not have samples from other times of the year, but it is significant that the population appears to be comprised largely of one- and two-year-old animals. Further, the ten specimens 300 mm or greater in total length are female. This may imply some size dimorphism, also suggested by Parker (1936) among males and females, or, possibly, longer-lived females which have the possibility of bearing a number of litters during their lives.

In conclusion, some characteristics of caecilian reproductive biology are shared by other amphibians and, indeed, by other vertebrates. However, the above discussion suggests that a number of aspects of reproductive biology involving morphology, physiology, and development are uniquely caecilian. There is rather little to indicate an "amphibian pattern" except at the most general level, and any discussion of amphibian reproductive biology must consider the variation shown in all three amphibian orders. Work in progress on the analysis of means of internal fertilization, egg retention and the acquisition of non-placental viviparity, the status of free-living larvae, direct development, and viviparity in various species, and species-specific characteristics of adult and fetal

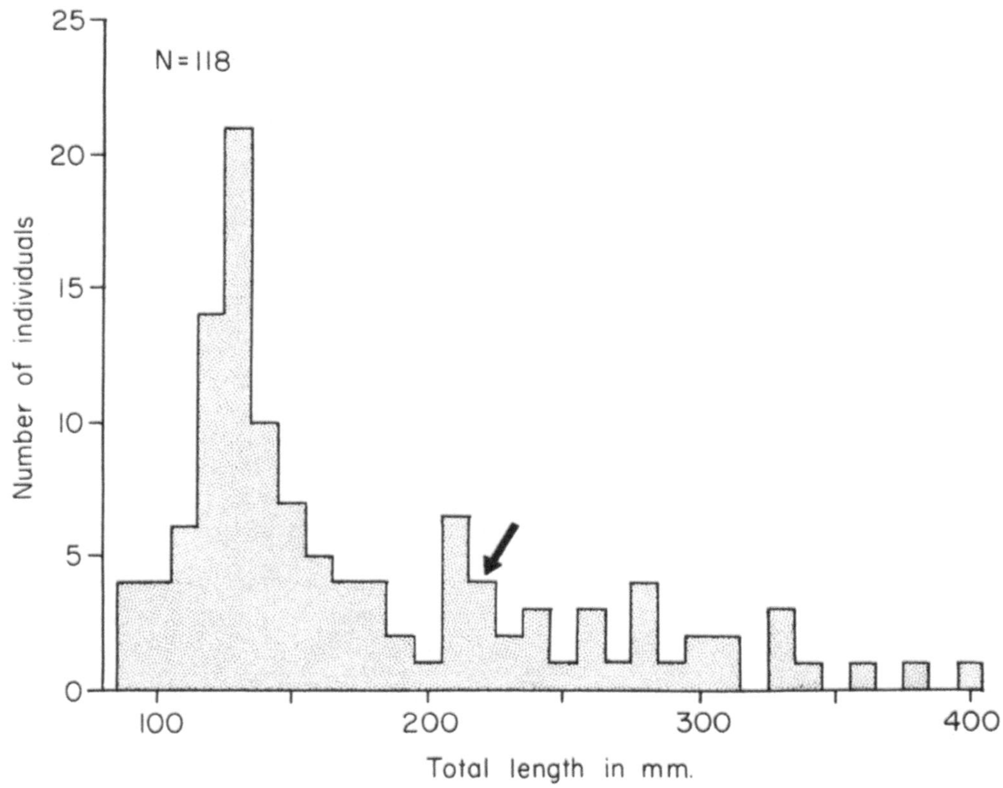

Fig. 11. Profile of a population of *Geotrypetes seraphini* from Tafo, Ghana. The smallest specimens collected are near size at birth, therefore the first peak may represent the youngest age class. 62% of the population is composed of immature juveniles. The second peak is composed of mature specimens. The arrow indicates the size of the smallest known pregnant female in the population, indicating that age at first reproduction may be as early as two years. Age classes cannot be distinguished for large specimens. All specimens 300 mm or greater are female.

morphology in lineages in the amphibian order Gymnophiona may provide considerable insight into patterns of evolution, especially the acquisition of viviparity, a problem of general significance to biologists.

ACKNOWLEDGEMENTS

I thank the many museums and curators cited earlier (Wake 1968, 1970, 1972, 1976) for their loans of specimens, and many colleagues for sending me for examination specimens that they had collected. These studies would not have been possible without their kind cooperation. Addye Brown, Department of Anatomy, University of Chicago, Penny Hermes, Cindy Hillery, and Gloria Wurst, Department of Zoology and Museum of Vertebrate Zoology, University of California, Berkeley, prepared histological materials. Gene M. Christman, Museum of Vertebrate Zoology, prepared figure 11; Teriann Asami typed the manuscript. I thank David B. Wake, Museum of Vertebrate Zoology, for critically reviewing the manuscript.

Portions of this work were supported by National Science Foundation grants GB 6432X and GB 17112 to D. B. Wake, and by funds to the author from the Committee on Research of the University of California, Berkeley, and from Biomedical Sciences Support Grant RR-7006 from the General Research Support Branch, Division of Research Resources, Bureau of Health Professions Education and Manpower Training, National Institutes of Health. The assistance of the Electron Microscope Laboratory of the University of California, Berkeley, in preparing figure 8 is gratefully acknowledged. Some recent aspects of this work were facilitated by a grant from the American Philosophical Society's Johnson Fund.

LITERATURE CITED

Barrio, A. 1969. Observaciónes sobre Chthonerpeton indistinctum (Gymnophiona, Caecilidae) y su reproduccion. Physis 28:449-503.

Bloom, W. and D. W. Fawcett. 1968. A textbook of histology. Ninth Ed. W. B. Saunders Co., Philadelphia, Pennsylvania.

Bond, A. N. 1960. An analysis of the response of salamander gills to changes in the oxygen concentration of the medium. Dev. Biol. 2:1-20.

Brauer, A. 1897. Beiträge zur Kenntniss der Entwicklungsgeschichte und der Anatomie der Gymnophionen. Zool. Jahrb. Anat. 10:389-472.

Brauer, A. 1899. Beiträge zur Kenntniss der Entwicklung und Anatomie der Gymnophionen. II. Entwicklung der aussern Form. Zool. Jahrb. Anat. 12:477-508.

Brauer, A. 1902. Entw. Anat. der Gymnophionen. III. Entw. der Excretionsorgane. Zool. Jahrb. Abt. Anat. 16:1-176.

Carroll, R. L. and P. J. Currie. 1975. Microsaurs as possible apodan ancestors. Zool. J. Linn. Soc. 57:229-247.

Corben, C. J., G. J. Ingram and M. J. Tyler. 1974. Gastric brooding: unique form of parental care in an Australian frog. Science 186:946-947.

Gallien, L. 1959. Endocrine basis for reproductive adaptations in Amphibia. IN: A. Gorbman (ed.), Comparative endocrinology. Wiley and Sons, Inc., New York, New York, pp. 479-487.

Gans, C. 1961. The first record of egg laying in Siphonops paulensis Boettger. Copeia 1961:490-491.

Goeldi, E. A. 1899. Über die Entwicklung von Siphonops annulatus. Zool. Jahrb. Syst. 12:170-173.

Joly, J. 1961. Le cycle sexuel biennial chez la femelle de Salamandra salamandra quadri-virgata dans les hautes-Pyrénées. C. R. Acad. Sci. Paris 252:3145-3147.

Joly, J. and B. Picheral. 1972. Ultrastructure, histochimie, et physiologie du follicule pre-ovulatoire et du corps jaune de l'urodele ovo-vivipare Salamandra salamandra (L.) Gen. Comp. Endo. 18:235-259.

Lamotte, M., P. Rey and M. Vogeli. 1964. Recherches sur l'ovaire de Nectophrynoides occidentalis, batracien anoure vivipare. Arch. Anat. Mic. Morph. Exp. 53:179-244.

Lamotte, M. and F. Xavier. 1972. Rechèrches sur le devèloppement embryonnaire de Nectophrynoides occidentalis Angel, amphibien anoure vivipare. I. Les principaux traits morphologiques et biometriques du developpement. Ann. Emb. et Morph. 5:315-340.

Largen, M. J., P. A. Morris and D. W. Yalden. 1972. Observations on the caecilian Geotrypetes grandisonae Taylor (Amphibia: Gymnophiona) from Ethiopia. Monitore Zool. Italiano, NS Supplemento IV, No. 8:185-205.

Lawson, R. 1963. The anatomy of Hypogeophis rostratus Cuvier (Amphibia, Apoda) Part I. The skin and skeleton. Proc. Univ. Durham Phil. Soc. 13:254-273.

Lawson, R. 1965a. The teeth of Hypogeophis rostratus, (Amphibia, Apoda), and tooth structure in the Amphibia. Proc. Zool. Soc. London 145:321-325.

Lawson, R. 1965b. The development and replacement of teeth of Hypogeophis rostratus (Amphibia, Apoda). J. Zool. 147:352-362.

Lawson, R. 1966a. The anatomy of the heart of Hypogeophis rostratus (Amphibia, Apoda) and its possible mode of action. J. Zool. London 149:320-336.

Lawson, R. 1966b. The development of the centrum of Hypogeophis rostratus (Amphibia, Apoda) with special reference to the notochordal (intravertebral) cartilage. J. Morph. 118:137-148.

Lodge, P. D. B. and C. L. Smith. 1960. Hormonal control of secretion in the oviduct of the Amphibia. Nature 185:774-775.

Lofts, B. 1974. Reproduction. IN: B. Lofts (ed.), Physiology of the Amphibia, vol. 2. Academic Press, New York, New York, pp. 107-218.

Marcus, H. 1908. Beiträge zur Kenntnis der Gymnophionen. I. Über das Schlundspaltengebeit. Arch. f. mik. Anat. 71:695-744.

Marcus, H. 1909. Beitrage zur Kenntnis der Gymnophionen. III. Über Entwicklungsgeschichte des Kopfes. Morph. Jahrb. 40:105-183.

Marcus, H. 1920. Beitrage zur Kenntnis der Gymnophionen. VIII. Über die Zähne and Korrelation ihrer Zahl mit dem Alter. D. Zahnheilkunde. 30:145-157.

Marcus, H. 1922. Der Kehlkopf bei Hypogeophis. Anat. Anz. 1:188-202.

Marcus, H. 1939. Beitrag zur Kenntnis der Gymnophionen. Über Keimbahn, Keimdrusen, Fettkörper u. Urogenitalverbindung bei Hypogeophis. Bio-Morphosis 1:360-384.

Nelsen, O. E. 1953. Comparative embryology of the vertebrates. Constable and Co., London.

Norris, H. and S. P. Hughes. 1918. The cranial and anterior spinal nerves of the caecilian amphibians. J. Morph. 31:489-560.

Parker, H. W. 1936. The caecilians of the Mamfe Division, Cameroons. Proc. Zool. Soc. London 1936:135-163.

Parker, H. W. 1956. Viviparous caecilians and amphibian phylogeny. Nature 178:250-252.

Parker, H. W. 1958. Caecilians of the Seychelles Islands with description of a new subspecies. Copeia 1958:71-76.

Parker, H. W. and E. R. Dunn. 1964. Dentitional metamorphosis in the Amphibia. Copeia 1964:75-85.

Peters, W. 1874. Über die Entwicklung der Caecilien und besonders der Caecilia compressicauda. Monats. Aka. der Wiss. Berlin 1874:45-49.

Peters, W. 1874. Über die Entwicklung der Caecilien. Monats. Aka. der Wiss. Berlin 1875:483-493.

Ramaswami, L. S. 1954. External gills of Gegenophis embryos. Anat. Anz. 101:120-122.

Salthe, S. N. and J. S. Mecham. 1974. Reproductive and courtship patterns. IN: B. Lofts (Ed.), Physiology of the Amphibia, vol. 2. Academic Press, New York, New York, pp. 309-521.

Sanderson, I. T. 1937. Animal treasure. Pyramid Books, New York, New York.

Sarasin, P. and F. Sarasin. 1887-90. Ergebnisse naturwissenschaftlichen. Forschungen auf Ceylon in den Jahren 1884-1886. Zur Entwicklungsgeschichte u. anat. der Ceylonische Blindwuhle Ichthyophis glutinosus. C. W. Kreidel's Verlag, Wiesbaden.

Semon, R. 1892. Studien über den Bauplan des Urogenitalsystem der Wirbeltiere. Dargelegt an der Entwicklung dieses Organsystems. Jena Zeitschr. Naturwiss. 26:80-203.

Seshachar, B. L. 1936. The spermatogenesis of Ichthyophis glutinosus L. The spermatogonia and their division. Z. Forsch. 24: 662.

Seshachar, B. L. 1937. Spermatogenesis of Ichthyophis glutinosus II. The meiotic divisions. Z. Forsch. 27:133-158.

Seshachar, B. L. 1939. The apodan sperm. Curr. Sci. Bangalore 9: 464-465.

Seshachar, B. L. 1942a. Stages in the spermatogenesis of Siphonops annulatus and Dermophis gregorii (Boul.). Proc. Ind. Aca. Sci. 15 B:263-277.

Seshachar, B. L. 1942b. Eggs and embryos of Gegenophis carnosus. Curr. Sci. 11:439.

Seshachar, B. L. 1943. Spermatogenesis of Ichthyophis glutinosus Lin. III. Spermatelosis. Proc. Nat. Inst. Sci. India 9:271-286.

Seshachar, B. L. 1945. Spermatelosis in Uraeotyphlus narayani Sesh. and Gegenophis carnosus Bedd. Proc. Nat. Inst. Sci. India 11:366-340.

Spengel, J. W. 1876. Das Urogenitalsystem der Amphibian. I. Theil, Der anatomische Bau des Urogenitalsystem. Arbeiten aus dem Zoologzootom. Inst. in Wurzburg 3:61-114.

Taylor, E. H. 1960. On the caecilian species Ichthyophis glutinosus and Ichthyophis monochrous with description of related species. Univ. Kansas Sci. Bull. 50:37-120.

Taylor, E. H. 1965. New Asiatic and African caecilians with a redescription of certain other species. Univ. Kansas Sci. Bull. 46:253-302.

Taylor, E. H. 1968. The caecilians of the world: A taxonomic review. Univ. Kansas Press, Lawrence, Kansas.

Taylor, E. H. 1969. A new family of African Gymnophiona. Univ. Kansas Sci. Bull. 48:297-305.

Taylor, E. H. 1970. The lateral-line sensory system in the caecilian family Ichthyophiidae (Amphibia: Gymnophiona). Univ. Kansas Sci. Bull. 48:861-868.

Tonutti, E. 1931. Beitrag zur Kenntnis der Gymnophionen. XV. Das Genitalsystem. Morph. Jahrb. 68:151-292.

Tonutti, E. 1932. Vergleichende morphologische Studien über Enddarm and Kopulationsorgane. Morph. Jahrb. 70:101-130.

Tonutti, E. 1933. Beitrag zur Kenntnis der Gymnophionen. XIX. Kopulationsorgane bis weiteren Gymnophionenarten. Morph. Jahrb. 72:155-211.

Vilter, V. and A. Vilter. 1960. Sur la gestation de la salamandre noire des Alpes, Salamandra atra Laur. C. R. Soc. Biol. Paris 154:290-291.

Vilter, V. and A. Vilter. 1964. Sur l'évolution des corps jaunes ovariens chez Salamandra atra Laur. des Alpes Vaudoises. C. R. Soc. Biol. Paris 158:467-461.

Vilter, V. and A. Lugano. 1959. Rechèrches sur le détèrminisme interne et externe du corps jaune gestatif chez le crapaud vivipare du Mont Nimba, le Nectophrynoides occidentalis de la Haute Guinee. C. R. Soc. Biol. Paris. 153:294-297.

Wake, M. H. 1967. Gill structure in the caecilian genus Gymnopis. Bull. So. Calif. Acad. Sci. 66 :109-116.

Wake, M. H. 1968. Evolutionary morphology of the caecilian urogenital system. Part I. The gonads and fat bodies. J. Morph. 126:291-332.

Wake, M. H. 1969. Gill ontogeny in Gymnopis. Copeia 1969:183-184.

Wake. M. H. 1970. Evolutionary morphology of the caecilian urogenital system. Part II. The kidneys and urogenital ducts. Acta Anatomica 75:321-358.

Wake, M. H. 1972. Evolutionary morphology of the caecilian urogenital system. Part IV. The cloaca. J. Morph. 136:353-366.

Wake, M. H. 1974. The comparative morphology of the caecilian lung. Anat. Rec. 178:483.

Wake, M. H. 1976. The development and replacement of teeth in viviparous caecilians. J. Morph. 148:33-64.

Wake, M. H. 1977. Fetal maintenance and its evolutionary significance in the Amphibia: Gymnophiona. J. Herp. (in press).

Wiedersheim, R. 1879. Die Anatomie der Gymnophionen. Verlag von Gustav Fischer, Jena.

Xavier, F., M. Zuber-Vogeli and Y. LeQuang Trong. 1970. Rechèrches sur l'activité endocrine de l'ovaire de Nectophrynoides occidentalis Angel (Amphibien anoure vivipare). Gen. Comp. Endo. 15:425-431.

ASPECTS OF THE REPRODUCTIVE BIOLOGY OF JAPANESE ANURANS

Toshijiro Kawamura and Midori Nishioka

Laboratory for Amphibian Biology, Faculty of Science, Hiroshima University, Higashisenda-cho, Hiroshima, Japan

Seventeen anurans occur on the four main islands of Japan and on the Islands of Tsushima situated between Japan and Korea. Two of these, *Rana nigromaculata* and *R. brevipoda*, are pond frogs allied to *Rana esculenta*.[1] Six other species, *Rana japonica*, *R. ornativentris*, *R. tagoi*, *R. chensinensis*, *R. tsushimensis* and *R. dyboswkii*, are brown frogs allied to *Rana temporaria*, although *Rana tagoi* differs from the others in many respects. In addition to these species, two small, dark-colored frogs, *Rana rugosa* and *R. limnocharis*, and a large American frog, *Rana catesbeiana*, are widely distributed in Japan. In other genera *Hyla arborea japonica*, *Bufo bufo japonicus*, *Bufo torrenticola* and three species of *Rhacophorus* occur. Although it is generally believed that *Bombina orientalis* occurs on Tsushima, no one has recently found it there. During the last ten years, we have made a number of studies of reproductive biology using most of the Japanese anurans and some foreign ones.

1. The sex of gynogenetic diploids

A fundamental problem in amphibian reproductive biology is to determine which is heterogametic in each species. As sex is essentially determined by genetic factors in all the species hitherto examined, these species seem to have functional sex chromosomes. However, visibly distinct sex chromosomes have seldom been observed in amphibians. The presumption that the male is heterogametic in *Rana nigromaculata* was made on the basis of the sex of partheno-

[1]See Berger's paper, this volume, for a redefinition of the *R. esculenta* complex in Europe.

genetically produced frogs (Kawamura 1939). Of 23 frogs, 16 were females with normal ovaries, two were females with degenerating ovaries, four were transforming into males and one was a male with an almost typical testis. If females are heterogametic in this species, one-half of the parthenogenetically produced frogs should be males. Male heterogamety was also presumed by Moriwaki (1959) in *Rana japonica* on the basis of the sex of the offspring of a parthenogenetically produced male frog, although seven of his 16 mature parthenogenetically produced frogs were males. As all the offspring produced by mating the parthenogenetically produced male with a normal female were females, it is certain that this male was a genetic female.

The diploidy of the frogs parthenogenetically produced by Kawamura (1939) was probably due to duplication of the haploid chromosomes during the first cleavage mitosis without an accompanying cytoplasmic division. In contrast, those obtained by Moriwaki (1959) were probably derived from eggs whose pronucleus fused with the second polar body nucleus retained in the egg as an effect of refrigeration. In both experiments, only a small number of parthenogenetically produced diploids resulted. It was later found that abundant gynogenetic diploids could be readily produced. Using this method, we obtained many interesting strains of *Rana nigromaculata*, *R. brevipoda*, *R. japonica*, *R. tsushimensis*, *R. rugosa*, *Hyla arborea japonica* and Korean *Bombina orientalis*. Our method for diploid gynogenesis will be briefly described. Unfertilized oviductal eggs are inseminated with UV-irradiated sperm from a different species. The eggs are refrigerated at 0 ~ 3° C for 1 ~ 4 hours or heat-shocked at 36 ~ 38° C for 3 ~ 8 minutes about 20 minutes after insemination. If a different species, spermatozoa of which could penetrate easily into the eggs, is not available, conspecific males are used. In this case, it is necessary to test the effectiveness of UV-irradiation in control eggs which are not refrigerated or heat-shocked. The eggs of brown frog species are usually heat-shocked, as they are much less affected by low temperatures, while those of the other species are generally refrigerated.

Throughout this paper (except where noted otherwise) juvenile hermaphrodites whose gonads were transforming from ovaries into testes are counted as males as they would have presumably matured into males.

a. *Rana nigromaculata*

Gynogenetic diploids were abundantly produced from 26 field-collected *Rana nigromaculata* in four years (Table 1). Eggs were inseminated with irradiated sperm of *Rana pipiens* or *R. brevipoda*. While 1286 (86%) of 1500 eggs became metamorphosed frogs in the control series, only 1463 (30%) of 4877 eggs did so in the experi-

Table 1. Sex of gynogenetically produced diploids using eggs of Rana nigromaculata. Experimental series were inseminated with UV-irradiated sperm of male R. pipiens or R. brevipoda. The number of juvenile hermaphrodites counted as males is in parentheses.

	Year or age (years)	Series	No. of female parents	No. of frogs whose sex was examined: Total	♀	♂	%(♂)
By year	1964	Cont.	3	201	102	99	49.3
		Exp.	3	252	249	3(2)	1.2
	1965	Cont.	7	312	157	155	49.7
		Exp.	7	413	359	54(6)	13.1
	1966	Cont.	6	331	164	167	50.5
		Exp.	6	518	511	7(2)	1.4
	1968	Cont.	10	440	225	215	48.9
		Exp.	10	279	268	11(1)	3.9
	Total	Cont.	26	1284	648	636	49.5
		Exp.	26	1462	1387	75(11)	5.1
By age at death	0 ~ 1	Cont.		147	79	68	46.3
		Exp.		598	578	20(11)	3.3
	1 ~ 2	Cont.		553	279	274	49.5
		Exp.		395	383	12	3.0
	2+	Cont.		584	290	294	50.3
		Exp.		469	426	43	9.2
	Total	Cont.		1284	648	636	49.5
		Exp.		1462	1387	75(11)	5.1

mental series. In the course of rearing, one frog in the experimental and two frogs in the control series were lost without being sexed or prior to metamorphosis. When the results for the four years were totalled, 50% of the 1284 frogs in the control series were males, but only 5% of the 1462 frogs in the experimental series were males and hermaphrodites. The gynogenetic diploids were inferior in viability. In the experimental series, 59% of the metamorphosed frogs lived for one year or more, while 88% did so in the control series.

Of the 26 females used in producing diploids gynogenetically, 25 were similar to each other in the sex ratio of offspring in the experimental series as well as in the control. One female (No. 11) used in 1965, however, was quite different from the others in this respect, although she produced nearly an equal number of males and females in the control series. Of 18 gynogenetic frogs from this female which died within one year after fertilization, 14 were females, three were hermaphrodites and one was a male; and of 44 frogs that lived for one year or more, 43 were males and one was a female. In other words, 47 of the 75 males and hermaphrodites produced from 26 female parents came from a single female. Moreover, all 41 diploids produced gynogenetically from this female that lived for two years or more were males. In the experimental series of this female (No. 11), 83% of 128 eggs cleaved normally. These eggs were almost the highest in viability among those of the 26 experimental series; 62 (48%) of the total eggs metamorphosed normally, 34% became one-year-old frogs, and 32% lived for two years or more. However, these frogs grew very slowly after metamorphosis and were extremely stunted even at two years.

If the offspring of female No. 11 were excluded from those of 26 females, there remained 1372 females, 8 juvenile hermaphrodites and 20 males among 1400 gynogenetic diploids in total. Two percent of these were males and hermaphrodites.

b. Rana brevipoda

Diploids were produced gynogenetically from 19 females in four years (Table 2). In the 1966 experimental series, 82 of the metamorphosed frogs died within one year after fertilization; 79 of these were females, two were transforming into males and one was a male. All the other 62 frogs that lived for one year or more were females except a single male. All seven hermaphrodites for 1972 were found among the frogs which died by the age of one year.

In 1968, second-generation gynogenetically produced diploids were obtained from three female diploids that were gynogenetically produced in 1966 (Table 2). In 1970, third-generation gynogenetically produced diploids were repeatedly produced from three second-generation female diploids produced gynogenetically from a gynogenetically produced diploid in 1968. In both years some eggs from the gynogenetically produced females were inseminated with sperm of males collected from the field and were used as controls.

Among the 1798 gynogetically produced diploids produced in four years, there were 1773 females, 11 hermaphrodites and 14 males; thus, 1.4% were males and hermaphrodites. The gynogenetically produced frogs were inferior in viability. In the experimental series of 1966, 1968 and 1970, 43, 58 and 46% respectively of the metamor-

Table 2. Sex of gynogenetically produced diploids using eggs of Rana brevipoda. Eggs in the experimental series were inseminated with UV-irradiated sperm of Rana pipiens or R. nigromaculata. GD1 females were produced gynogenetically from wild-caught females, while GD2 females were produced gynogenetically from female GD1. The number of juvenile hermaphrodites counted as males is presented in parentheses.

		Female parents		% metamorphosed	No. of metamorphosed frogs whose sex was examined			
Year	Series	Origin	No.		Total	♀	♂	%(♂)
1966	Cont.	Wild	3	78	145	74	71	49.0
	Exp.	Wild	3	29	144	140	4(2)	2.8
1968	Cont.	GD1	3	81	175	86	89(2)	50.9
	Exp.	GD1	3	24	182	181	1	0.5
1970	Cont.	GD2	3	71	88	41	47(1)	53.4
	Exp.	GD2	3	51	186	181	5(2)	2.7
1972	Cont.	Wild	10	88	1068	533	535	50.1
	Exp.	Wild	10	29	1286	1271	15(7)	1.2
Total	Cont.		19	80	1476	734	742(3)	50.3
	Exp.		19	33	1798	1773	25(11)	1.4

phosed frogs lived more than one year, while 71, 66 and 92%, respectively, did so in the control series.

c. Rana japonica, R. tsushimensis and R. rugosa

Of the six brown frog species found in Japan, Rana japonica and R. tsushimensis were used to produce diploids gynogenetically. Hybrids between these two species are inviable. Diploids were produced gynogenetically from seven female R. japonica and three female R. tsushimensis (Table 3). The eggs of one species were activated with UV-irradiated spermatozoa of the two species. In the experimental series of R. japonica, 784 (38%) of 2048 eggs developed into metamorphosed frogs, as did 679 (80%) of 844 control eggs. A few of these were lost without their sex being determined; of the remainder, 8% of the gynogenetically produced frogs were males and hermaphrodites, and 49% of the controls were males. The percentage of males and hermaphrodites among diploids gynogenetically produced by activation with irradiated spermatozoa of R. tsushimensis was

Table 3. Sex of gynogenetically produced diploids using eggs of R. japonica, R. tsushimensis and R. rugosa. Eggs in the experimental series were inseminated with UV-irradiated sperm of R. japonica (1), R. tsushimensis (2) or R. rugosa (3). The number of juvenile hermaphrodites counted as males is presented in parentheses. The Rana rugosa experiments were done by Yano in the authors' laboratory.

Species	Series	No. of female parents	No. of frogs whose sex was examined: Total	♀	♂	%(♂)
R. japonica	Cont. 1	5	462	235	227	49.1
	Exp. 1(1)	5	445	401	44(15)	9.9
	Cont. 2	2	200	101	99	49.5
	Exp. 2(1)	2	179	169	10(5)	5.6
	Exp. 2(2)	2	157	149	8(4)	5.1
	Total Cont.	7	662	336	326	49.2
	Exp.	7	781	719	62(24)	7.9
R. tsushimensis	Cont. 1	3	135	67	68	50.4
	Exp. 1(1)	3	231	199	32(12)	13.9
	Exp. 1(2)	2	151	140	11(5)	7.3
	Total Cont.	3	135	67	68	50.4
	Exp.	3	382	339	43(17)	11.3
R. rugosa	Cont. (3)	10	392	206	186	47.4
	Exp. (3)	10	414	388	26(7)	6.3

similar to that obtained by irradiated spermatozoa of R. japonica (Table 3).

In the experimental series of Rana tsushimensis, 382 (33%) of 1143 eggs became metamorphosed frogs as did 167 (78%) of 215 in the control. The percentage of males including hermaphrodites in diploids gynogenetically produced by activation with irradiated spermatozoa of R. japonica was considerably higher than those from activation with spermatozoa of R. tsushimensis. Fourteen percent of the former and 7% of the latter frogs were males or hermaphrodites; in total, 11% were males or hermaphrodites.

Diploids from ten female Rana rugosa were produced gynogenetically in our laboratory by Yano. Of 5661 eggs refrigerated at 0 ~ 2° C for two hours after insemination with conspecific irradiated sperm, 866 (15%) developed into metamorphosed frogs. In the control series, 599 (70%) of 860 eggs became frogs. Immediately after metamorphosis, 414 frogs in the experimental series as well as 392 in the control were killed to examine their sex (Table 3). In the experimental series, 6% were males and hermaphrodites, whereas 47% of the control series were males.

d. Hyla arborea japonica and Bombina orientalis

Diploids from three female Hyla arborea japonica were produced gynogenetically in our laboratory by Nishiguchi (Table 4). Of 1498 eggs refrigerated after insemination with UV-irradiated sperm of the same subspecies, 108 (7%) developed into metamorphosed frogs. One of the latter was lost without the sex being determined. Of the remainder, only one was a male and 106 were females; no juvenile hermaphrodites were found. In the control series, 191 (70%) of 272 eggs became metamorphosed frogs. Of 184 frogs sexed at metamorphosis, 46% were males. Thirty-eight of the gynogenetic females had extremely degenerated ovaries; however, these showed no sign of change into testes. The single male in the experimental series seemed to derive from an egg inseminated with a spermatozoon on which UV-irradiation was unexpectedly ineffective, as sex-reversal from a female into a male does not seem to occur in this form (see below).

Table 4. Sex of gynogenetically produced diploids using eggs of Hyla arborea japonica and Bombina orientalis. Eggs of H. a. japonica and those of B. orientalis were inseminated with UV-irradiated sperm of H. a. japonica and B. orientalis, respectively. No juvenile hermaphrodites were found. The Hyla arborea japonica and Bombina orientalis experiments were done by Nishiguchi and Ueda, respectively, in the authors' laboratory.

Species	Series	No. of female parents	No. of frogs whose sex was examined			
			Total	♀	♂	%(♂)
H. a. japonica	Cont.	3	184	99	85	46.2
	Exp.	3	107	106	1	0.9
B. orientalis	Cont.	14	427	208	219	51.3
	Exp.	14	129	127	2	1.6

Diploids in Bombina orientalis were produced gynogenetically in our laboratory by Ueda (1971) from 14 females (Table 4). Of 90 gynogenetically produced frogs which died or were killed shortly after metamorphosis, 89 were females and one was a male, while there was an equal number of males and females among 266 control frogs. At sexual maturity, 38 frogs were females and one was a male in the experimental series, while 75 frogs were females and 86 were males in the control. Four mature females produced from one of the 14 females in the experimental series had degenerated ovaries. The two males found in the experimental series seemed to derive from eggs inseminated with spermatozoa on which UV-irradiation was ineffective, as their testes were completely normal and the mature male had no traces of Müllerian ducts.

2. Effects of sex hormones

a. Rana nigromaculata

That sex reversal of genetic females into males can be induced by injection of the androgen testosterone propionate (Enarmon, Teikokuzoki, Inc.) and that females are almost exclusively produced from masculinized females by mating with normal females has been reported in Rana japonica by Kawamura and Yokota (1959). The same phenomena were also observed in Rana nigromaculata. Each of 300 tadpoles was injected with 0.25 mg of testosterone propionate dissolved in 0.01 cc of oil solvent at the stage of the appearance of the first-toe rudiment. Many experimental tadpoles died, and only 170 completed metamorphosis, while 96 of 100 control tadpoles metamorphosed normally. Within one month after metamorphosis, 71 frogs in the experimental series died or were killed. Of these, only nine were females with normal ovaries, 18 were females with underdeveloped ovaries, 13 were hermaphrodites and 31 were males with normal testes. Of 45 control frogs, 21 were females with normal ovaries, three were females with underdeveloped ovaries, two were hermaphrodites and 19 were males with normal testes. In the experimental series, 64 frogs died before sexual maturity. Three of them were females with normal ovaries, 11 were females with underdeveloped ovaries and 50 were males with normal testes, while 15 of 30 control frogs were females with normal ovaries and 15 were males with normal testes. The remaining frogs in the experimental and control series were sexually mature at the age of two years. In the experimental series, six mature frogs were females and 29 were males, while 10 were females and 11 were males among controls.

Five experimental males and five control males were mated with normal females collected from the field in 1968. The offspring were killed and sexed one month after metamorphosis (Table 5). In the controls, males and females of nearly the same number were obtained from each male parent. Three of the five male parents in

Table 5. Sex ratios in the offspring of five males derived from androgen-injected tadpoles of Rana nigromaculata.

Series	Parents: Female No.	Parents: Male No.	No. of metamorphosed frogs	No. of offspring whose sex was examined one month after metamorphosis: Total	♀	♂	⚥
Control	Wild 1	Wild 1	45	45	22	23	-
	Wild 2	Wild 2	62	62	32	30	-
	Wild 3	Wild 3	73	71	35	36	-
	Wild 4	Wild 4	54	54	27	27	-
	Wild 5	Wild 5	60	60	29	31	-
Experimental	Wild 1	Androgen 1	123	113	56	57	-
	Wild 2	Androgen 2	96	89	46	43	-
	Wild 3	Androgen 3	105	102	99	0	3
	Wild 4	Androgen 4	110	105	52	53	-
	Wild 5	Androgen 5	127	114	103	0	11

the experimental series produced also an almost equal number of males and females, but the other two produced no males. One produced 99 females and three hermaphrodites, the other 103 females and 11 hermaphrodites. Thus, the latter two males were sex-reversed genetic females.

b. Hyla arborea japonica

The effects of sex hormones on the sex differentiation of Hyla arborea japonica were reported by Takahashi (1958, 1959). Some of the genetic males raised in an aqueous solution of estradiol turned into females, but no genetic females kept in aquarium water with methyltestosterone changed their sex, although the cortices of their ovaries were somewhat suppressed. Such a feminization of genetic males in the Japanese tree-frog was ascertained by Abe in our laboratory. He injected estradiol propionate (Ovahormon Dept., Teikoku-zoki Inc.) in 179 tadpoles and examined the gonads one month after metamorphosis; 44 of 52 were females, two were hermaphrodites and six were males. In the control series, 37 of 80 frogs were females and 43 were males. However, 22 of the females in the experimental series had underdeveloped ovaries, while only two had such ovaries in the control series.

In 1975, we carried out the same experiments as those done by Abe. The tadpoles used for the experiments were those which developed from eggs laid by two females. The offspring of female No. 1

were divided into three groups. Each tadpole belonging to one group was injected with 0.25 mg testosterone propionate at the stage of appearance of the first-toe rudiment. Each of those belonging to another group was injected with 0.05 mg estradiol dipropionate (Ovahormon Dept., Teikokuzoki Inc.) at the same stage. The remaining group was reared as uninjected controls. Within one month after metamorphosis, all the frogs in the experimental and control series died or were killed and their sex was examined (Table 6). In the control series, 46% were females. In the series given testosterone propionate, 38% were females; in contrast, 64 (97%) of 66 frogs derived from estrogen-injected tadpoles were females.

The sex of the progeny of female No. 2 was very interesting. The tadpoles were divided into five groups. In two of these groups, testosterone propionate or estradiol dipropionate was injected into tadpoles at the stage of appearance of the first-toe rudiments. In two other groups, the above steroid hormones were injected twice, first at the stage of appearance of the first-toe rudiments and next within a week after the completion of metamorphosis. The tadpoles of the remaining group were reared as uninjected controls. Of 128 control frogs that died or were killed within ten days after metamorphosis, 23% were females (Table 6). Of 36 control frogs at sexual maturity only 25% were females. Thus, about 25% of the

Table 6. Effect of sex hormones on sex ratios in *Hyla arborea japonica*. Testosterone propionate (T) or estradiol dipropionate (E) was injected once (T1, E1) or twice (T2, E2).

Series	No. of tad-poles	No. of metamor-phosed frogs	No. of juvenile frogs					No. of mature frogs			
			♀N	♀U	⚥	♂	%♀	♀N	♀U	♂	%♀
Control 1	40	37	16	1	0	20	45.9				
Exp. 1, T1	80	61	16	2	3	27	37.5				
Exp. 1, E1	100	70	29	35	2	0	97.0				
Control 2	200	179	29	1	3	95	23.4	9	0	27	25.0
Exp. 2, T1	350	161	17	9	7	82	22.6	16	1	27	38.6
Exp. 2, T2		130	36	1	2	45	44.0	13(7)*	0	23	36.1
Exp. 2, E1	450	235	118	23	21	0	87.0	40	2	21	66.7
Exp. 2, E2		130	57	8	0	0	100	37	2	24	61.9

N = normal ovaries; U = underdeveloped ovaries.
*seven females had male secondary sexual characters.

control frogs were females. This seemed to show that the female parent (No. 2) was a sex-reversed male (XY).

In the experimental series in which androgen was injected only once at the tadpole stage, 23% of 115 frogs that died or were killed within ten days after metamorphosis were females mostly with normal ovaries. At the sexually mature stage, 39% of 44 frogs in the same experimental series were females. In the experimental series in which estrogen was injected only once at the tadpole stage, 87% of 162 frogs that died or were killed within ten days after metamorphosis were females mostly with normal ovaries. At sexual maturity, 67% of 63 frogs in the same experimental series were females.

In the experimental series in which androgen was injected twice, 44% of 84 frogs that died within one month after metamorphosis were females. At the sexually mature stage, 36% of 36 frogs in the same experimental series were females. Seven of these had large vocal sacs, a male secondary sexual character, although they had normal ovaries. This sex-reversal in appearance was probably attributable to the second injection of testosterone propionate. In the experimental series in which estrogen was injected twice, all the frogs that died within one month after metamorphosis were females. However, 62% of 63 frogs were females at sexual maturity.

These results clearly indicate that reversal of genetic males into females occurs by administration of estrogen in Hyla arborea japonica. Most of the sex-reversed males keep their phenotypic sex up to sexual maturity, but others seem to return to their genetic sex. However, sex reversal of genetic females into males does not seem to occur by administration of androgen.

c. Bombina orientalis

Ueda (1977) injected testosterone propionate or estradiol benzoate into tadpoles of Bombina orientalis. Moreover, he injected androgen together with a pituitary suspension from Rana nigromaculata in order to increase the effects of this sex hormone. He also raised tadpoles of B. orientalis in aquarium water with the same androgen or estrogen. In another experiment, he transplanted a piece of the testis or ovary taken from a young Bombina into the coelom of each tadpole.

No sex reveral occurred in any cases. Males and females of nearly equal number were found at the stage immediately after metamorphosis in all the experimental series. Accordingly, it is believed that the genetic sex-differentiating mechanism is very stable in Bombina orientalis.

3. YY males in the field

As sex reversal of genetic males into females occurs in Hyla arborea japonica by administration of estrogens, it is probable that there are some sex-reversed females in the field. One such female has actually been discovered. If genetic males are heterogametic (XY) for sex chromosomes or sex-determining genes, YY males should also be observed in the field as the result of crossing between a normal male (XY) and a feminized male (XY), provided that YY males survive in the field.

Of 14 mature albino tree-frogs used for mating experiments during the past ten years, three males were considered to be YY (Nishioka and Ueda 1977). Two of the three were collected from Kake, Hiroshima Prefecture, while the other was from Yamaguchi, Yamaguchi Prefecture. The two collecting stations are about 100 km apart and separated by mountains. These male albinos were collected at the tadpole stage and reared to sexual maturity. The offspring produced from them by mating with wild-type females collected from the field were all males. Some of the male offspring were mated with female albinos as well as with wild-type females. As the result, males and females were produced in nearly a 1:1 ratio in all the matings. From these mating experiments, it was confirmed that the three male albinos collected from Kake and Yamaguchi were YY, and at the same time that the males were heterogametic in Hyla arborea japonica.

4. The sex of haploids and polyploids

The sex of 18 haploid frogs and metamorphosing tadpoles of Rana nigromaculata has been reported by Miyada (1960). Fourteen of them were females, one was a hermaphrodite and three were males. Many haploid Rana rugosa were produced by Yano in our laboratory. Eggs were activated with UV-irradiated sperm of the same species. The haploidy of the tadpoles developed from pseudofertilized eggs was confirmed by the tail-tip method. The haploid Rana rugosa were much superior in viability to the haploid Rana nigromaculata. Of 2691 eggs, 2519 (94%) cleaved normally following insemination with UV-irradiated spermatozoa, 132 (5%) attained the metamorphosing stage (stage XX) and 94 (3.5%) completed metamorphosis. When the sex of 108 of these metamorphosing or metamorphosed frogs was examined, it was found that five were females with normal ovaries, 25 were transforming into males, three were males with rudimentary testes and 75 were males with normal testes. When the juvenile hermaphrodites were counted as males, 95% of the total frogs were males. As the normal males are heterogametic in Rana rugosa as well as in Rana nigromaculata, the haploids produced by pseudofertilization are genetic females, having an X-chromosome. Thus, the haploid males are sex-reversed females.

Six haploids produced by Yano (1977) lived for five months or more. Two of them, both males, died at the age of about five months. Another haploid, an immature female, died at the age of one year. The remaining three haploids were mature females at the age of one year and two of them laid many eggs. Almost all the eggs obtained from one haploid female cleaved normally after insemination with sperm from a normal male, but the cleaved eggs died simultaneously before the gastrula stage.

We have studied the sex and reproductive ability of polyploid Japanese amphibians since 1940. Based on the sex of triploids, there are two kinds of anuran species. In one of them, nearly equal numbers of males and females are produced if the eggs are not overmatured, as in Rana nigromaculata (Kawamura 1941), while in the other, almost all the triploids become males, as in Rana pipiens (Humphrey et al. 1950) and Rana japonica (Kawamura and Tokunaga 1952). In Rana rugosa, the sex of 200 triploids which had been produced from fertilized eggs by the refrigeration method was examined by Yano (1975) within one month after metamorphosis. It was found that seven were transforming into males and 193 were males with normal testes; no females were observed. Unlike Rana rugosa, Nishiguchi found that 18 of 27 triploid Hyla arborea japonica produced in our laboratory from fertilized eggs by the refrigeration method were females with normal ovaries; the other nine were males with normal testes; there were nearly an equal number of males and females in the control series. Sex reversal from females into males does not seem to occur in the Japanese tree-frog, although some XXY individuals may be differentiated as females. The sex of triploids in Bombina orientalis was examined by Ueda (1971). Triploids were produced from fertilized eggs by the refrigeration method. Of 80 triploids, 38 were females and 42 were males, while 66 were females and 65 were males among the 131 controls. Since nearly equal numbers of males and females occur in triploids and, moreover, gynogenetically produced diploids are all females, the males of this species are believed to be heterogametic.

While triploid females of Bombina orientalis were quite sterile, some of the triploid males obtained by Ueda were slightly fertile and produced abnormal embryos when mated with diploid females, just as in Rana nigromaculata (Kawamura 1951).

We have found autotetraploids and amphidiploids in two groups of Japanese frogs (Kawamura and Nishioka 1960, 1963, 1967). Both males and females are found among autotetraploids of Rana nigromaculata as well as among allotetraploid hybrids between R. nigromaculate and R. brevipoda. The allotetraploid male and female hybrids are relatively fertile compared to diploid hybrids, in which males are almost completely sterile. Autotetraploid males are also fertile to some extent. Unlike triploid females, autotetraploid females can produce mature eggs. Both males and females are found among the allotetraploid hybrids between female Rana japonica and male R. ornativentris; some of the allotetraploid males are slightly fertile. In these respects, the allotetraploids in brown frogs are quite different from the diploid hybrids, which are all sterile males.

5. Effects of irradiation

Effects of irradiation with X-rays (X) or neutrons (N) on the sex of offspring were examined in Rana japonica and R. nigromaculata. First-generation offspring were produced from irradiated spermatozoa or unfertilized eggs by fertilizing with untreated gametes. Second-generation offspring were produced by mating males and females of the first-generation with frogs collected from the field. Third- and fourth-generation offspring were produced by the same procedure.

a. Rana japonica

Spermatozoa and unfertilized eggs were exposed to 150 ~ 250 rads of X-rays or 100 ~ 250 rads of neutrons. The sex of frogs in each of the four generations was examined about one month after metamorphosis as well as at the sexually mature stage (Table 7). Although there were always a few juvenile hermaphrodites with gonads transforming from ovaries into testes, they were counted as males, since they would ultimately become males.

In the first generation control series, 49 (46%) of the 106 juvenile frogs and 47 (51%) of the 92 mature ones were males. Four experimental groups were produced using X-irradiated spermatozoa (SX), X-irradiated eggs (EX), neutron-irradiated spermatozoa (SN) and neutron-irradiated eggs (EN). In each group 41 ~ 44% of 180 ~ 303 juvenile frogs and 54 ~ 61% of 23 ~ 123 mature frogs were males. Irradiation of gametes seems to delay differentiation of the gonads of some genetic males of the first-generation offspring and later to give rise to sex reversal in some genetic females.

Each group of experimental animals was divided into two in the subsequent generations; one included individuals produced from male offspring by mating with females collected from the field, while the other included those produced from female offspring by mating with males collected from the field. In the second and third generations, the juvenile frogs whose sex was determined were produced from several males or females in each series (Table 7). Mature frogs in the same series as that of the juvenile frogs were, however, produced from only one of the males or females. Accordingly, the sex ratio of mature frogs is not comparable with that of juveniles in the same series (Table 7).

Table 7. The number and percentage of male frogs in the first, second, third and fourth generation offspring derived from irradiated spermatozoa or eggs in *Rana japonica*. Each of the second to fourth generations was derived from males (m) in the previous generation by mating with normal females or from females (f) by mating with normal males. The sex was examined at the young or sexually mature stage.

Group of Series	1st				2nd				3rd				4th	
	Young		Mature		Young		Mature		Young		Mature		Young	
	Total	Male	Total	Male	Total	Male	Total	Male	Total	Male	Total	Male	Total	Male
Control	106	49(46.2%)	92	47(51.1%)	1078	540(50.1%)	131	67(51.1%)	896	445(49.7%)	75	38(50.7%)	1416	706(49.9%)
SX (m)	303	132(43.6%)	123	74(60.2%)	1284	650(50.6%)	89	47(52.8%)	629	252(40.1%)	26	17(65.4%)	1231	655(53.2%)
(f)					440	220(50.0%)	80	64(80.0%)	549	301(54.8%)	59	38(64.4%)	1508	819(54.3%)
EX (m)	180	76(42.2%)	23	14(60.9%)	881	537(61.0%)	48	23(47.9%)	500	267(53.4%)	28	18(64.3%)	1261	664(52.7%)
(f)					514	250(48.6%)	24	11(45.8%)	725	400(55.2%)	40	26(65.0%)	1055	589(55.8%)
SN (m)	301	131(43.5%)	72	39(54.2%)	1832	703(38.4%)	111	57(51.4%)	548	228(41.6%)	25	18(72.0%)	1492	766(51.3%)
(f)					437	259(59.3%)	30	22(73.3%)	569	300(52.7%)	68	37(54.4%)	1326	696(52.5%)
EN (m)	219	90(41.1%)	26	14(53.8%)	766	382(50.0%)	43	30(69.8%)	482	278(57.7%)	23	14(60.9%)	1347	706(52.4%)
(f)					529	291(55.0%)	27	13(48.1%)	771	407(52.8%)	35	18(51.4%)	1091	566(51.9%)

SX; derived from spermatozoa exposed to 150 or 250 rads of X-rays
EX; derived from eggs exposed to 150 or 250 rads of X-rays
SN; derived from spermatozoa exposed to 100 or 150 rads of neutrons
EN; derived from eggs exposed to 150 or 250 rads of neutrons

In the second generation, 38% of 1832 juveniles were males in one of the eight groups' series (Table 7). This 38% was of the total from seven male first-generation offspring which were derived from neutron-irradiated spermatozoa. Specifically, 11% and 31% of juveniles produced from two of the seven males were males, while nearly an equal number of males and females were produced from each of the other five males. In each of the three groups of experimental series in which 70 ~ 80% of mature frogs were males (Table 7), there was approximately an equal number of males and females about one month after metamorphosis. Accordingly, it is believed that female sex reversal occurred late in these series.

In the third generation, there were two groups of experimental series in which juvenile males were comparatively few, 40 and 42% (Table 7). The 40% was of the total from nine male second-generation offspring derived from X-irradiated spermatozoa. One of the nine males produced 50 females and 24 (32%) males, another produced 65 females and two males, and still another produced 47 females but no males. The remaining males produced nearly an equal number of males and females. The group of series in which 42% of the juveniles were males included frogs from eight male second-generation offspring derived from neutron-irradiated spermatozoa. One of these eight males produced 92 females and three males, while the other males produced nearly an equal number of males and females. When the sex of mature frogs was examined, males were remarkably more numerous than females in six of the eight groups of experimental series; 61 ~ 65% of 23 ~ 59 mature frogs were males. In the two other groups of experimental series, 51% of 35 and 54% of 68 mature frogs were males.

In the fourth generation, 51 ~ 56% of 1055 ~ 1508 juvenile frogs were males in the eight groups of experimental series, while 50% of 1416 juvenile frogs were males in the control series.

It is clear that irradiation affected the sex of first- to fourth-generation offspring. Although the effect of irradiation delayed the differentiation of gonads in some genetic males of the first-generation offspring and resulted in retaining an ovarian structure for an unusually long period, eventually it gave rise to sex reversal in some genetic females in each of the four generations. Accordingly, it seems certain that some of the males used to produce their offspring were sex-reversed females which produced only genetic females. The fact that there were more males than females in most groups of the experimental series in each generation seems to show that the effects of irradiation on sex differentiation of genetic females continue from generation to generation.

b. Rana nigromaculata

Irradiation of gametes with X-rays or neutrons in Rana nigromaculata gives nearly the same effect on sex differentiation of individuals derived from the gametes as in Rana japonica. Spermatozoa or unfertilized eggs were exposed to 90 ~ 240 rads of X-rays or 50 ~ 130 rads of neutrons and then fertilized with untreated gametes to produce the first-generation offspring. Second-generation offspring were produced from males or females of the first-generation offspring by mating with normal frogs collected from the field. The third- and fourth-generation offspring were produced in the same way (Table 8). The sex of frogs in the first generation was principally determined about three months after metamorphosis, while sex in the second, third and fourth generations was analyzed about one month after metamorphosis. Juvenile hermaphrodites transforming from females into males were counted as males. The sex of a comparatively small number of frogs was examined about five months after metamorphosis or at the sexually mature stage. Unlike Rana japonica, the sex ratio examined about one month after metamorphosis scarcely differed from that examined four months later or at the sexually mature stage.

First generation males were slightly greater in number than females in all the 12 experimental series except one, while the sex ratio in the control series was 50:50. In an exceptional series (EX-200; Table 8), 40 (41%) of 98 frogs were males. This low percentage may be attributed to the fact that male offspring were more adversely affected by the large dose of X-rays than females, although the frogs examined were far fewer than those in the other series.

Second generation males, obtained by mating male first-generation offspring with normal females from the field, were remarkably fewer than females in most experimental series; 29 ~ 42% of 152 ~ 548 frogs were males in nine of the 12 series, while 49% of 1539 were males in the control series (Table 8). In contrast, second generation males, obtained by mating female first-generation offspring with normal males from the field, were almost equal to or slightly more numerous than females in ten of the 12 experimental series. The sex ratio in the control series was 50:50. In the remaining two experimental series, 43% of 1160 and 40% of 802 frogs were males. The 43% was found in the series derived from spermatozoa exposed to 50 rads of neutrons. This series was produced from ten females of which one produced 96 females and 14 males and another 79 females and ten males. Each of the other eight females produced nearly an equal number of males and females. The experimental series in which the 40% were males was derived from spermatozoa exposed to 130 rads of neutrons. The frogs in this series were produced from eight females; one of the latter produced 123 females

Table 8. The number and percentage of male frogs in the first, second, third and fourth generation offspring derived from irradiated spermatozoa or eggs in *Rana nigromaculata*. The sex of frogs was determined one or three months after metamorphosis.

Series	1st		2nd				3rd				4th			
			From ♂		From ♀		From ♂,♂		From ♀,♀		From ♂,♂,♂		From ♂,♂,♀	
	Total	Male	Total	Male	Total	Male	Total	Male	Total	Male	Total	Male	Total	Male
Control	916	461(50.3%)	1539	760(49.4%)	653	328(50.2%)	441	218(49.3%)	492	239(48.6%)	452	222(49.1%)	675	338(50.1%)
SX-90	483	271(56.1%)	487	142(29.2%)	872	532(61.0%)	302	169(56.0%)	261	133(51.0%)	736	374(50.8%)	460	233(50.7%)
SX-170	355	203(57.2%)	292	120(41.1%)	586	343(58.5%)	184	94(51.1%)	240	122(50.8%)	755	382(50.6%)	505	259(51.3%)
SX-240	147	85(57.8%)	269	133(49.4%)	109	52(47.7%)	50	28(56.0%)	246	133(54.1%)	592	300(50.7%)	274	147(53.6%)
EX-90	485	273(56.3%)	853	413(48.4%)	616	363(58.9%)	172	87(50.6%)	355	184(51.8%)	720	367(51.0%)	410	229(55.9%)
EX-145	290	184(63.4%)	986	451(45.7%)	401	209(52.1%)	491	246(50.1%)	382	200(52.4%)	754	385(51.1%)	632	331(52.4%)
Ex-200	98	40(40.8%)	152	47(30.9%)	83	39(47.0%)	716	386(53.9%)	15	8(53.3%)	600	303(50.5%)	214	115(53.7%)
SN-50	485	269(55.5%)	182	52(28.6%)	1160	501(43.2%)	236	125(53.0%)	160	85(53.1%)	630	312(49.5%)	470	239(50.9%)
SN-90	260	143(55.0%)	488	191(39.1%)	579	301(52.0%)	207	115(55.6%)	293	158(53.9%)	333	170(51.1%)	361	190(52.6%)
SN-130	102	57(55.9%)	548	228(41.6%)	802	323(40.3%)	529	299(56.5%)	324	185(57.1%)	605	325(53.7%)	529	274(51.8%)
EN-50	553	307(55.5%)	421	169(40.1%)	831	421(50.7%)	218	127(58.3%)	290	149(51.4%)	475	241(50.7%)	220	112(50.9%)
EN-90	321	187(58.3%)	492	205(41.7%)	303	146(48.2%)	138	81(58.7%)	300	170(56.7%)	626	319(51.0%)	386	199(51.6%)
EN-130	178	98(55.1%)	404	127(31.4%)	144	75(52.1%)	208	107(51.4%)	250	132(52.8%)	513	260(50.7%)	535	283(52.9%)
Mean	3757	2117(56.3%)	5574	2278(40.9%)	6486	3305(51.0%)	3451	1864(54.0%)	3116	1659(53.2%)	7339	3738(50.9%)	4996	2611(52.3%)

Sx-90, 170, 240; derived from spermatozoa exposed to 90, 170 or 240 rads of X-rays
EX-90, 145, 200; derived from eggs exposed to 90, 145 or 200 rads of X-rays
SN-50, 90, 130; derived from spermatozoa exposed to 50, 90 or 130 rads of neutrons
EN-50, 90, 130; derived from eggs exposed to 50, 90 or 130 rads of neutrons

and 76 males, and another produced 111 females and 40 males. The above four females, which produced numerous females in the two experimental series, were peculiar in that the males among their offspring were almost inviable; very few survived to sexual maturity. In three experimental series, 59 ~ 61% of 586 ~ 872 frogs were males. It is very probable that sex reversal occurred in some genetic females as an effect of irradiation given to spermatozoa or unfertilized eggs.

In the third generation, four groups of offspring were produced in each experimental series; two of them are presented in table 8. One group contained frogs produced from male first- and second-generation offspring, while the other contained frogs which came from female first- and second-generation offspring. In the control series of the two groups, 49% of 441 or 492 frogs were males. In eight of the 12 experimental series of the first group, 53 ~ 59% of 50 ~ 716 frogs were males, while 50 ~ 51% of 172 ~ 491 frogs were males in the other four experimental series. In seven of the 12 experimental series of the second group, 53 ~ 57% of 15 ~ 324 frogs were males, while 51 ~ 52% of 240 ~ 382 frogs were males in the other five experimental series. Thus, it is probable that sex reversal occurred in some genetic females in the third generation.

In the fourth generation, two groups of offspring were produced in each experimental series. One group contained frogs which came from male first-, second- and third-generation offspring, while the other contained frogs which came from male first-, male second- and female third-generation offspring. In one of the 12 experimental series belonging to the former group, 54% of 605 frogs were males, while 50 ~ 51% of 333 ~ 755 frogs were males in the other eleven series. In the control series of this group, 49% of 452 frogs were males. In the other group, 53 ~ 56% of 214 ~ 535 frogs were males in five of the 12 experimental series, while 51 ~ 52% of 220 ~ 632 frogs were males in the remaining seven experimental series. In the control series of this group, the sex ratio was 50:50. Accordingly, sex reversal of some genetic females seems to occur even in the fourth-generation offspring.

In order to clarify the relationship of irradiation of gametes with sex reversal in the offspring, gynogentically produced diploids were produced from eggs of female first- and second-generation offspring in each experimental series (Table 9). When diploids were produced gynogentically from eggs of females raised from X-irradiated spermatozoa or eggs, 42 ~ 51% of 147 ~ 200 frogs, 48% on the average, were males in four experimental series. On the other hand, 14 ~ 21% of 158 ~ 274 frogs, 18% on the average, were males in series in which diploids were produced gynogenetically from eggs of females raised from neutron-irradiated spermatozoa or eggs. In the control series in which diploids were produced gynogenetically from eggs of females raised from non-irradiated spermatozoa and eggs, only three percent were males.

When third-generation offspring were produced from males of these gynogenetically produced diploids by mating them with normal females collected from the field (Table 9), 20 ~ 54% of 109 ~ 197 frogs, 41% on the average, were males in series from X-irradiated grandparental spermatozoa or eggs. In similar series from neutron-irradiated grandparental spermatozoa or eggs, 10 ~ 33% of 61 ~ 220 frogs were males. In the control series in which males of the control gynogenetically produced diploids were mated with normal females, 13% of 75 frogs were males. When female gynogenetically produced diploids were mated with male diploids gynogenetically produced, 42 ~ 59% of 34 ~ 264 frogs, 47% on the average, were males in series from X-irradiated grandparental spermatozoa or eggs. Only 27 ~ 30% of 60 ~ 141 frogs, 28% on the average, were males in series from neutron-irradiated grandparental spermatozoa or eggs. In the control series in which matings were made between males and females of the control gynogenetically produced diploids, only 3% of 111 frogs were males.

Second-generation diploids were produced gynogenetically from eggs of some of the gynogenetically produced females (Table 9); 55 ~ 59% of 33 ~ 221 frogs, 57% on the average, were males in series from X-irradiated great-grandparental spermatozoa or eggs, while 26% of 66 frogs and 36% of 58 frogs, 31% on the average, were males in two series from neutron-irradiated great-grandparental spermatozoa or eggs. In the control series in which second-generation diploids were produced gynogenetically from eggs of some of the control gynogenetically produced females, no males were produced.

The gynogenetically produced diploids as well as the progeny produced by gynogenetically produced males were all genetic females, since the female is homogametic in <u>Rana nigromaculata</u>. Accordingly, the male gynogenetically produced diploids and their male offspring were all sex-reversed females. These experiments show that X-irradiation of gametes produced sex reversal in females more frequently than neutron-irradiation. The larger the dose of X-irradiation, the more frequently the sex reversal occurred.

6. Reproductive isolation

a. Pond frogs

We have presented an outline of isolating mechanisms among different pond frog species (1973, 1975). Up to the present, Japanese <u>Rana nigromaculata</u> and <u>R. brevipoda</u>, Korean <u>Rana nigromaculata</u> and <u>R. plancyi chosenica</u>, Formosan <u>Rana plancyi fukienensis</u>, and European <u>Rana esculenta</u> (northern France), <u>R. lessonae</u> (Luxem-

Table 9. The number and percentage of male frogs in the second- and third-generations produced by diploid gynogenesis of eggs of first-generation females raised from irradiated spermatozoa or eggs in Rana nigromaculata. Diploids GD1 were produced gynogenetically from females of the first-generation offspring, while diploids GD2 were produced gynogenetically from female GD1.

Series	1st		2nd		3rd					
	Wild		GD1		Wild♀ x ♂ GD1		♀GD1 x ♂ GD1		GD2	
Parents	Total	Male	Total	Male	Total	Male	Total	Male	Total	Male
Control	916	461(50.3%)	356	10(2.8%)	75	10(13.3%)	111	3(2.7%)	55	0
SX-90	483	271(56.1%)	152	64(42.1%)	109	22(20.2%)	264	126(47.7%)	130	77(59.2%)
SX-170	355	203(57.2%)	200	102(51.0%)	114	48(42.1%)	80	41(51.3%)	33	18(54.5%)
EX-90	485	272(56.1%)	147	72(49.0%)	197	82(41.6%)	164	69(42.1%)	221	124(56.1%)
EX-145	290	184(63.4%)	153	76(49.7)	142	77(54.2%)	34	20(58.8%		
SN-50	485	269(55.5%)	274	46(16.8%)	111	11(9.9%)	141	38(27.0%)		
SN-130	102	57(55.9%)	266	53(19.9%)	180	59(32.8%)	60	16(26.7%)	58	21(36.2%)
EN-50	553	307(55.5%)	175	24(13.7%)	61	19(31.1%)	135	40(29.6%)		
EN-130	178	98(55.1%)	158	33(20.9%)	220	48(21.8%)	130	37(28.5%)	66	17(25.8%)

SX-90, 170; derived from spermatozoa exposed to 90 or 170 rads of X-rays
EX-90, 145; derived from eggs exposed to 90 or 145 rads of X-rays
SN-50, 130; derived from spermatozoa exposed to 50 or 130 rads of neutrons
EN-50, 130; derived from eggs exposed to 50 or 130 rads of neutrons

bourg), R. ridibunda ridibunda (Turkey) and R. ridibunda perezi (southern France) were used for crossing experiments. No pair of pond-frog species are completely isolated from each other by gametic isolation, although incomplete gametic isolation is usually found between Far Eastern and European species. All the different species examined hitherto are completely or incompletely isolated from one another by hybrid inviability or hybrid sterility (Tables 10 and 11).

The two Japanese pond frog species, Rana nigromaculata and R. brevipoda, are not isolated from each other by gametic isolation nor by hybrid inviability; usually, however, they have seasonal, ecological and behavioral differences that reduce hybridization among them. The males of reciprocal hybrids between them are almost completely sterile, differing from females which are fertile to a large extent. According to Moriya (1960), male and female hybrids between female R. brevipoda brevipoda and male R. brevipoda porosa are almost completely fertile, although male hybrids are somewhat abnormal in spermatogenesis. In Japan, R. nigromaculata is widely distributed on three of the four main islands, Honshu, Shikoku and Kyushu, although it is not found in the Kanto and Sendai plains nor in most of the areas along the Shinano River (Fig. 1). Rana brevipoda porosa occurs in these areas where R. nigromaculata is not found. In the Tokai and Kinki districts and in the areas along the shores of the eastern half of the Inland Sea (Fukuyama to Nagoya, Fig. 1), R. nigromaculata is sympatric with R. brevipoda brevipoda. Natural hybridization between the two species is rather frequent in local areas near the lower and upper Shinano River. At these localities, the range of R. brevipoda porosa comes in contact with that of R. nigromaculata. These localities are peculiar in that they are the coldest districts where R. brevipoda occurs; the area around the lower Shinano River is covered with heavy snow until late spring, the areas along the upper Shinano River are probably the highest above sea-level. While R. nigromaculata spawns earlier than R. brevipoda in other sympatric areas, the two species spawn almost simultaneously in these two colder regions. Accordingly, natural hybridization occurs most frequently in these districts.

It seems probable that R. brevipoda porosa is a stabilized population in which introgression of genes from R. nigromaculata has occurred since antiquity and reached such an extent that we can not distinguish the population from interspecific hybrids by outward appearance, and that this newly established population dislodged the R. nigromaculata from most of the areas along the Shinano River and then from the Kanto and Sendai plains. This introgression extends slightly to the population of R. brevipoda distributed in nearly half of the range where they are sympatric with R. nigromaculata, although it diminishes gradually with distance from the Kanto plain.

One of the Korean pond-frog subspecies, R. plancyi chosenica, is not isolated from Korean and Japanese R. nigromaculata by gametic

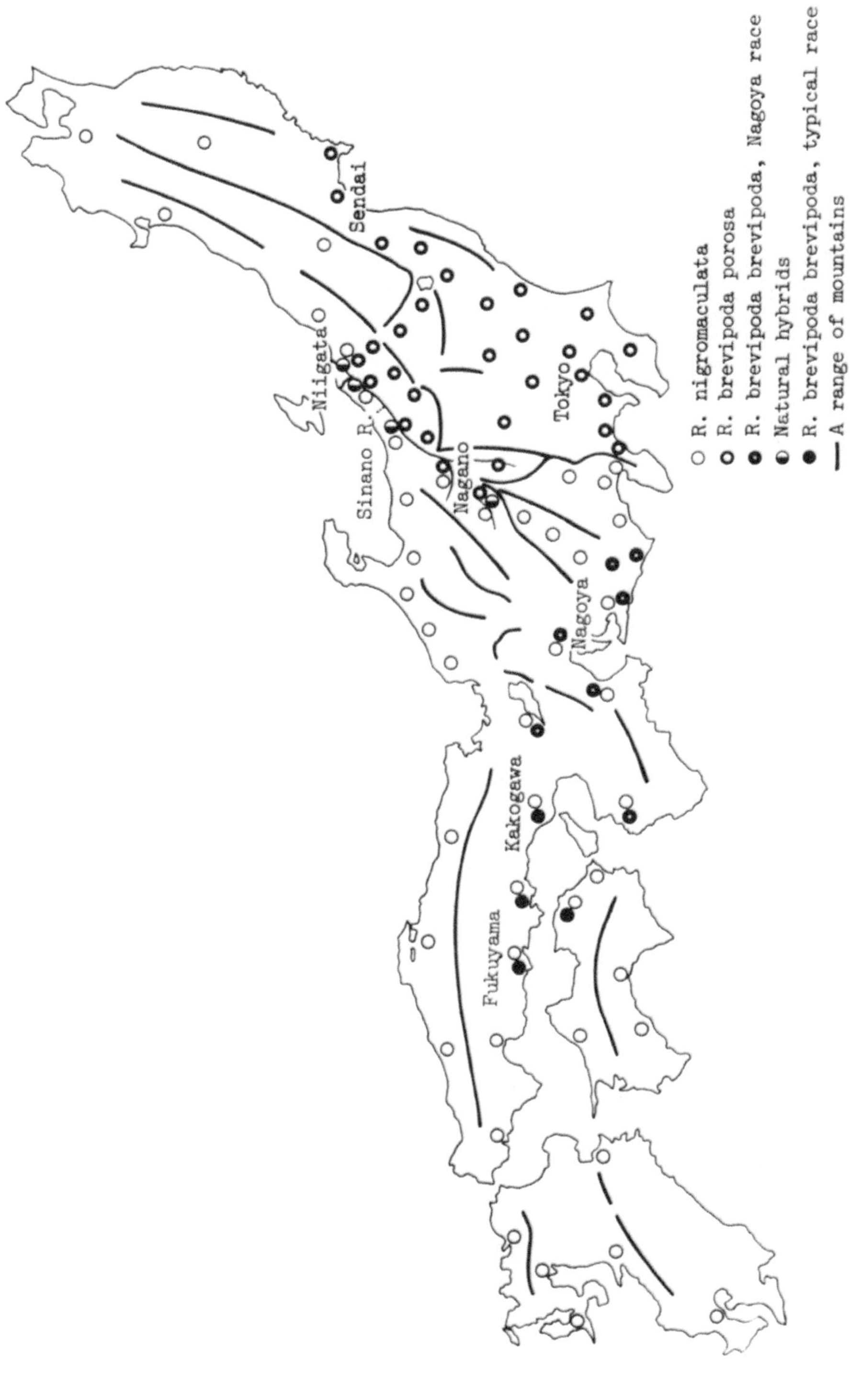

Fig. 1. Distribution of the pond frog group in Japan.

isolation nor by hybrid inviability. The males of reciprocal hybrids between these two species, however, are sterile to a slightly smaller extent than those of reciprocal hybrids between R. nigromaculata and R. brevipoda. The female hybrids are fertile to the same extent as those of the latter. The alliance between R. plancyi chosenica and R. brevipoda seems less intimate than that of the former with R. nigromaculata, since most of the females as well as the males of reciprocal hybrids between R. plancyi chosenica and R. brevipoda are completely sterile (Table 10).

Males of the Formosan subspecies, R. plancyi fukienensis, were mated with females of the two Japanese species. Most of the eggs of female R. nigromaculata are not fertilized with sperm of R. plancyi fukienensis, although almost all the fertilized eggs are viable and become metamorphosed frogs. Mature hybrids are nearly the same in reproductive capacity as reciprocal hybrids between R. nigromaculata and R. brevipoda or between R. nigromaculata and R. plancyi

Table 10. Isolating mechanisms between different pond frog species or subspecies distributed in Japan, Korea and Formosa. In gametic isolation, "incomplete" indicates that less than 50% of the eggs are fertilized. In hybrid inviability, "incomplete" indicates that more than 75% of fertilized eggs die in embryonic or tadpole stages.

Species or subspecies ♀	♂	Isolating mechanism: Gametic isolation	Hybrid inviability	Hybrid sterility ♀	♂
nigromaculata	brevipoda	No	No	No	Almost C
brevipoda	nigromaculata	No	No	No	Almost C
nigromaculata	chosenica	No	No	No	Almost C
chosenica	nigromaculata	No	No	No	Almost C
brevipoda	chosenica	No	No or C	Almost C	Almost C
chosenica	brevipoda	No	No	Almost C	Almost C
nigromaculata	fukienensis	I	No	No	Almost C
brevipoda	fukienensis	No	I or C		C
chosenica	fukienensis	No	I	No	Almost C
fukienensis	chosenica	No	I	No	Almost C

brevipoda, Rana brevipoda from Okayama Pref., Honshu
chosenica, Rana plancyi chosenica from Korea
fukienensis, Rana plancyi fukienensis from Formosa
nigromaculata, Rana nigromaculata from Hiroshima, Honshu

C, Complete I, Incomplete

chosenica. Male hybrids are almost completely sterile, while female hybrids are fertile to a large extent. In contrast, most of the eggs of female R. brevipoda are fertilized with sperm of R. plancyi fukienensis. The hybrids are barely viable, however. Two mature hybrids obtained hitherto were both completely sterile males.

Reciprocal crosses were made between R. plancyi chosenica and R. plancyi fukienensis. Although these two subspecies are not isolated by gametic isolation, reciprocal hybrids are mostly inviable in embryonic or tadpole stages. A small number of tadpoles complete metamorphosis and attain sexual maturity. While the hybrids of female R. plancyi chosenica and male R. plancyi fukienensis grow rapidly after metamorphosis and become large frogs, the reciprocal ones grow slowly and become dwarfs. Mature females of reciprocal hybrids spawn normally. However, mature males are probably almost completely sterile; the spermatozoa contained in the seminal tubules of their testes are distinctly fewer than those of the control males and mostly abnormal in shape and size.

The Japanese and Korean species are not isolated from two European species, Rana lessonae and R. ridibunda perezi, by gametic isolation (Table 11). While R. nigromaculata is completely isolated from R. lessonae and R. ridibunda perezi by hybrid inviability, R. brevipoda is incompletely isolated from the latter two. However, male and female hybrids between R. brevipoda and R. lessonae or R. ridibunda perezi are completely sterile. Although R. plancyi chosenica is not always completely isolated from R. ridibunda perezi by hybrid inviability, all the males and females of reciprocal hybrids are completely sterile. While there is no hybrid inviability between female R. plancyi chosenica and male R. lessonae, all or most of the reciprocal hybrids are inviable. Mature hybrids between R. plancyi chosenica and R. lessonae are incompletely sterile. When crosses were made between two female R. lessonae and a male R. plancyi fukienensis, all the hybrids were inviable, although most of the eggs were fertilized. We wish to add that the two European species R. lessonae and R. ridibunda perezi are either completely or contrarily scarcely isolated from each other by hybrid inviability, and male and female hybrids are incompletely sterile (Table 11).

b. Brown frogs

Eight brown-frog species, Rana japonica, R. ornativentris, R. chensinensis, R. tsushimensis, R. dybowskii, R. amurensis coreana, R. temporaria and R. arvalis, were used for crossing experiments in our laboratory. The first two were collected from Hiroshima Prefecture, R. chensinensis from Hokkaido, R. tsushimensis from Tsushima Island, R. dybowskii from Tsushima Island and Korea, R. amurensis coreana from Korea and the last two from Europe. Fifteen reciprocal combinations were made among the six Japanese and Korean species to examine the existence of isolating mechanisms (Table 12). It was found that each of the six species is completely isolated from

Table 11. Isolating mechanisms between Far Eastern and European pond frogs. In gametic isolation, "Incomplete" indicates that less than 50% of the total eggs are fertilized. In hybrid inviability, "incomplete" indicates that more than 75% of fertilized eggs die in embryonic or tadpole stages. In hybrid sterility, "incomplete" indicates that less than 25% of mature frogs are somewhat fertile.

Species or subspecies		Isolating mechanism			
		Gametic isolation	Hybrid inviability	Hybrid sterility	
♀	♂			♀	♂
nigromaculata	lessonae	No	C	-	-
lessonae	nigromaculata	No	C	-	-
nigromaculata	perezi	No	C	-	-
perezi	nigromaculata	No	C	-	-
brevipoda	lessonae	No	I	C	C
lessonae	brevipoda	No	I	C	C
brevipoda	perezi	No	I	C	C
perezi	brevipoda	No	I	C	C
chosenica	perezi	No	No or C	C	C
perezi	chosenica	No	I	C	C
chosenica	lessonae	No	No	I	I
lessonae	chosenica	No	I or C	I	I
lessonae	fukienensis	No	C	-	-
lessonae	perezi	No	No or C	I	I
perezi	lessonae	No	No or C	I	I

brevipoda, *Rana brevipoda* from Okayama Pref., Honshu
chosenica, *Rana plancyi chosenica* from Korea
fukienensis, *Rana plancyi fukienensis* from Formosa
nigromaculata, *Rana nigromaculata* from Hiroshima, Honshu
lessonae, *Rana lessonae* from Luxembourg
perezi, *Rana ridibunda* perezi from southern France

C, Complete I, Incomplete

all the others by gametic isolation, hybrid inviability or hybrid sterility. Some of these results have been reported by Kawamura and his collaborators since 1943.

Table 12. Isolating mechanisms between different brown frog species distributed in Japan and Korea. In gametic isolation, "incomplete" indicates that fewer than 50% of the total eggs are fertilized. In hybrid inviability, "incomplete" indicates that more than 75% of fertilized eggs die in embryonic or tadpole stages.

Species		Isolating mechanism			
		Gametic isolation	Hybrid inviability	Hybrid sterility	
♀	♂			♀	♂
japonica	ornativentris	No	No	-	C
ornativentris	japonica	C	-	-	-
japonica	chensinensis	No	No	-	C
chensinensis	japonica	I	I	-	C
japonica	tsushimensis	No	C	-	-
tsushimensis	japonica	No	C	-	-
japonica	dybowskii	No	C	-	-
dybowskii	japonica	No	I	-	C
japonica	coreana	No	C	-	-
coreana	japonica	No	C	-	-
ornativentris	chensinensis	I	I	-	C
chensinensis	ornativentris	No	I	-	C
ornativentris	tsushimensis	C	-	-	-
tsushimensis	ornativentris	No	C	-	-
ornativentris	dybowskii	C	-	-	-
dybowskii	ornativentris	No	No	-	C
ornativentris	coreana	C	-	-	-
coreana	ornativentris	No	C	-	-
chensinensis	tsushimensis	I	C	-	-
tsushimensis	chensinensis	No	C	-	-
chensinensis	dybowskii	No	C	-	-
dybowskii	chensinensis	No	I	-	C
chensinensis	coreana	C	-	-	-
coreana	chensinensis	No	C	-	-
tsushimensis	dybowskii	No	C	-	-
dybowskii	tsushimensis	No	C	-	-
tsushimensis	coreana	No	C	-	-
coreana	tsushimensis	No	C	-	-
dybowskii	coreana	No	C	-	-
coreana	dybowskii	No	C	-	-

C, Complete I, Incomplete

chensinensis, *Rana chensinensis* from Hokkaido
coreana, *Rana amurensis coreana* from Korea
dybowskii, *Rana dybowskii* from Tsushima Island
japonica, *Rana japonica* from Hiroshima, Honshu
ornativentris, *Rana ornativentris* from Hiroshima Pref., Honshu
tsushimensis, *Rana tsushimensis* from Tsushima Island

Eggs of R. ornativentris can not be fertilized by sperm of four species, that is, R. japonica, R. tsushimensis, R. dybowskii and R. amurensis coreana. Although some of the eggs of R. ornativentris are fertilized by sperm of R. chensinensis, most of the fertilized eggs are inviable at embryonic or tadpole stages. All the remaining R. ornativentris hybrids become completely sterile males. Eggs of R. chensinensis are not fertilized by sperm of R. amurensis coreana, while some of the eggs are fertilized by sperm of R. japonica and R. tsushimensis. While all the fertilized eggs between female R. chensinensis and male R. tsushimensis are inviable in embryonic or tadpole stages, some of the fertilized eggs between female R. chensinensis and male R. japonica develop into frogs. All these are completely sterile males, however.

In all of the other 22 combinations (Table 12) eggs of each species are mostly fertilized by sperm of the other. However, in 16 of these combinations, all the hybrids die in embryonic or tadpole stages. In the other six combinations, some or many of the fertilized eggs develop into sterile male frogs. It is remarkable that metamorphosed frogs are only produced in eight of the 30 combinations made among the six species, and that all the frogs become completely sterile males (Table 12).

Crossing experiments were performed between five Japanese species, R. japonica, R. ornativentris, R. chensinensis, R. tsushimensis and R. dybowskii, and two European species, R. arvalis and R. temporaria (Table 13). The Japanese species are completely isolated from the European species by gametic isolation, hybrid inviability or hybrid sterility, as also observed between different species distributed in Japan or Korea. Eggs of R. temporaria are not fertilized by sperm of R. japonica, R. ornativentris, R. chensinensis, R. tsushimensis, nor European R. arvalis. Eggs of R. arvalis are easily fertilized by sperm of R. japonica and R. ornativentris and some are fertilized by sperm of R. chensinensis, R. tsushimensis and R. dybowskii. On the other hand, eggs of R. japonica, R. chensinensis and R. tsushimensis are more or less fertilizable by sperm of R. arvalis and R. temporaria. A few of the eggs of R. dybowskii and none of the eggs of R. ornativentris are fertilized by sperm of R. arvalis.

The hybrids produced from crosses between female R. arvalis and males of R. dybowskii, R. chensinensis or R.tsushimensis, between female R. tsushimensis and males of R. arvalis or R. temporaria, between female R. chensinensis and male R. temporaria and between female R. dybowskii and male R. arvalis are inviable at embryonic or tadpole stages. Although the hybrids from crosses between female R. japonica and males of R. arvalis or R. temporaria, between female R. arvalis and males of R. japonica and R. ornativentris, between female R. ornativentris and male R. temporaria and between

Table 13. Isolating mechanisms between Japanese and European brown frogs. In gametic isolation, "incomplete" indicates that less than 50% of the total eggs are fertilized. In hybrid inviability, "incomplete" indicates that more than 75% of fertilized eggs die in embryonic or tadpole stages.

Species ♀	Species ♂	Isolating mechanism: Gametic isolation	Hybrid inviability	Hybrid sterility ♀	Hybrid sterility ♂
japonica	arvalis	No	No	-	C
arvalis	japonica	No	I	-	C
japonica	temporaria	No	No	-	C
temporaria	japonica	C	-	-	-
ornativentris	arvalis	C	-	-	-
arvalis	ornativentris	No	No	-	C
ornativentris	temporaria	No	I	-	C
temporaria	ornativentris	C	-	-	-
chensinensis	arvalis	I	I	-	C
arvalis	chensinensis	I	C	-	-
chensinensis	temporaria	No	C	-	-
temporaria	chensinensis	C	-	-	-
tsushimensis	arvalis	No	C	-	-
arvalis	tsushimensis	I	C	-	-
tsushimensis	temporaria	I	C	-	-
temporaria	tsushimensis	C	-	-	-
dybowskii	arvalis	I	C	-	-
arvalis	dybowskii	I	C	-	-
temporaria	dybowskii	C	-	-	-
arvalis	temporaria	I	No	-	C
temporaria	arvalis	C	-	-	-

C, Complete I, Incomplete

arvalis, Rana arvalis from East Germany
chensinensis, Rana chensinensis from Hokkaido
dybowskii, Rana dybowskii from Tsushima
japonica, Rana japonica from Hiroshima, Honshu
ornativentris, Rana ornativentris from Hiroshima Pref., Honshu
temporaria, Rana temporaria from Southern Switzerland
tsushimensis, Rana tsushimensis from Tsushima Island

female R. chensinensis and male R. arvalis are more or less viable and all metamorphose into completely sterile males, as do those between female R. arvalis and male R. temporaria.

Recently, Kuramoto (1974) reported on the results of artificial hybridization between Rana longicrus from Formosa and R. okinavana from the Ryukyu Islands, and between these two species and five brown frog species from Kyushu and Tsushima Island. Of 22 combinations, six produced hybrids which were viable through metamorphosis. The gonads of these hybrids were observed immediately after metamorphosis. According to him, the testes of hybrids were rather normal in size and structure, while the ovaries were all underdeveloped. There were females in relatively high frequencies in many kinds of hybrids. However, it is our assumption that all the viable hybrids produced by Kuramoto (1974) would become completely sterile males like those described above if they were reared to sexual maturity.

c. Hyla, Rhacophorus and Bombina

Crosses between Japanese and European tree-frogs were made for the first time by Daito (1968) in our laboratory. Eggs of two Hyla arborea japonica were inseminated by sperm of a male Hyla arborea sarda from Corsica. Although these two subspecies were not isolated by gametic isolation nor by hybrid inviability, the hybrids seemed to be completely sterile. Afterwards, we (Kawamura and Nishioka 1972) conducted hybridization experiments between European Hyla arborea arborea and three populations of Hyla arborea japonica as well as among the latter three populations. These populations of Hyla arborea japonica were collected from Hiroshima, Tsushima Island and Korea, while the Hyla arborea arborea were from Luxembourg. The relationship between H. arborea japonica and H. arborea arborea is nearly the same as that reported by Daito. The male and female hybrids between female H. arborea japonica from the three districts and male H. arborea arborea from Luxembourg are completely sterile. Among the three populations of H. arborea japonica, there is a slight degree of hybrid inviability as well as of hybrid sterility.

Two Japanese Rhacophorus species, R. schlegelii and R. arboreus, were crossed by Nakane (1953, 1954). These two species are completely isolated from each other by gametic isolation or hybrid inviability. Kuramoto (1974) has briefly reported that no viable hybrids were obtained from five combinations made among five Rhacophorus species collected from Kyushu and southwestern islands of Japan. According to him, eggs either were not fertilized or else died at the blastula stage.

Korean Bombina orientalis are completely isolated from European Bombina variegata by hybrid sterility (Ueda, 1977). Both males and females of reciprocal hybrids had extremely degenerated gonads.

Conclusions

In all the Japanese anurans used for our experiments, the male is considered to be the heterogametic sex. This presumption was made for the first time on the basis of the sex of parthenogenetically produced frogs in Rana nigromaculata (Kawamura 1939) and of the progeny of sex-reversed females in Rana japonica (Kawamura and Yokota 1959). Thereafter, the male heterogamety in these two species has been ascertained by producing diploids gynogenetically, sex reversed females and polyploids. By similar experiments, male heterogamety was established also in Rana brevipoda and R. rugosa. These four Rana species are very similar to each other in that they are semi-differentiated in sex determination, that sex reversal always occurs from a female into a male, and that the genetic sex is not altered by sex reversal.

Hyla arborea japonica and Korean Bombina orientalis are the same as the above four Rana species in that the male is heterogametic. The tree-frog is peculiar, however, in that sex reversal occurs easily from a male into a female. In this respect, the tree-frog is like Xenopus laevis (Chang and Witschi 1956), although the female is the heterogametic sex in Xenopus laevis. In Hyla arborea japonica, gynogenetically produced diploids are all females. Although this fact is simply explained by female homogamety, it is not inconsistent with female heterogamety, if the separation of W and Z chromosomes occurs at the second meiotic division. If the latter is true, all the triploid tree-frogs produced by refrigeration of fertilized eggs should be females and should be WZZ in chromosome constitution. Both males and females are found among triploids, however, so it appears that the male is the heterogametic sex in Hyla arborea japonica. This male heterogamety was supported by our voluminous mating experiments. The male heterogamety in Bombina orientalis is also clarified on the basis of the sex of gynogenetically produced diploids as well as of triploids, as it is in Hyla. In Bombina orientalis, however, sex reversal does not occur in gynogenetically produced diploids and triploids, or by administration of sex hormones.

An XY female and three YY males of Hyla arborea japonica were found among the animals from the field. The XY female is a sex-reversed male, while the YY males probably result from matings of sex-reversed males (XY) and normal males (XY). Although all three YY males were albinos, it is very probable that many sex-reversed males as well as YY males with normal appearance would be found in the field. It is also assumed that there are sex-reversed females

(XX) of Rana species in the field, since sex reversal from a female into a male frequently occurs by various experimental means, such as overripening of unfertilized eggs, raising of tadpoles at high temperatures or administration of androgens.

A female Rana nigromaculata was peculiar in that the gynogenetic diploids produced from this female were all dwarf males at the age of two years. This extraordinary female was clearly heterozygous for chromosomal or gene mutations which give rise to sex reversal as well as a dwarf body in the homozygous state. While such a powerful chromosomal or gene mutation is rarely found in natural populations of frogs, some recessive chromosomal or gene mutations which give rise to sex reversal in a small percentage of homozygotes seem to occur in nearly all natural populations, as shown by experiments producing diploids gynogenetically.

Chromosomal or gene mutations causing sex reversal are abundantly induced by irradiation of gametes with X-rays or neutrons in Rana nigromaculata and R. japonica. Some of them are dominant, differing from the spontaneous ones found in natural populations. Both dominant and recessive mutations frequently appear after a long latent period, that is, one, two or even three generations. As sex-reversed frogs appearing in the first to fourth generation derived from irradiated gametes are always accompanied by individuals which have some chromosome aberrations and various morphological abnormalities and are lethal in embryonic or tadpole stages (1977a, 1977b), the sex reversal seems to be a by-product of a kind of abnormal metabolism which produces morphological abnormalities. Although the nature of the abnormal metabolism causing sex reversal is not yet known, such abnormal metabolism seems to be brought about by many kinds of chromosomal or gene mutations. In this respect, the spontaneous chromosomal or gene mutations producing sex reversal are probably the same as those induced by irradiation of gametes. They differ from the latter only in being all recessive.

Sex reversal found in haploids and polyploids as well as in interspecific hybrids seems also to be a by-product of a kind of abnormal metabolism caused by abnormal combinations of chromosomes and cytoplasm. Sex reversal occurs most frequently in brown frogs. This seems to indicate that the sex-differentiating mechanism in brown frogs is less stable than the mechanism in pond frogs, although sex reversal from a male into a female is firmly resisted in all the Rana species. In Hyla arborea japonica, sex reversal from a female into a male is strictly resisted. Bombina orientalis is the most stable among the anuran species used for our experiments in sex-differentiating mechanism.

The degree of alliance of one species to another in phylogeny seems most exactly reflected in the degree of hybrid sterility, while the existence of gametic isolation or hybrid inviability is not always indicative of a remote relationship of the two species. We often see examples in which gametic isolation or hybrid inviability is complete in a combination of two species, while it is incomplete or it does not exist in the reciprocal combination. When hybrid inviability is complete, we usually can not ascertain the phylogenic relationship of the two species.

REFERENCES

Chang, C. Y., and E. Witschi. 1956. Genetic control and hormonal reversal of sex differentiation in Xenopus. Proc. Soc. Exptl. Biol. Med. 93:140-144.

Daito, Y. 1968. Intersubspecific hybrids between Japanese and European tree-frogs. Zool. Mag. (Tokyo) 77:117-127.

Humphrey, R. R., R. Briggs and G. Fankhauser. 1950. Sex differentiation in triploid Rana pipiens larvae and the subsequent reversal of females to males. J. Exp. Zool. 115:399-428.

Kobayashi, M. 1962a. Studies on reproductive isolation mechanisms in brown frogs. I. Development and inviability of hybrids. J. Sci. Hiroshima Univ., Ser. B, Div. 1. 20:147-156.

Kobayashi, M. 1962b. II. Hybrid sterility. J. Sci. Hiroshima Univ., Ser. B, Div. 1. 20:157-179.

Kawamura, T. 1939. Artificial parthenogenesis in the frog. II. The sex of parthenogenetic frogs. J. Sci. Hiroshima Univ., Ser. B, Div. 1. 7:39-86.

Kawamura, T. 1941. The sex of triploid frogs in Rana nigromaculata. Zool. Mag. (Tokyo) 53:334-347.

Kawamura, T. 1943. Studies on hybridization in amphibians. I. The species hybrid of Rana japonica Guenther x Rana temporaria Linné. Zool. Mag. (Tokyo) 55:315-330.

Kawamura, T. 1950. II. Interspecific hybrids in red-colored frogs. J. Sci. Hiroshima Univ., Ser. B, Div. 1. 11:61-70.

Kawamura, T. 1951. The offspring of triploid males of the frog, Rana nigromaculata. J. Sci. Hiroshima Univ., Ser. B, Div. 1. 12:11-20.

Kawamura, T. 1962. On the names of some Japanese frogs. J. Sci. Hiroshima Univ., Ser. B, Div. 1. 20:181-193.

Kawamura, T., and M. Kobayashi. 1959. Studies on hybridization in amphibians. VI. Reciprocal hybrids between Rana temporaria temporaria L. and Rana temporaria ornativentris Werner. J. Sci. Hiroshima Univ., Ser. B, Div. 1. 18:1-15.

Kawamura, T., and M. Kobayashi. 1960. VII. Hybrids between Japanese and European brown frogs. J. Sci. Hiroshima Univ., Ser. B, Div. 1. 18:221-238.

Kawamura, T., and M. Nishioka. 1960. Amphidiploid frogs produced by artificial means. J. Sci. Hiroshima Univ., Ser. B, Div. 1. 18:195-220.

Kawamura, T., and M. Nishioka. 1962. Hybridization between European and Japanese Rana temporaria temporaria. Zool. Mag. (Tokyo) 71:395.

Kawamura, T., and M. Nishioka. 1963a. On the physiological isolation of Rana tsushimensis. Zool. Mag. (Tokyo) 72:257.

Kawamura, T., and M. Nishioka. 1963b. Reproductive capacity of an amphidiploid male produced by nuclear transplantation in amphibians. J. Sci. Hiroshima Univ., Ser. B, Div. 1. 21:1-13.

Kawamura, T., and M. Nishioka. 1963c. Autotetraploids and the production of allotetraploids and diploid nucleo-cytoplasmic hybrids in pond frogs. J. Sci. Hiroshima Univ., Ser. B, Div. 1. 21:85-106.

Kawamura, T., and M. Nishioka. 1967. On the sex and reproductive capacity of tetraploids in amphibians. Gunma Symposia on Endocrinology 4:25-40.

Kawamura, T., and M. Nishioka. 1973. Superiority of anuran amphibians as experimental materials. Exp. animals 22(suppl.): 115-126.

Kawamura, T., and M. Nishioka. 1975. On the pond frogs in the palearctic region, with special reference to the isolating mechanisms between different species. Proc. Jap. Soc. Syst. Zool. 11:61-78.

Kawamura, T., and M. Nishioka. 1977. On the abnormalities found in the offspring of Rana nigromaculata produced from irradiated eggs or sperm. Sci. Rep. Lab. Amphibian Biol., Hiroshima Univ. 2: (in press).

Kawamura, T., M. Nishioka and M. Kuramoto. 1972. Interspecific hybrids between Japanese and European pond frogs. Sci. Rep.

Lab. Amphibian Biol., Hiroshima Univ. 1:277-301.

Kawamura, T., M. Nishioka and Y. Myorei. 1963. Reproductive capacity of autotetraploid males in brown frogs. J. Sci. Hiroshima Univ., Ser. B, Div. 1. 21:15-24.

Kawamura, T., M. Nishioka and H. Ueda. 1972a. Isolating mechanisms in tree-frogs distributed in Japan, Korea and Europe. Zool. Mag. (Tokyo) 81:329-330.

Kawamura, T., M. Nishioka and H. Ueda. 1972b. Reproduction of the oriental fire-bellied toad, Bombina orientalis, in the laboratory, with special reference to the superiority of this species as an experimental animal. Sci. Rep. Lab. Amphibian Biol., Hiroshima Univ. 1:303-317.

Kawamura, T., and C. Tokunaga. 1952. The sex of triploid frogs, Rana japonica Guenther. J. Sci. Hiroshima Univ., Ser. B, Div. 1. 13:121-128.

Kawamura, T., and R. Yokota. 1959. The offspring of sex-reversed females of Rana japonica Guenther. J. Sci. Hiroshima Univ., Ser. B, Div. 1. 18:31-38.

Kuramoto, M. 1974a. Experimental hybridization between the brown frogs of Taiwan, the Ryukyu Islands and Japan. Copeia. 1974: 815-822.

Kuramoto, M. 1974b. Miscellaneous hybridization experiments among the frogs. Jap. J. Herp. 5:85-90.

Miyada, S. 1960. Studies on haploid frogs. J. Sci. Hiroshima Univ., Ser. B, Div. 1. 19:1-56.

Moriwaki, T. 1959. Studies on matured parthenogenetic frogs. I. The development and reproductive ability. J. Sci. Hiroshima Univ., Ser. B, Div. 1. 17:13-32.

Moriya, K. 1951. On the isolating mechanisms between two subspecies of the pond frog, Rana nigromaculata. J. Sci. Hiroshima Univ., Ser. B, Div. 1. 12:47-56.

Moriya, K. 1954. Studies on the five races of the Japanese pond frog, Rana nigromaculata Hallowell. I. Differences in the morphological characters. J. Sci. Hiroshima Univ., Ser. B, Div. 1. 15:1-21.

Moriya, K. 1959a. II. Differences in character of development. J. Sci. Hiroshima Univ., Ser. B, Div. 1. 18:109-124.

Moriya, K. 1959b. III. Sterility in interracial hybrids. J. Sci. Hiroshima Univ., Ser. B, Div. 1. 18:125-156.

Nakane, K. 1953. Isolation between Rhacophorus schlegelii schlegelii and R. schlegelii arborea. Zool. Mag. (Tokyo) 62:76.

Nakane, K. 1954. Isolation between Rhacophorus schlegelii schlegelii and R. schlegelii arborea. II. Zool. Mag. (Tokyo) 63: 433-434.

Nishioka, M. 1963. Studies on amphidiploids and diplo-tetraploid hybrids in brown frogs. J. Sci. Hiroshima Univ., Ser. B, Div. 1. 21:25-64.

Nishioka, M. 1971a. Abnormal combinations of the nucleus and cytoplasm and their effects in amphibians. Symposia for Cell Biology 22:189-203.

Nishioka, M. 1971b. On the offspring of frogs produced by diploid gynogenesis. Japan. J. Genetics 46:433.

Nishioka, M. 1974. Sex reversal in the offspring of Rana nigromaculata derived from irradiated eggs or sperm. Zool. Mag. (Tokyo) 83:476.

Nishioka, M. 1977. Abnormalities found in the offspring of Rana japonica produced from irradiated eggs or sperm. Sci. Rep. Lab. Amphibian Biol., Hiroshima Univ. 2: (in press).

Nishioka, M., and H. Ueda. 1976. Sex reversal and existence of YY males in Hyla arborea japonica. Japan. J. Genetics 51:(in press).

Nishioka, M., and H. Ueda. 1977. Genetic and ultramicroscopic studies on ten stocks of albinos in Hyla arborea japonica. Sci. Rep. Lab. Amphibian Biol., Hiroshima Univ. 2:(in press).

Takahashi, H. 1958. Gonadal reaction in tree-frog larvae (Hyla arborea japonica Guenther) to the androgen. J. Fac. Sci. Hokkaido Univ., Ser. VI, Zool. 14:92-99.

Takahashi, H. 1959. Partial feminization of larval gonads of Hyla arborea japonica Guenther induced by treatment with estradiol. J. Fac. Sci. Hokkaido Univ., Ser. VI, Zool. 14:210-221.

Ueda, H. 1971. On the diploid gynogenesis in Bombina orientalis. Japan. J. Genetics 46:446.

Ueda, H. 1977. Interspecific hybrids between Bombina orientalis

(Boulenger) and _Bombina variegata_ (L.). _Sci. Rep. Lab. Amphibian. Biol., Hiroshima Univ._ _2_:(in press).

Yano, K. 1975. Sex of triploids and gynogenetic haploids and diploids in _Rana rugosa_. _Zool. Mag._ (Tokyo) _84_:507.

Yano, K. 1977. Mature haploids and their reproductive capacity in _Rana rugosa_. _Sci. Rep. Lab. Amphibian. Biol., Hiroshima Univ._ _2_:(in press).

THE EVOLUTION OF COURTSHIP BEHAVIOR IN NEW WORLD SALAMANDERS WITH SOME COMMENTS ON OLD WORLD SALAMANDRIDS

Stevan J. Arnold

Department of Biology, University of Chicago
Chicago, Illinois 60637

Salamander courtship has been the subject of several recent reviews (Joly 1966; Organ and Organ 1968; Salthe 1967; Salthe and Mecham 1974). These reviews introduce the reader to the extensive descriptive literature on salamander courtship and provide some interesting speculations about the phylogeny of sexual behavior. Since macroevolution of salamander courtship has been reviewed recently, I will focus, instead, on the sexual dynamics of populations and on the selective pressures that stabilize and direct the course of behavioral evolution.

Innovation in techniques for sperm transfer and female persuasion are two major themes in the evolution of salamander courtship. Salamanders show a great diversity in modes of sperm transfer and an even greater diversity of behaviors which seem to represent tactics for female persuasion. At least two families (cryptobranchids and hynobiids) practice external fertilization while the remaining families, accounting for about 90% of the extant species, transfer sperm indirectly via spermatophores attached to the substrate. In these species fertilization is internal. Internal fertilization evolved from external fertilization in salamanders (Dunn 1923; Salthe 1967) but the details of this crucial transition are still poorly understood. The present review makes no contribution to this interesting problem. Instead, this paper focuses on species with indirect sperm transfer. Perhaps the most important realization in resolving sexual diversity in these salamanders is that indirect sperm transfer is intrinsically difficult. The failure rate of spermatophores in particular species ranges from about 50-70%. Part of the difficulty is the problem of finding a small sperm-bearing structure placed at some distance from the female. Many solutions are possible to this intrinsic problem, and the variety of sperm transfer modes in salamanders probably represents a suite of alternative solutions to the problem of spermatophore failure. Other groups with indirect sperm transfer, e.g., terrestrial arthropods, show comparable diversity in modes of transfer (Schaller 1971). The significance of tactics for female persuasion will be discussed in later sections.

Although the major features of courtship evolution in salamanders are known, we are still struggling to understand why this evolution has occurred. We can arrange salamander behaviors in evolutionary series, but the transitions remain a mystery. In most instances we have reached an impasse because we do not understand the function of particular behaviors. While there is a great need for experimental analysis of function, such work is time consuming and only a few taxa will ever be thoroughly analyzed. We need to know where this work should be focused. Comparative studies serve this role by revealing dramatic evolutionary transitions. One of the goals of this review is to point out promising areas for experimental work. Experimental studies can also guide comparative work since important evolutionary transitions may have subtle causes.

There is a considerable body of descriptive work on the courtship of New World salamanders but the record is very incomplete. Courtship behavior for most species has been described only once and often only fragments of courtship are reported. The available reports often lack the detail necessary for comparative purposes; additionally, little attention has been given to temporal and sequential variability in courtship. For many taxa there is no information. We know nothing about courtship in the following genera: Phaeognathus, Leurognathus, Batrachoseps, Lineatriton, Oedipina, Thorius, Parvimolge, Gyrinophilus, Stereochilus, Typhlomolge, Typhlotriton, Haideotriton, Rhyacosiredon, Pseudobranchus and Siren. The published accounts of courtship in Amphiuma (Baker et al. 1947) and Necturus (Bishop 1926) are so incomplete that no meaningful interpretations are possible. In addition, there is still much to be learned about the best known species.

In this review I will focus on a few topics that have yielded promising results and neglect many equally important issues. In particular, I will consider relative rates of behavioral evolution, the use of courtship glands, the sexual resources of males and interactions between sexual rivals, since they raise important evolutionary questions. Consideration of evolutionary rates provides a useful focus. Some aspects of courtship are relatively ancient and conservative. For these aspects long term preservation is the salient issue. Other aspects of courtship evolve rapidly and our attention is directed to the causes of behavioral diversity. The study of courtship glands, a second topic, promises an interpretation of many sexual dimorphisms in morphology as well as of many otherwise inexplicable courtship activities. Finally, we may be able to generate testable hypotheses for courtship evolution by considering the sexual resources of males and contests between sexual rivals.

RATES OF BEHAVIORAL EVOLUTION

The phyletic distribution of courtship behaviors suggests that some behaviors have had a very conservative evolutionary history, while other behaviors have evolved relatively rapidly. In particular, behaviors directly concerned with sperm transfer seem to evolve slowly, whereas behaviors that occur at the onset of courtship seem prone to relatively rapid evolution. The following sections survey courtship at various taxonomic levels. The supposition guiding this survey is that characters which are constant at a particular taxonomic level have evolved slower than characters which are variable at the same taxonomic level.

Geographic Variation within Species

The only thorough investigations of geographic variation in salamander courtship are Kawamura and Sawada's (1959) and Sawada's (1963) studies of the Asiatic newt Cynops pyrrhogaster. F_1 hybrids among the geographic races of this newt are completely viable (Sawada 1963). Nevertheless, there are sufficient geographic differences in male and female sexual behavior to cause partial or complete sexual isolation between most pairs of races. Males execute a tail-fanning display in front of the female as a precursor to spermatophore deposition. The position assumed by the male during his display and the duration of the display vary geographically. Females show geographic variation in the kind of response given to the male's display (some populations bite the male, while others simply nudge him) and in propensity to respond to various kinds of male display. Matings between newts from different populations fail for a variety of reasons: the female does not respond to the male's display; the male fails to recognize critical cues from the female; the tail-nudging walk, during which sperm transfer is accomplished, is uncoordinated because the partners walk at different speeds; or the female fails to accompany the male during the tail-nudging walk. Sexual coordination within populations of Cynops pyrrhogaster is apparently a consequence of strong selection for male behavior suited to local female behavior and selection for female behavior suited to local male behavior.

Although these preliminary courtship activities show geographic variation in Cynops pyrrhogaster, postures used during sperm transfer do not vary (Sawada 1963).

There are no comparable studies for North American salamanders. Arnold (1976) did not find differences in courtship behavior among four of the geographic isolates of Plethodon jordani recognized by Highton (1970). Geographic variation in courtship has been proposed in each of two Ambystoma species, but both cases are suspect. Licht (1969) observed a clasping behavior in Ambystoma gracile from British Columbia that was not reported in Knudsen's (1960) account of Washington animals and proposed geographic variation as one possible explanation. I recently observed courtship behavior in animals from Knudsen's population and found that individual males used both the clasping posture described by Knudsen and the one described by Licht. Courtship behavior in Ambystoma texanum has been described for animals from Michigan (Arnold 1972); Jackson Co., Illinois (Garton 1972) and McLean Co., Illinois (Wyman 1971). Male clasping of females was observed in the last population but not in the other two. Possible hybridization between A. texanum and A. jeffersonianum is currently being investigated (F. Downs pers. comm.). Since A. jeffersonianum consistently clasps (Uzzell 1964, 1969), while at least some populations of A. texanum do not, the systematics of these two forms must be resolved before a case can be made for geographic variation in A. texanum courtship.

Variation among Congeneric Species

Courtship behavior has been observed in eight species of the genus Plethodon (Arnold 1972, 1976; Organ 1958, 1960a, 1960b): P. caddoensis, P. cinereus, P. glutinosus, P. jordani, P. richmondi, P. ouachitae, P. welleri, and P. yonahlossee. Males of each species execute a unique combination of actions during the initial stages of courtship. These actions include tail undulations, rasping motions with the premaxillary teeth, and a peculiar action of the limbs called "foot dance" (Organ 1958). The process of sperm transfer has been observed in only five of these species (P. jordani, P. glutinosus, P. ouachitae, P. welleri, and P. yonahlossee). In contrast to the diversity of behaviors in the early stages of courtship, sperm transfer is strikingly uniform. Male P. yonahlossee execute a slightly different action while the female retrieves sperm, but the five species seem otherwise identical.

The most pronounced differences among species in the genus Ambystoma are in the presence and mode of clasping. Clasping may be absent (A. annulatum, A. dumerili, A. maculatum, A. opacum, A. talpoideum), males may clasp the female's pectoral region with the forelimbs (A. jeffersonianum, A. laterale, A. macrodactylum), or males may use this same clasping mode and then shuffle forward and clasp with the hindlimbs as in A. gracile (Anderson 1961; Arnold 1972, 1976; Brandon 1970; Noble and Brady 1933; Shoop 1960; Spotila and Beumer 1970; Storez 1969; Uzzell 1969). This genus also shows modest variation in the mode of sperm transfer. In species with clasping, the female generally follows the male closely during spermatophore deposition and sperm transfer, whereas the female finds spermatophores without closely following the male in some nonclasping species (e.g., A. maculatum, A. opacum).

The genus Triturus shows species differences in the preliminary courtship display (Halliday, this volume). The form of execution of apparently homologous display actions differs among species and, in addition, some species incorporate unique actions into their displays. The nature of transitions between display actions also varies in the genus. Some species revert to the preliminary display between successive spermatophore depositions, but other species do not. In contrast, the postures and actions related to sperm transfer are remarkably constant.

Behaviors Unique to Particular Genera

A peculiar tail-wagging display (Fig. 1) is unique to the monotypic ambystomatid genus Rhyacotriton. The display is performed by sexually active males prior to spermatophore deposition. It may be executed in close proximity to the female or at a distance of several centimeters from her. The function of the display is unknown.

Male Desmognathus execute a unique forelimb action while approaching females during courtship (Organ 1961a, 1961b). Both forelimbs are simultaneously swung forward in an action that resembles the dolphin or butterfly swimming stroke. The action is performed repeatedly as the male creeps towards the female prior to the tail-straddling walk. This forelimb action has been observed in each of the four species of Desmognathus for which

Fig. 1. The tail-wagging display of Rhyacotriton olympicus. The male wags just the tip of his tail. A female is facing away from the male, and only her tail (stippled) is indicated. From a photograph.

courtship observations are available but not in any other plethodontid genus.

Behaviors Unique to Particular Families

Plethodontids engage in a tail-straddling walk (Fig. 2) immediately before spermatophore deposition. This behavior was first observed in Eurycea bislineata (Noble 1929) and later reported in a variety of plethodontids (Noble and Brady 1930). Subsequently, tail-straddling walks have been observed in nine genera distributed in both subfamilies and all recognized tribes (Wake 1966) of the family. The behavior is apparently an obligatory precursor to sperm transfer in plethodontids and is not found in any other family. In the one plethodontid for which quantitative information is available, the constancy of male and female positions during the walk is the result of regulatory behavior by the male (Arnold 1976). As a consequence of this behavior, spermatophores are invariably deposited immediately in front of the female's head. The positions assumed by the male and female during sperm pick-up are likewise very uniform in the family Plethodontidae (Fig. 3). A minor variation on this posture is seen in three genera (Eurycea, Ensatina and Aneides). In these genera the male massages the female's sacrum with his tail as she picks up the sperm mass (Stebbins 1949; Arnold 1972).

The wide distribution of the tail-straddling walk among the diverse tribes of plethodontids, and its absence in all other families, indicates that the behavior arose very early in the history of the family. Wake (1966) argued that the divergence of two subfamilies (desmognathines and Plethodontines) occurred in the Mesozoic. Since the tail-straddling walk is found in both subfamilies, this behavioral trait is probablv at least 70 million years old.

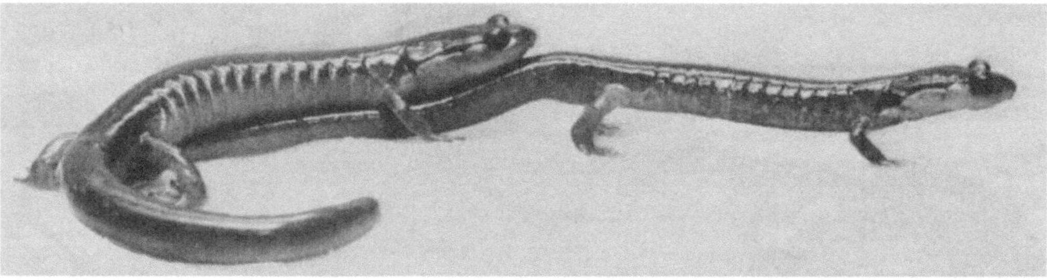

Fig. 2. The tail-straddling walk of Plethodon jordani. The female (left) is straddling the male's tail. (From Arnold 1976.) Reproduced with permission of Verlag Paul Parey, Berlin and Hamburg; copyright 1976.

Fig. 3. The postures assumed during sperm pick-up in *Pseudotriton ruber* (above) and *Plethodon jordani* (below). In each case the female (stippled) has inserted the male's spermatophore into her cloaca. From photographs.

Salamandrids show a unique behavior during sperm transfer. Males in five genera displace the sacrum laterally in an arc by pivoting about the contralateral forelimb. The action is executed immediately after spermatophore deposition. In *Notophthalmus*, *Taricha* and *Triturus*, the male displaces his sacrum through a 90° arc. The behavior apparently helps the female find the spermatophore, since the male blocks her path at a distance of one body length in front of the spermatophore (Arnold 1972; Halliday 1974b). Male *Pleurodeles* and *Salamandra* also displace the sacrum laterally after spermatophore deposition, but the extent of the movement differs from that in the first three genera. In *Pleurodeles*, the male swings his sacrum through an arc of 180° or more by pivoting about the contralateral forelimb. In *Salamandra*, the male swings only 45°, and this is accomplished simply by flexing the vertebral column laterally. The effect of lateral displacement of the sacrum is not to block the female's path in these two genera. The *Salamandra* male holds the female on his dorsum during spermatophore deposition by clasping her forelimbs with his. Lateral displacement of the sacrum allows the female's vent to drop down onto the spermatophore. In *Pleurodeles*, sperm transfer is accomplished as the pair pivots about interlocked forelimbs while facing each other head to head. Lateral displacement of the sacrum is absent in *Cynops* and *Euproctus*. *Cynops* males march forward on a straight path after spermatophore deposition. *Euproctus* males embrace the female during deposition and transfer and pass the spermatophore directly to the female with the hindfeet. Thus lateral displacement of the sacrum after spermatophore deposition is unique to salamandrids but found in only five of the seven genera so far observed.

The peculiar sacral movement of salamandrids is probably very old. Modern Taricha show little osteological modification from their Oligocene ancestor, Paleotaricha (Wake and Özeti 1969). If Triturus and Notophthalmus diverged from Taricha during or before the Oligocene, then the sperm transfer behavior shared by these genera is at least 25 million years old.

Behaviors Shared by Families

Salamanders in four families (ambystomatids, Plethodontids, salamandrids, and the proteid genus Proteus) characteristically deposit spermatophores just in front of the female's snout. In these families the female finds and orients to the spermatophore by touch as she crawls over the spermatophore and it contacts her ventral surface. Spermatophore deposition in front of the female's snout probably represents an adaptive optimum dictated by female behavior. An unrestrained female is most likely to find a spermatophore that is close to her and in her line of travel. Departures from the typical site of spermatophore deposition are of two kinds: (1) In species that restrain the female during spermatophore deposition (Euproctus, Pleurodeles and Salamandra), the site of deposition is stereotyped in each genus but the site varies according to the type of restraint imposed on the female. Eurpoctus, Pleurodeles and Salamandra have apparently achieved new adaptive optima in deposition site as a consequence of innovations in the ability of the males to restrain females during deposition and sperm transfer; (2) In species that court and deposit spermatophores in polyandrous aggregations (some species of Ambystoma), the site of deposition is highly variable within populations. In such species, competition among males apparently places a premium on rapid deposition and this precludes stereotyped deposition in front of the female's snout (Arnold 1976).

Other aspects of sperm transfer are strikingly similar in plethodontids, ambystomatids, salamandrids and Proteus, suggesting that the basic mechanism of sperm transfer is ancient. The following characteristics, in addition to site of spermatophore deposition, are common to these four families: (1) The spermatophore consists of a gelatinous base, attached to the substrate, and an apical sperm mass; (2) The male faces away from the female during spermatophore deposition; (3) The female orients to glands in the male's cloaca (ambystomatids, Fig. 4; salamandrids and Proteus) or

Fig. 4. The tail-nudging walk of Ambystoma mexicanum. The male exposes his cloacal papillae by elevating his tail. The female (stippled) nudges these papillae as she follows him during sperm transfer. From a 16 mm movie.

to the dorsum of the male's tail base (plethodontids) just before spermatophore deposition and as the male departs from the spermatophore; (4) The male responds to contact with his cloaca or tail just before and during sperm transfer. Thus the shared syndrome of traits involved with sperm transfer includes the morphology of the spermatophore, the site of deposition, the position assumed by the sexes and the cues used during sperm transfer. Whether one adopts the view that these shared traits are homologous or that two or more families have independently evolved the same syndrome of attributes, one must conclude that behaviors directly concerned with sperm transfer evolve very slowly compared with the preliminary behaviors of courtship.

THE USE OF COURTSHIP GLANDS AND EPIDERMAL SECRETIONS

Exocrine secretions appear to serve three functions during courtship: identification of species and sex, orientation during the locomotory phases of courtship, and persuasion (Tinbergen 1953) of the female prior to spermatophore deposition. None of these supposed functions has been established with conclusive experimentation. Instead the appeal of each function rests on simple observational data and preliminary experimentation. The observational record is, however, extensive enough to reveal promising avenues for critical experimentation.

Identification of Species and Sex

Epidermal secretions may play a role in species and sex identification in ambystomatids and plethodontids, but specialized male courtship glands do not appear to serve these functions. Capacity for sex and species identification has been established for several species by the simple technique of housing males with only conspecific males or with only heterospecific females. Under these conditions, males usually fail to court and deposit spermatophores (Arnold 1976; Noble and Brady 1930). Since in all salamanders the male contacts the female's epidermis with his snout before spermatophore deposition, chemoreception has been proposed as a mechanism for sex and species identification (Arnold 1976; Noble and Brady 1930; Organ and Organ 1968). The case for chemoreception as a major sensory modality for species and sex identification in ambystomatids and plethodontids remains circumstantial, however, Visual cues may also be important for species and sex identification in plethodontids (Organ 1960b; Organ and Organ 1968).

The only direct evidence for chemoreception as a mechanism for sex and species identification in North American salamanders comes from Twitty's (1955, 1961a) work with the salamandred genus *Taricha*. Twitty (1955) found that male *T. rivularis* were attracted to sponges soaked with skin secretions from conspecific females, but they ignored control sponges that lacked secretions. In another field experiment, females from four populations (*T. rivularis*, *T. granulosa*, *T. torosa torosa* and *T. t. sierrae*) were anchored side by side in a stream naturally inhabited by *T. rivularis* and *T. granulosa* (Twitty 1961a:454). Native male *T. rivularis* showed a strong response to conspecific females and female *T. t. sierrae* and a weak response to *T. granulosa* and *T. t. torosa* females. Twitty rarely found *T. rivularis* males clasping sympatric *T. granulosa* females in nature (Twitty 1955, 1961a). Twitty and Davis (1964) found that *T. rivularis* males would occasionally clasp sympatric *T. granulosa* females if skin secretions from conspecific females were introduced into the aquarium. Such heterospecific clasping was ephemeral and never resulted in complete courtship. Blinded T. rivularis males were also able to discriminate between conspecific and heterospecific females. Although chemoreception is strongly implicated in preventing hybridization between T. rivularis and T. granulosa, the precise role of this sensory modality has not been elucidated. Experiments with anosmic males might provide additional clarification. It is not known whether females play any role in species recognition. The problem is particularly interesting since artificially produced hybrids between T. rivularis and T. granulosa are viable and fertile, yet hybridization is very rare in nature (Twitty 1961b).

Cedrini and Fasolo (1971) monitored the olfactory bulb in spinal female Triturus cristatus and reported discrimination between skin secretions from males and females.

Orientation during Courtship

Salamandrids (Cynops, Notophthalmus, Taricha and Triturus), ambystomatids (some Ambystoma spp.) and, apparently, Proteus (Briegleb 1962) engage in a tail-nudging walk before spermatophore deposition. During such walks chemoreception appears to be a major channel of communication to the female. During the walk the male's elevated tail exposes cloacal papillae to the female behind him (Fig. 4). In Triturus these papillae are the exit points of the male's abdominal glands (Zeller 1905), but the identity of comparable papillae in Ambystoma has not been established (Arnold 1976). Delivery of secretions might be accomplished by physical contact, since the female repeatedly nudges the papillae with her snout, or via water currents produced by the male's tail undulations (Noble 1927b). The function of the various cloacal glands has not been tested by behavioral experimentation. Recent electrophysiological studies, however, indicate that extracts of the male's abdominal gland elicit strong responses in the olfactory bulb of female Triturus (Cedrini and Fasolo 1971).

Plethodontid females may orient by chemoreception during the tail-straddling walk that precedes spermatophore deposition. Male Eurycea bislineata possess a glandular bump on the dorsum of the tail base, and the female's snout rests on this bump during the tail-straddling walk (Noble 1929). In most plethodontids there is no visible structure in this region, but histological work might be focused here with profit.

Persuasion

Male salamanders possess a diverse array of epidermal glands that are repeatedly applied to the female during preliminary courtship. Noble (1927a) called these "hedonic" glands, but the name has long since served its purpose. Whatever function these glands have, we will never know if they are hedonic (pleasure-giving). Since the function of male glands is still unresolved, they should be called "courtship" glands and particular glands should be named in reference to their anatomical position (e.g., mental gland, submandibular gland, genial gland).

The behavioral contexts in which courtship glands are employed provide some clues as to their function. Recent behavioral observations leave the problem of function largely unresolved but they do suggest a priming rather than triggering function for courtship glands. In particular, the contexts in which *Notophthalmus* use genial glands and plethodontids use mental glands have not been generally recognized so emphasis will be placed on these salamanders.

The use of cephalic glands in salamandrid courtship. Male *Notophthalmus* possess 3-4 glandular pits on each side of the head posterior to the eye (Cope 1889; Hilton 1902). These glands may be called "genial" glands in reference to their position on the cheek. They are applied to the female's nares as the male clasps her dorsum (Fig. 5). This clasping posture is commonly maintained for 45 minutes or more and during this time the genial glands are repeatedly applied to the female's nares. The complex temporal structure of the clasping display has been recognized since the earliest descriptions of *Notophthalmus* courtship (Zeller 1890; Jordan 1891; Humphries 1955); additional details were given by Arnold (1972).

The courtship of *Notophthalmus* presents an enigma that may eventually lead to an elucidation of genial gland function. In some courtships the male never uses his genial glands. Instead, the male deletes the entire clasping display and goes directly to spermatophore deposition and sperm transfer. Such abbreviated courtship occurs if the female actively approaches the male; he will then delete capture and clasping and immediately begin spermatophore deposition (Arnold 1972; Humphries 1955). The male captures the female and subjects her to the lengthy clasping display only if she attempts to escape from him when he first encounters her.

The short-circuit of courtship can be explained, in part, by the results of Rogoff's (1927) experiments to determine the effect of genial gland secretions on the female. In the first experiment, delivery of genial gland secretions to the female was enhanced. Secretion of the glands was increased by injecting males with pilocarpine hydrochloride, and the nares of the female were held against the gland apertures. Such females subsequently followed the male, but control females did not. Delivery of secretions was blocked in two separate experiments by occluding the genial gland apertures of the male and by occluding the female's nares. Both manipulations

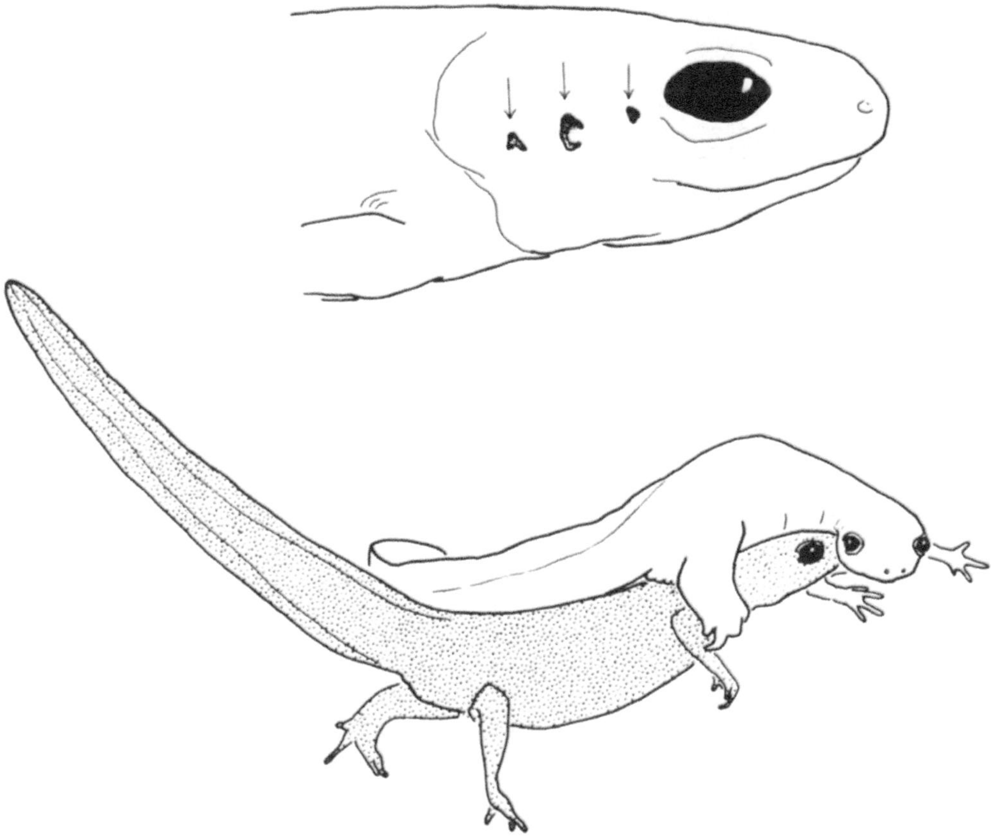

Fig. 5. Application of the male's genial glands during the clasping display of Notophthalmus viridescens. The three apertures of the male's genial glands are located on his cheek (above). During the clasping display (below), the male positions the female's head with his forelimb and presses his genial glands against her nares. From FMNH 85546 and a 16 mm movie.

had the same result: the females failed to follow the male. Rogoff's results were reported in a brief note, and the experiments could profitably be repeated. The experiments suggest that genial gland secretions enhance female responsiveness during the tail-nudging walk that precedes spermatophore deposition.

Why is glandular application deleted in some courtships? Perhaps the male adjusts his courtship to the sexual responsiveness of his partner, deleting glandular application if the female already is responsive at the onset of courtship and attempting to make the female responsive by repeated application of his genial glands if she is not. This working hypothesis is complex and a complete test will require several experiments. In particular we suppose that females differ in sexual responsiveness, and that males can assay and change female sexual responsiveness.

A comparison of Notophthalmus with Taricha presents a new problem. In Notophthalmus, the clasping by the males is facultative, but in Taricha clasping is an obligatory precursor to spermatophore deposition. The basis of this difference is not understood. Male Taricha clasp the female's dorsum with both the forelimbs and hindlimbs. From this position the male repeatedly shifts forward and rubs a submandibular gland over the female's nares (Fig. 6). The duration of the clasping display is even longer than in Notophthalmus and may continue for several hours (Arnold 1972; Davis and Twitty 1964; Smith 1941). During this period the submandibular gland is applied a few times each minute.

The use of mental glands and premaxillary teeth in plethodontid courtship. Male plethodontids develop a disc-shaped protuberance on the chin. This protuberance is called the mental gland (Noble 1929) and is absent in females. The mental gland is composed of a dense cluster of exocrine glands, which release their secretory product on the surface of the chin or mentum. Sever (1976) recently reviewed the histology of these glands. The secretory product appears to be a glyco- or mucoprotein (Lanza 1959; Sever 1975). The relative development of the gland corresponds to courtship season. In seasonal breeders, the gland hypertrophies during the courtship season, but in aseasonal breeders the gland remains in a constant state of development (Houck, this volume). Development and maintenance of the gland is apparently mediated by gonadal hormones. Testosterone injection can induce mental gland development in female Eurycea quadridigitata in just a few weeks (Sever 1976).

A Dunn (1926) pointed out, some plethodontids have extraordinary premaxillary teeth, which protrude forward in males and actually pierce the lip (Fig. 7). The morphology and development of

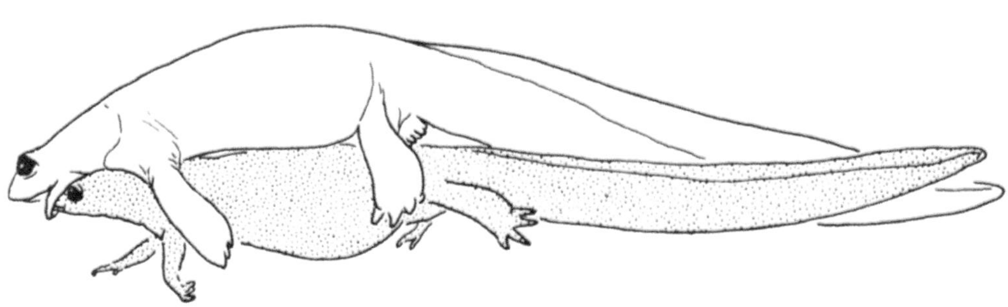

Fig. 6. A male Taricha granulosa (above) rubbing his submandibular gland on the female's snout. From a 16 mm movie.

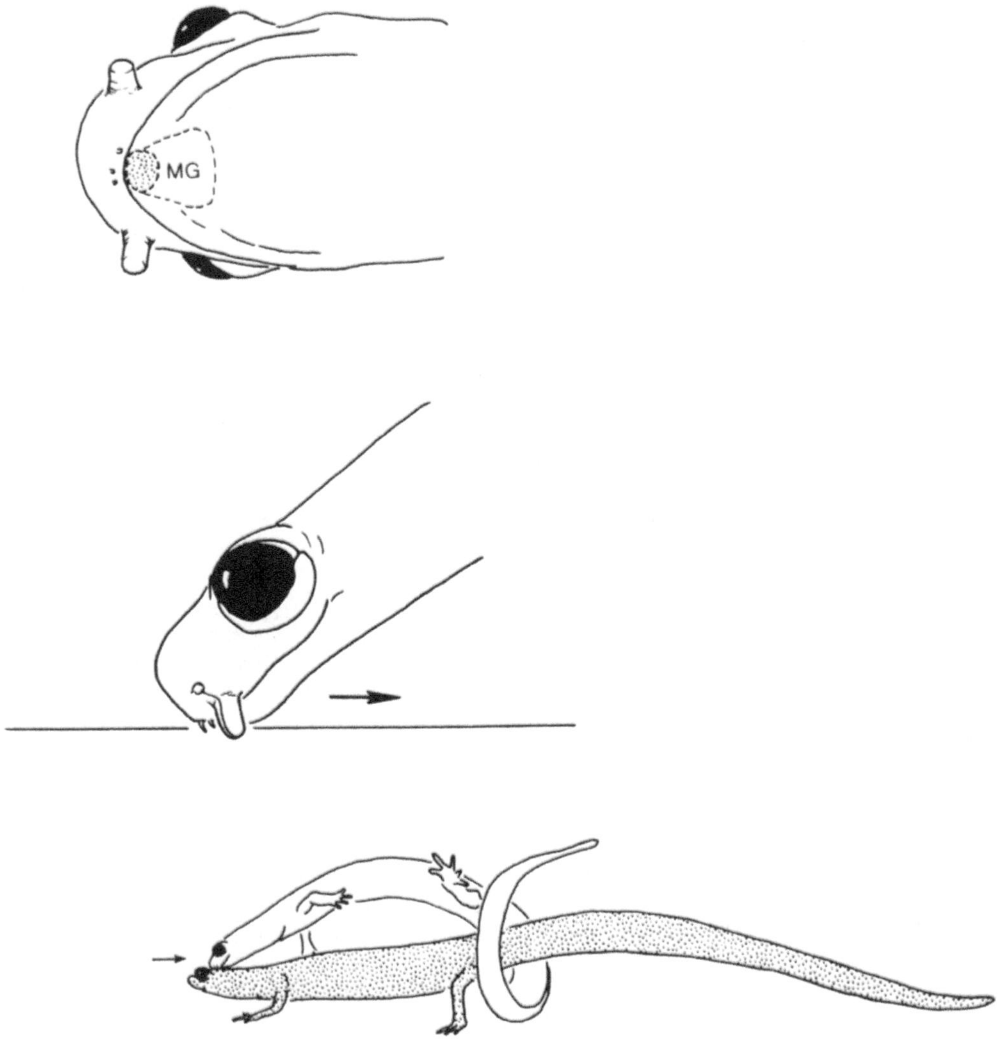

Fig. 7. Delivery of mental gland secretions in Eurycea bislineata. The male's premaxillary teeth pierce his lip (upper figure). The approximate extent of the mental gland (MG) is indicated with a dotted line. The region of secretory release is indicated with stippling. During courtship (middle and lower figures) the male appears to introduce mental gland secretions into the female's circulation by abrading her skin with his premaxillary teeth. In the lower figures, the male has just pulled his premaxillary teeth and mental gland over the female's head in the direction indicated by the arrow. From FMNH 90632 and a 16 mm movie of a pair courting underwater.

premaxillary teeth has been studied in two sexually dimorphic genera (Desmognathus and Eurycea). Adult males differ from females in kind and number of premaxillary teeth (Noble 1929; Noble and Pope 1928; Stewart 1958). The male's teeth are monocuspid rather than bicuspid as in females. They are longer in males and protrude anteriad to the symphysis of the lower jaw, sometimes piercing the lip. Eurycea bislineata males show a seasonal replacement of premaxillary teeth; the specialized monocuspid teeth are maintained only during the winter courtship season (Stewart 1958). Adult male E. bislineata have fewer premaxillary teeth than females or immature males (Stewart 1958). Sexual dimorphism in premaxillary teeth is apparently mediated by gonadal hormones. The specialized teeth of Desmognathus are replaced by bicuspid teeth in castrated males, and ovariectomized females grow male teeth when impanted with testes (Noble and Pope 1929). When female Eurycea quadridigitata are injected with testosterone ethanate, they develop protruding premaxillary teeth as well as mental glands and nasolabial cirri (Sever 1976 and pers. comm.).

In some plethodontids, males appear to deliver mental gland secretions by "vaccinating" the female with the premaxillary teeth (Fig. 7). This function for premaxillary teeth was first proposed by Organ (1961a). During courtship Desmognathus males press the mental gland against the female's dorsum and then fling themselves away from the female in a snapping motion. Organ (1961a) proposed that the premaxillary teeth abrade the female's skin during this snapping action, and that mental gland secretions are introduced directly into her superficial circulation. The morphology of the male's snout certainly suggests this interpretation. The Desmognathus mental gland is situated on a platform anterior to the mandibular symphysis (Organ 1961; Sever 1976). Since the male moves posteriad during the snapping action his mental gland will rub the female's dorsum and the protruding premaxillary teeth might immediately abrade the same site. The male arches his head downward at the onset of snapping, sometimes lifting his forefeet off the substrate, so that only his mentum and premaxilla touch the female. The positioning probably would enhance the proposed effect. The snapping action that Organ (1961a) observed in Desmognathus apparently is widespread in plethodontids. I have observed a similar action in Eurycea bislineata, Plethodon cinereus, P. richmondi, Bolitoglossa flavimembris, B. occidentalis, and B. subpalmata as well as in Desmognathus fuscus and D. monticola. Salthe and Salthe (1964:574) observed this action in Pseudoeurycea belli. In Bolitoglossa occidentalis females, scratches could be observed on the dorsum of the trunk after courtship, but in other species scratches were not visible to the unaided eye. A second action, similar to snapping, may also have the effect of "vaccinating" females with mental gland secretions. In this action, which may be called "pulling," the male applies his mentum to the female's dorsum in a succession of quick strokes without lifting his head from her body. A male Eurycea bislineata in the act of pulling is illustrated in Fig. 7. Pulling has been observed in the same species that employ snapping, as well as in Aneides lugubris and Hydromantes platycephalus; in male Hydromantes the maxillary as well as premaxillary teeth protrude. All of these species have protruding premaxillary teeth and mental glands; snapping and pulling were not observed in species that lack mental glands and protruding premaxillary teeth (Pseudotriton ruber and Ensatina eschscholtzi). Both actions are usually performed with the male perpendicular to the female, but occasionally they are executed with the male parallel to and on top of the female (Fig. 7). Both actions may be executed anywhere on the female's dorsum but most commonly on the female's trunk. Snapping may occur during preliminary courtship as well as during the tail-straddling walk, but pulling was observed only in the former context. The male performs snapping during the tail-straddling walk by turning back towards the female and applying his mentum to her dorsum. Females sometimes react to snapping or pulling by arching the back or by bolting forward slightly, but snapping is often executed without disrupting the tail-straddling walk.

Some species deliver mental gland secretions by slapping the mental gland on the female's snout (Fig. 8). This mode of application has been ogserved only in large species of the genus Plethodon (P. jordani, P. glutinosus, and P. yonahlossee; Arnold 1976; Organ 1960a). Arnold (1976) gave a complete account of the action in Plethodon jordani. Presumably the mental gland secretions are taken in through the nares, perhaps via the nasolabial grooves. Species of Plethodon exploit one or the other of the two different modes for delivery of mental gland secretions. Variations in mental glands and premaxillary teeth in this genus coincide with the two delivery modes. Species that employ the snapping and pulling modes (P. cinereus and P. richmondi) have relatively small mental glands situated forward on the mentum just behind the mandibular symphysis and the protruding premaxillary teeth. The premaxillary teeth of these species are specialized; the anterior cusp is recurved (Highton 1962; Noble 1927a) and ideally suited for a rasping function. Species that slap the mental gland on the female's snout have relatively large mental glands situated near the center of the chin (Fig. 8). The premaxillary teeth are simple (Highton 1962) and do not protrude.

In light of the fact that premaxillary teeth are used in conjunction with the mental gland during courtship, it is not surprising that the premaxillary teeth act as a unit separate from the maxillary teeth during ontogeny (Noble 1927a). Many curious features of premaxillary tooth morphology will probably be understood as our knowledge of plethodontid courtship increases. The most extreme condition is seen in the genus Oedipina (Wake 1966). In species

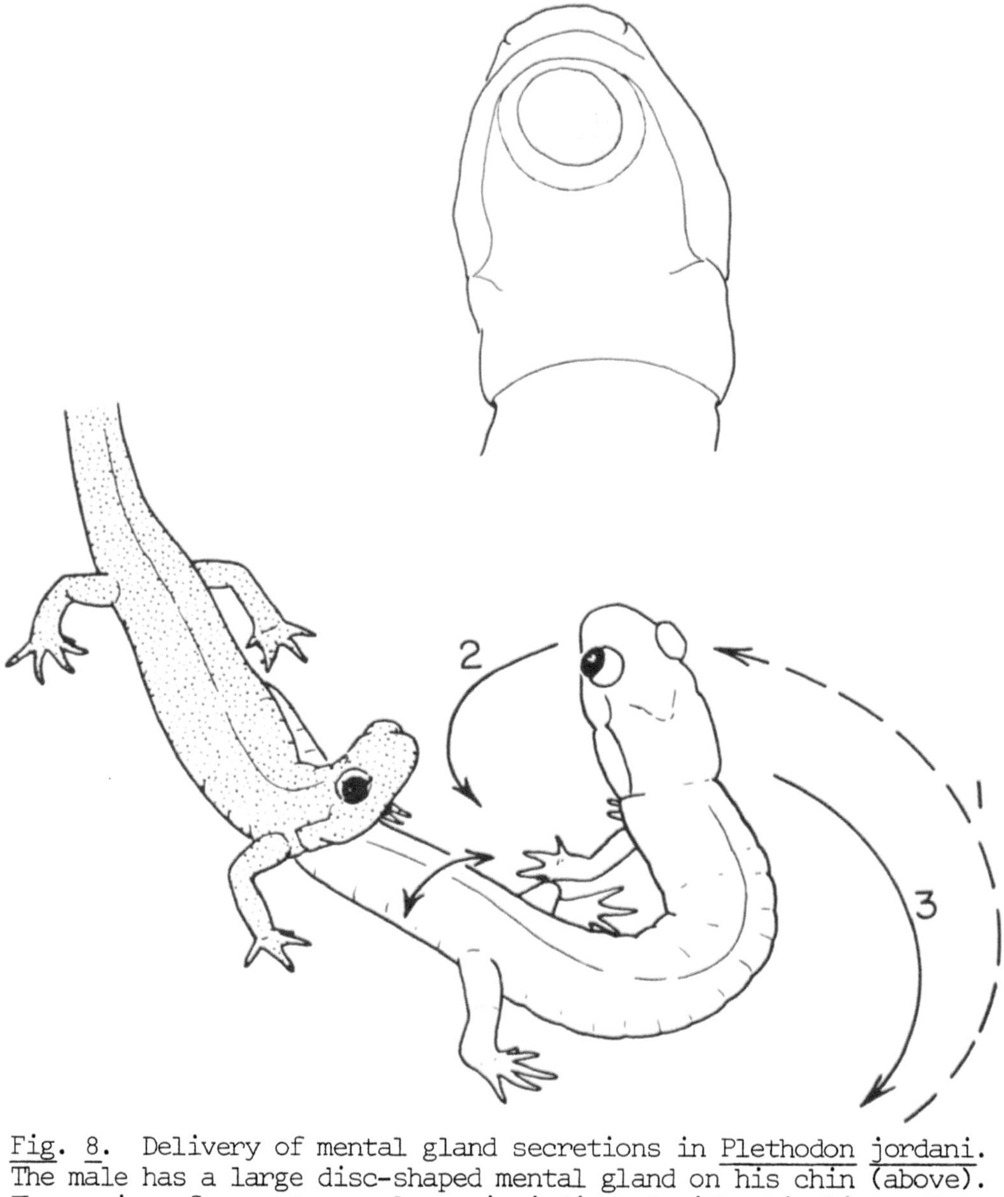

Fig. 8. Delivery of mental gland secretions in *Plethodon jordani*. The male has a large disc-shaped mental gland on his chin (above). The region of secretory release is indicated with stippling. During the tail-straddling walk (below), the male slaps his mental gland on the female's nares. From FMNH 47748 and a photograph of a courting pair.

such as *O. carablanca* and *O. elongatus*, maxillary teeth are completely lacking, but males retain two premaxillary teeth that protrude and pierce the lip (Brame 1968). In other species (O. parvipes) the pars dentalis of the premaxilla is so reduced that it houses only a single tooth (Wake 1966), while a few maxillary teeth are retained (Brame 1968). Is the retention of a single premaxillary tooth in O. parvipes related to its utility in courtship?

While it is clear that mental gland secretions are delivered to the female during courtship, the function of these secretions remains an open question. The context in which the mental gland is used suggests that its secretions have a priming effect on the female. Extensive courtship observations are available for only one plethodontid species, Plethodon jordani. During some courtships the male repeatedly slaps his mental gland on the female's head during a prolonged tail-straddling walk. In other courtships the tail-straddling walk is brief and slapping never occurs (Arnold 1976). Such courtship variability may reflect male response to differences in female receptivity. The pattern of repeated delivery of secretions to the female suggests that the secretion primes rather than triggers sexual responsiveness. Salthe and Salthe (1958) offer an alternate hypothesis. They suggest that mental gland application in Pseudoeurycea belli simply tests the receptivity of the female. Many other interpretations are possible. Mental gland secretions may be needed to induce ovulation in some females, for example. There is an obvious need for physiological studies of the effects of mental gland secretions. We need to know whether mental gland secretions increase the probability that the female will respond to the male's spermatophore.

THE SEXUAL RESOURCES OF THE MALE

One of the most important points to emerge from recent quantitative studies of salamander courtship is that male sexual resources are limited (Arnold 1976; Halliday 1976). This observation sets the stage for a new round of inquiry into the phenomenon of courtship evolution. If the male's sexual resources limit his genic representation in the next generation, then alternative patterns for the distribution of sexual resources could have profound effects on male reproductive success. For example, should the male expend his entire spermatophore supply early in the courtship season with a few females, or should he conserve spermatophores and court for a longer time? The problem is one of distribution.

Our goal is a testable theory for the distribution of sexual resources. There is not enough information, at present, to formalize a general theory, but we can characterize the task of the theory. The theory should make predictions about measurable variables that show dramatic variability across taxa. Furthermore, the variables should have a direct relation to fitness. Some aspects of courtship are easily measured, have a direct relation to particular components of male fitness, and differ among salamander taxa. From these standpoints, the most promising variables for a theory to work on are: (1) the male's temporal investment per spermatophore, (2) the number of spermatophores deposited per courtship, and (3) the temporal distribution of spermatophores among courtships. The following sections simply point out the need for a theory by surveying taxonomic differences in these variables. I will also sketch the outline of a theory by offering some preliminary interpretations of courtship differences.

Number of Spermatophores per Courtship

This aspect of courtship is rather uniform within families but strikingly different among families. For example, one spermatophore per courtship is an almost invariant rule in plethodontids. In Plethodon jordani one spermatophore was deposited in each of 49 courtships, and two were deposited in two courtships (Arnold 1976). A few courtships with spermatophore deposition have been observed in 17 other species of plethodontids and these provide no exceptions to the P. jordani pattern (Organ 1960a, 1960b, 1961a, 1961b; Stebbins 1949). In my own observations of Pseudotriton, Eurycea, Desmognathus, Plethodon, Aneides, Ensatina, Chiropterotriton, Pseudoeurycea, and Bolitoglossa, one spermatophore was the rule, two spermatophores were rare, and three spermatophores were never deposited in a single night. The deposition of a second spermatophore does not seem to depend on the success of the first spermatophore as Organ and Organ (1968) suggest. In many instances Plethodon jordani males terminated courtship after a single spermatophore deposition even though the spermatophore failed to inseminate the female (Arnold 1976).

In contrast to plethodontids, ambystomatid males are very fecund. A male Ambystoma maculatum can deposit 81 spermatophores in a single evening with one female. Determinations of spermatophore production by a single male courting a single female for one night are available for five species of Ambystoma (Table 1). Species means range from 12 to 40 spermatophores per night. Dicamptodon males also deposit numerous spermatophores per night of courtship (R. Nussbaum, pers. comm.). In contrast, Rhyacotriton males deposit only 1-3 spermatophores per night of courtship (Arnold 1972).

Salamandrids deposit relatively few spermatophores in a courtship, but more than plethodontids. Quantitative studies of Triturus yield a range of 1-4 spermatophores per courtship (Halliday 1974). Halliday's thorough studies of factors affecting spermatophore number per courtship in Triturus are discussed elsewhere in this volume. Literature reports for Euproctus, Notophthalmus, Cynops, and

Table 1. Number of Spermatophores Deposited by *Ambystoma* Males during a Single Evening of Courtship with One Female.

Species	Mean	Standard Deviation	Range	Sample Size	Reference
A. maculatum	40.4	18.8	10-81	29	Arnold (1976)
A. tigrinum	20.6	6.4	8-37	32	Arnold (1976)
A. dumerili	13.0	2.9	9-16	4	Brandon (1970)
A. laterale[1]	19.3		1-32	4	Uzzell (1969)
	23.6	9.2	13-34	9	pers. observ.
A. jeffersonianum	12.4		6-21	8	Uzzell (1969)

[1]The two estimates for this species are both based on animals from Livingston Co., Michigan.

Salamandra, as well as my own meager records for *Salamandra* and *Pleurodeles*, all fall in the range of 1-4 spermatophores per courtship (Bedriaga 1882a; Dähne 1926; Despax 1923; Häfeli 1971; Jordan 1891; Kawamura and Sawada 1959; Sawada 1963; Tsutsui 1931; and Zeller 1890). Gallien (1953), however, reported a maximum of 6-7 spermatophores for *Pleurodeles waltl*, and I have one record of five successive spermatophore depositions for *Cynops pyrrhogaster*. I intercepted *Notophthalmus viridescens* adults as they migrated to a breeding pond in early spring; consequently none of the males had courted. The mean number of spermatophores per courtship was 3.8 (standard deviation = 1.8, mode = 3, range = 1-8, n = 18 males). Thus, maximum daily spermatophore production by salamandrids is less than the mean production by *Ambystoma*.

Courtship Investment per Spermatophore

Males invest courtship time, metabolic energy and materials in each spermatophore. Courtship time is the easiest aspect of investment to measure and, given our current knowledge, the most suitable for taxonomic comparison. Courtship time per spermatophore, not the actual time for deposition, can be estimated by dividing the total time elapsed for a series of courtships by the total number of spermatophore depositions. The energetic investment in the spermatophore itself could be estimated by dry weight or, better yet, by bomb calorimetry. No such determinations are available for any salamander. Although courtship time is only part of the male's spermatophore investment, this variable shows considerable variation among salamander taxa.

Plethodontids invest a surprising amount of courtship time in each spermatophore. A sample of 60 courtships by Plethodon jordani yields an estimate of 56 minutes of courtship per spermatophore (Arnold 1976). Less extensive observations on other plethodontids suggest that this is not an exceptional estimate. Indeed, some plethodontids appear to make much larger temporal investments: in Ensatina the tail-straddling walk alone may last for 5-6 hours (Arnold 1972; Stebbins 1949).

Ambystomatid males invest a small amount of time in each spermatophore. Ambystoma maculatum invest 1.4 minutes and A. tigrinum 4.5 minutes per spermatophore (Arnold 1976). Ambystoma dumerili (Brandon 1970), A. opacum, A. laterale, A. texanum, A. mexicanum and A. gracile appear to deposit spermatophores at comparable rates. None of the Ambystoma species so far observed spend as much courtship time per spermatophore as plethodontids, but species that insert a tail-nudging walk between spermatophore depositions (A. tigrinum, A. dumerili, A. mexicanum, A. laterale) spend more time per spermatophore than species that lack a tail-nudging walk (A. maculatum, A. opacum and A. texanum).

The diversity of temporal structure in salamandrid courtship thwarts any simple comparison with plethodontids and ambystomatids. Salamandrid species differ in the kinds of courtship actions that are intercalated between spermatophore depositions. Species that intercalate more actions between successive spermatophores take a longer time but the actual durations of these actions have been timed only in the genus Triturus (Halliday 1974, 1976). The following discussion will simply focus on the sequence diversity in the family.

Most salamandrid males execute a display as a preliminary courtship activity before any spermatophores are deposited. During such displays glandular secretions from the male's cloaca or head are delivered in a variety of ways. The male may execute his display while clasping the female or while stationed in front of her. Some species intercalate this same display, or parts of it, between successive spermatophore depositions, but other species simply deposit successive spermatophores without repeating the preliminary courtship display. The preliminary displays of Cynops, Notophthalmus, and Triturus share many behavioral components and are probably homologous (Arnold 1972). The diversity of actions performed between successive spermatophore depositions increases in the series Cynops, Notophthalmus, and Triturus. Cynops males deposit spermatophores in rapid succession while marching forward in a straight line. Notophthalmus males execute a lateral displacement of the sacrum between spermatophore depositions and consequently the interval between spermatophore depositions is longer than in Cynops. In Triturus this same action is executed (called "creep on"; Halliday 1974), but, in addition, various components of the preliminary display (retreat and static displays) may be inserted between spermatophore depositions (Halliday, this volume). Taricha and Salamandra perform the preliminary display while clasping the female (Davis and Twitty 1964; Häfeli 1971; Joly 1966). Both genera repeat the clasping display between spermatophore depositions. Salamandra salamandra males actually release and recapture the female between successive spermatophore depositions despite the fact that the male does not release the female during sperm transfer (Arnold 1972). The simplest temporal structure in the family, that of Cynops, is comparable to the most elaborate structure in the ambystomatid genus Ambystoma, as in A. tigrinum.

A relationship between courtship investment and spermatophore success is important because it would help explain the otherwise perplexing elaborateness of salamander courtship. Elaborate, time-consuming courtship may be adaptive because it promotes the success of the male's spermatophores. Spermatophores can fail to inseminate the female for a variety of reasons: the female may have no interest in the spermatophore, the female may be interested but simply fails to find it, or she may find it but then fumble her attempt to retrieve sperm. An elaborate courtship can guard against these kinds of failure by diagnosing the female's interest before spermatophore deposition (and hence eliminating spermatophore deposition with nonresponsive females), by enhancing the female's responsiveness with courtship glands, and by coordinating locomotion during the sperm transfer process. The best test would relate particular kinds of failure to particular deficiencies in male courtship, and it should be conducted with a single population of salamanders. Although the ideal test has not been made, we can relate differences in spermatophore failure to courtship differences in two congeneric salamanders. Ambystoma species with simple, rapid courtship seem to have low success with each spermatophore because spermatophore placement is haphazard (Arnold 1976). Ambystoma maculatum males hardly pause between spermatophore depositions and place their spermatophores in a great diversity of sites. The male does not orient the female during spermatophore discovery, and the probability of the female finding a particular spermatophore is correspondingly low. In species with a tail-nudging walk such as A. tigrinum, there is much less variability in deposition site, and the female is more likely to find the spermatophore (Fig. 9).

Courtship investment per spermatophore may affect the probability of sperm transfer during a courtship. Species with greater investments per spermatophore seem to have a higher success rate per spermatophore. The available data are presented in Fig. 9. There is a suggestion of a trend in these data; it will be interesting to see if the relationship persists in the face of additional information.

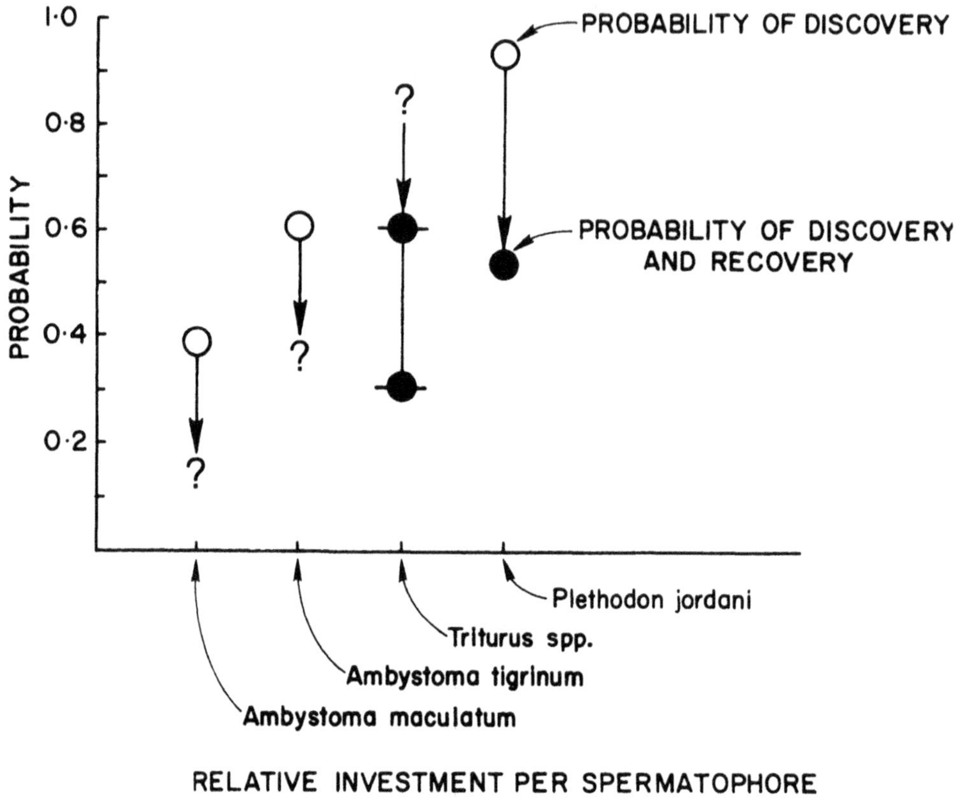

Fig. 9. The success of individual spermatophores as a function of the male's relative investment per spermatophore. Data on the probability of discovery by the female and the probability of sperm mass recovery were taken from Arnold (1976) and Halliday (1972). The range indicated for Triturus is for three species studied by Halliday.

For the moment we can consider only six species in three genera. The relative spermatophore investments of these species were ordered by the following considerations. Ambystoma tigrinum males spend more time per spermatophore than A. maculatum because they insert a tail-nudging walk between spermatophores. Triturus males intercalate parts of their preliminary display, as well as a tail-nudging walk, between spermatophore depositions. Plethodon jordani males engage the female in a very lengthy tail-straddling walk before spermatophore deposition. The same kind of relationship emerges from detailed studies of particular Triturus species (Halliday 1974). Spermatophores deposited in the later sequences of courtship, and hence after more courtship has been expended, are more successful.

The Total Spermatophore Supply and Its Distribution

The male's total spermatophore supply and his potential for daily production can be determined by the simple expedient of exposing him to a succession of females (Arnold 1976) or by inducing spermatophore depositions on successive days with artificial females (Halliday 1976). Three such determinations are graphed together in Fig. 10. The total spermatophore supply can be estimated for two species, but the sample sizes are very small in each case. The average total supply for five Ambystoma maculatum males was 77.0 spermatophores (standard deviation = 39.3, range = 28-137; Arnold 1976), and the average for five Triturus vulgaris males was 47.8 (standard deviation = 16.6, range = 27-73; Halliday 1976). Both of these are probably underestimates since the Ambystoma males were

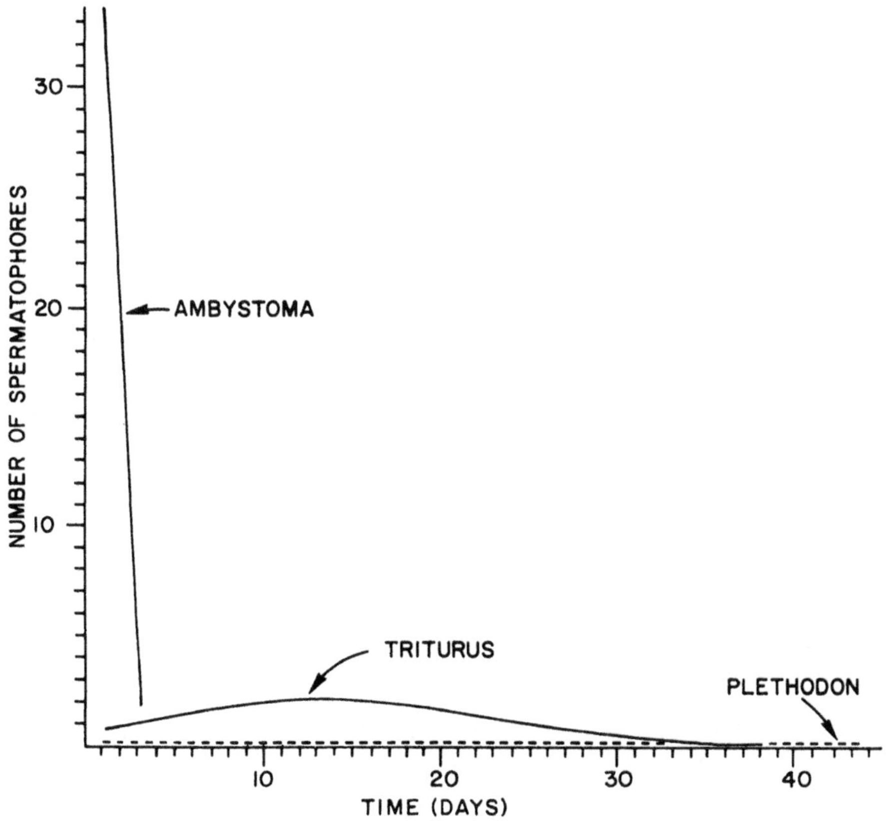

Fig. 10. Spermatophore production on successive days in three salamander genera. Data on A. maculatum and P. jordani are from Arnold (1976), and data on T. vulgaris are from Halliday (1976).

allowed to court virgin females for only three successive nights, and the Triturus males may have expended some of their supply in nature. The total supply of Plethodon jordani males could not be estimated for the latter reason. We will need many more determinations in order to detect individual, population, specific and generic differences in total spermatophore supply.

The finite spermatophore supply is distributed in very different ways by different genera (Fig. 10). The supply can be expended by depositing numerous spermatophores in a few nights of courtship as in Ambystoma, by depositing a few spermatophores each day for a few weeks as in Triturus, or by depositing one spermatophore each week for several weeks as in Plethodon. The particular pattern of distribution is probably related to the length of the courtship season. In Ambystoma there is a premium on early insemination of females during a short courtship season but, in Plethodon, females do not oviposit until well after the courtship season begins, and the period of active courtship lasts for several weeks (Arnold 1976). Thus the timing of oviposition in relation to courtship, and climatic constraints on the length of the courtship season, may have profound effects on the evolution of spermatophore distribution. If the courtship season is long, selection may favor males that conserve spermatophores and are thereby capable of inseminating females encountered late in the season. If the courtship season is short, selection may favor males that expend all of their spermatophores on the first few females they encounter; conservation is pointless.

INTERACTIONS BETWEEN RIVAL MALES

Competition between male salamanders for mating success is a neglected topic. Competition between males in nature has long been known (Blanchard 1930; Breder 1927; Ritter 1897; Tsutsui 1931), but the tactics employed during such competition can usually be appreciated only in the laboratory where close observation is possible. Only a few taxa have been observed closely, but even these present a surprising array of tactics. Interactions between rival males are fascinating in their own right, but they could also have a great impact on the evolution of courtship. The full extent of this impact, however, is unknown.

Contests between rivals are not trivial if sexual success is limited by the other sex. The relative sexual success of a contestant depends both on the success of rivals and on the individual's success. A particular behavior may evolve because it depresses the success of rivals or because it directly promotes sexual success. Contests between rivals for mating success have two components: an offensive component, "sexual interference," and a defensive component, "sexual defense" (Arnold 1976). In the ideal case, sexual interference depresses the sexual fitness of rivals by active usurpation of mating success. Sexual defense depresses the rival's success by defense of a limiting sexual resources such as a female.

Sexual interference and sexual defense take a variety of forms in salamanders (Arnold 1976). Sexual interference can be accomplished by stealing a female, disrupting the spermatophore depositions of rivals, by covering the spermatophores of rivals, or by duping rivals into unprofitable spermatophore depositions. Sexual defense can be accomplished by transporting females away from rivals or by chasing rivals away from females. Competition has been observed among ambystomatid and plethodontid males (Arnold 1976; Organ 1958, 1960a, 1960b, 1961a; Storez 1969) and in the salamandrid genera _Taricha_ and _Pleurodeles_ (also see Joly 1966, for remarks on _Salamandra_).

Interactions between Ambystomatid Males

Ambystoma maculatum males actively cover the spermatophores of rivals (Fig. 11). This behavior has an important consequence: the rival's sperm are rendered inaccessible to the female (Fig. 12). A male courting a female in the absence of rivals will also cover his own spermatophores. Is this behavior nonadaptive? Probably not, if we consider the typical social context for courtship in this species. _A. maculatum_ usually court in polyandrous aggregations: as many as 10-50 males may simultaneously vie for representation in the female's spermatheca. A male will cover any spermatophore he encounters in such aggregations, and, because of the abundance of rivals, he is more likely to encounter a rival's spermatophore than his own. One might suppose that spermatophore covering is advantageous only if the rival's spermatophores outnumber the male's spermatophores, but Arnold (1976) showed that spermatophore covering is advantageous regardless of relative abundance so long as at least one rival spermatophore is present. Thus spermatophore covering is an evolutionarily stable strategy in the sense of Maynard Smith (1974). The trait is advantageous when covering males are in low frequency, and a population of covering males cannot be invaded by males that do not cover. The universality of spermatophore covering in the genus _Ambystoma_ coincides with this theoretical prediction. Multiple spermatophores have been reported in almost every _Ambystoma_ species studied so far: _A. annulatum_ (Spotila and Beumer 1970), _A. jeffersonianum_ (Uzzell 1969), my observations on _A. opacum_, _A. maculatum_, _A. tigrinum_, _A. laterale_, _A. macrodactylum_, _A. gracile_, _A. texanum_, _A. mexicanum_; as well as literature reports for these species (Gasco 1881; Noble and Brady 1933; Smith 1911; Uzzel 1969).

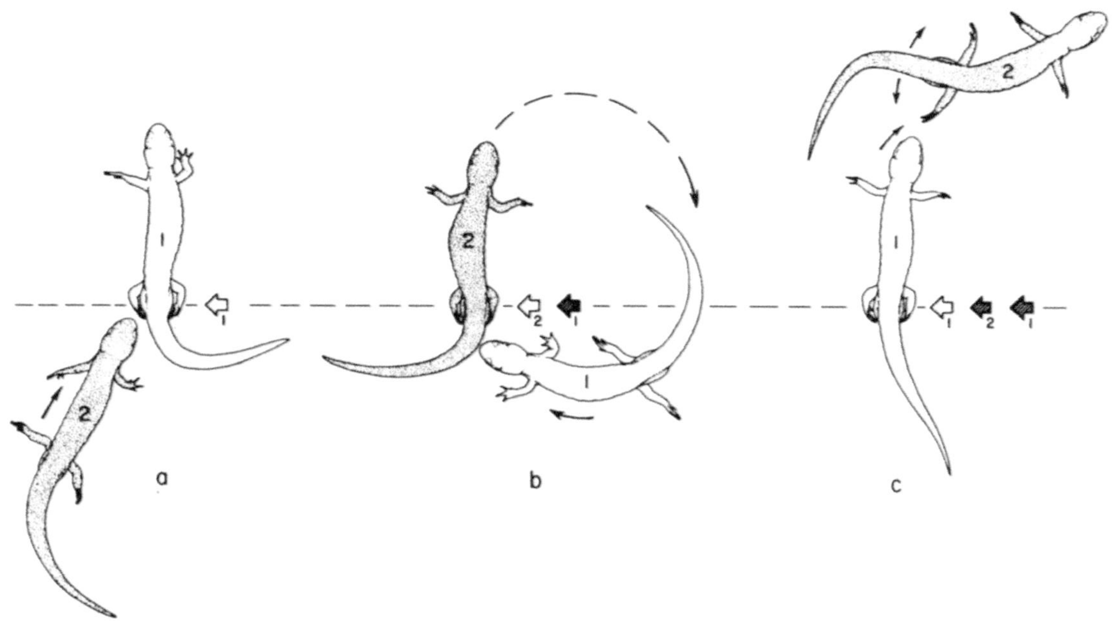

Fig. 11. Active spermatophore covering in *Ambystoma maculatum*. (a) Male 2 approaches as male 1 deposits a spermatophore (indicated with an open arrow 1). (b) Male 1 has circled around as male 2 now covers the spermatophore of male 1 with one of his own. (c) Male 1 has moved behind male 2 and now covers the spermatophore of male 2 with his spermatophore. The spermatophore is now a triple structure as in Fig. 12. (From Arnold 1976). Reproduced with permission of Verlag Paul Parey, Berlin and Hamburg, copyright 1976.

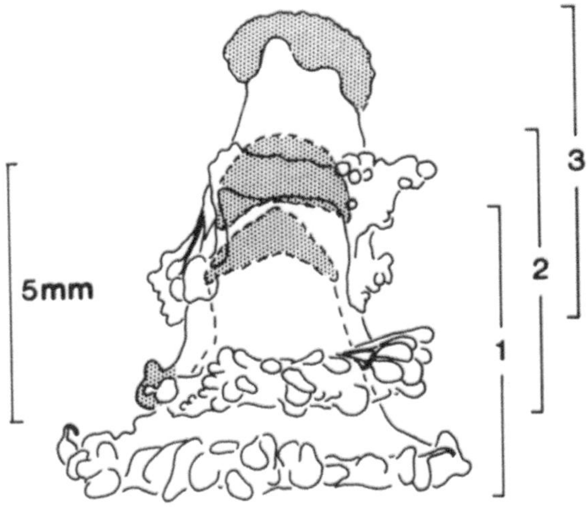

Fig. 12. A triple spermatophore of Ambystoma maculatum. Notice that the sperm masses (stippled) of the two lower spermatophores are completely enclosed by the spermatophore bases on top of them. (From Arnold 1976). Reproduced with permission of Verlag Paul Parey, Berlin and Hamburg, copyright 1976.

Spermatophore covering occurs in a slightly more elaborate context in some species. Ambystoma tigrinum males will actively intrude into the courtship of another male and dupe the rival into unprofitable spermatophore deposition (Fig. 13). Males guard against such sexual interference by transporting the female away from rivals before beginning the vulnerable process of sperm transfer (Arnold 1976). The male transports the female by shoving her for a distance of several body lengths. This shoving behavior almost invariably precedes epochs of spermatophore deposition even if the male is alone with the female. The same kind of shoving behavior has been observed in the closely related species A. mexicanum and A. dumerili (Arnold 1972; Brandon 1976; Gasco 1881), but interactions between males have not been studies.

Sexual monopoly of the female may be one of the main functions of clasping in Ambystoma. Storez (1969) found that A. laterale males could transport the female away from rivals while clasping her. This species clasps the female's dorsum in much the same posture as Taricha (Fig. 6), and the female is transported by vigorous swimming movements of the male's tail (Kumpf and Yeaton 1932). Storez (1969) suggested that swimming transportation is elicited only in the presence of rivals, but it can also be observed in solitary pairs (pers. observ.). A. gracile and A. macrodactylum males also are able to transport females during clasping (Knudsen 1960, pers. observ.). The nature and impact of male-male interactions is by no means clear for these three Ambystoma species since only a few observations are available for each of them. Storez (1969) noted that an intruding male A. laterale will sometimes try to clasp

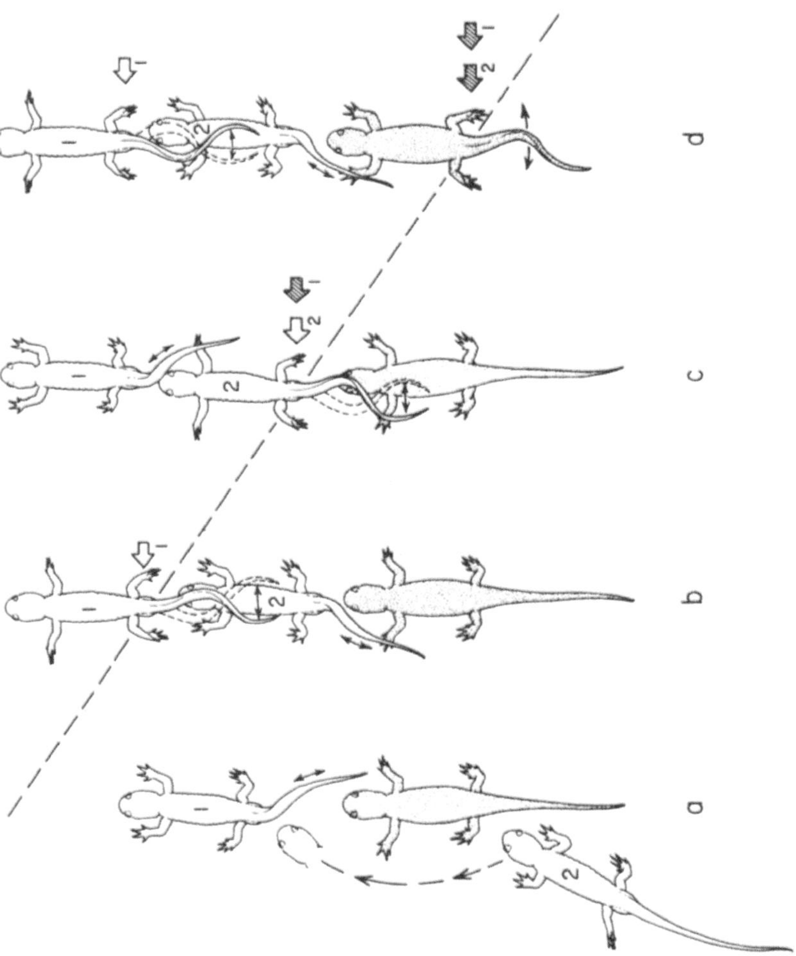

Fig. 13. Sexual interference in *Ambystoma tigrinum*. (a) Male 2 intrudes as male 1 leads the female (stippled) in a tail-nudging walk. (b) Male 2 has now moved between male 1 and the female and monitors the female's position by tapping with his tail. Male 1 deposits a spermatophore. (c) Male 2 covers male 1's spermatophore with one of his own. (d) The female attempts to retrieve sperm from the double spermatophore. Only the sperm of male 2 is accessible, however. Male 1 deposits another spermatophore. (From Arnold 1976). Reproduced with permission of Verlag Paul Parey, Berlin and Hamburg, copyright 1976.

a clasping pair, and Kumpf and Yeaton (1932) observed males clasping each other. I observed the latter behavior in A. gracile. A. gracile males also follow other males during spermatophore deposition, but I did not see spermatophore covering in this context. Detailed observations of male-male interactions in these species may help elucidate the significance of clasping.

A. jeffersonianum, A. laterale, A. macrodactylum and A. gracile males rub their chins on the female's snout during clasping (Anderson 1961; Kumpf and Yeaton 1932; Licht 1969; Mohr 1930; Storez 1969; Uzzell 1969). It is not known whether males possess a special gland on their chins. The behavior may simply stop females from moving forward so that the male can dismount and begin sperm transfer.

Aggressive interactions between sexually active males have been observed in the ambystomatid genus Rhyacotriton (Arnold 1972). When several males were housed with females, I frequently observed males pursuing and biting other males. I never saw interactions of this kind between males and females or between females. Preliminary courtship was observed on several occasions (Fig. 1) and spermatophores were discovered in the cages but spermatophore deposition was never observed.

Interactions between Salamandrid Males and the Significance of Clasping

In two clasping genera, Pleurodeles and Taricha, males grapple for the possession of females, but a clasping male sometimes can transport the female away from rivals. In Taricha the male clasps the dorsum of the female (Fig. 6) and transports her with vigorous swimming motions of his tail. In the field I have watched a male spirit the female away from rivals in this way. Female monopoly is not foolproof, however. Two males may attempt to clasp the same female in the laboratory, and the ensuing grappling contest may last for many minutes. "Balls" of newts, each composed of one female and several males, are sometimes found in the field.

Only laboratory observations are available for Pleurodeles. Males capture the female by rushing beneath her and clasping her forelimbs (Fig. 14). A clasping male can transport the female by swimming with his tail and walking with his hindlimbs. An intruding male sometimes succeeds in clasping the original male's hindlimbs or the distal portions of the female's forelimbs. If the intruder seizes the female's forelimbs, both males may grapple for possession of the female for several minutes, sometimes wrapping their tails around the female or the more dorsal contestant. Casual observations suggest that the original male has the competitive advantage in both Pleurodeles and Taricha.

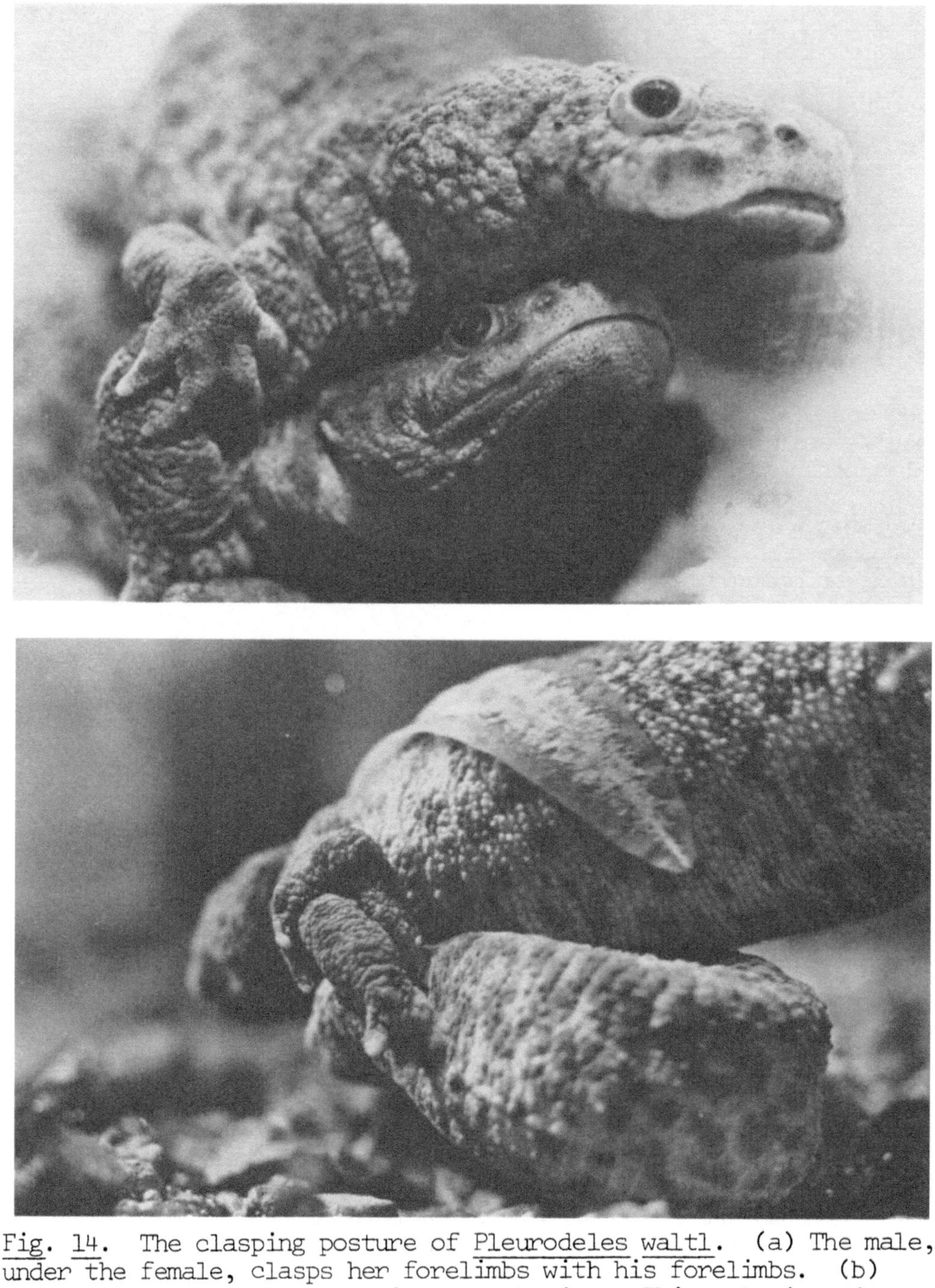

Fig. 14. The clasping posture of Pleurodeles waltl. (a) The male, under the female, clasps her forelimbs with his forelimbs. (b) The male's forelimb posture in lateral view. This male is using his tail as an adjunct to forelimb clasping by wrapping it around the female's body.

Sexual interference during sperm transfer has not been observed in Taricha, although it may occur. In Pleurodeles the male releases one of the female's forelimbs during sperm transfer (Bedriaga 1882b; Arnold 1972). On one occasion an intruding male succeeded in clasping the female's free forelimb and sperm transfer was disrupted by the resulting struggle. Intrusion during sperm transfer may place the two rivals in unbiased competition since their positions are symmetrical and the original male no longer has the advantage of a more stable posture. Males appear to be sensitive to the vulnerabilities of sperm transfer. A male will quickly reclasp the female's free limb and transport her away if any other animal contacts him during sperm transfer.

Clasping undoubtedly serves several functions in Pleurodeles and Taricha in addition to sexual monopoly. In both genera males are able to transport the female to the water surface for breathing. The necessity of breathing places a severe constraint on courtship in Triturus males which must abandon the female to breathe (Halliday 1976, 1977; Halliday and Sweatman 1976). Taricha and Pleurodeles can retain the female during breathing and are less likely to curtail or hasten courtship because of oxygen debt. Relative freedom from the constraints of oxygen debt is not a universal advantage of clasping; male Notophthalmus are unable to transport the female to the surface.

Clasping also enables males of some taxa to apply courtship glands to a restrained female. Male Notophthalmus apply genial glands and Taricha apply submandibular glands. Pleurodeles males do not rub special epidermal glands on the female but the male's everted cloaca is sometimes applied to her snout during clasping.

Clasping probably increases the coordination of partners during sperm transfer. In taxa which clasp the female's dorsum (Notophthalmus, Taricha) the male dismounts directly over the female's head and places his spermatophore in front of her snouth. Clasping may facilitate this alignment. In Euproctus and Salamandra clasping facilitates the placement of spermatophores near the female's vent. Pleurodeles males constrain the female's movement during sperm transfer by clasping one of her forelimbs.

In modern salamandrids clasping is maintained because it serves several functions: sexual defense, sexual interference, freedom from breathing constraints, facilitation of gland application and coordination during sperm transfer. Clasping takes many forms and may be polyphyletic. In order to elucidate the selective pressures responsible for the origin and diversification of clasping modes particular attention should be directed to conflicts between the various postulated functions of clasping. For example, the peculiar clasping mode of Notophthalmus seems to facilitate application of cephalic glands while precluding breathing at the water surface.

Interactions between Plethodontid Males

Two kinds of male-male interactions have been observed in plethodontid salamanders: homosexual courtships in which one male behaves as a pseudofemale, and aggressive encounters in which sexual rivals bite and pursue each other. In homosexual courtships, one male elicits a tail-straddling walk and dupes the other male into an unprofitable spermatophore deposition. Such interaction has been observed in Pseudotriton ruber (Organ and Organ 1968), Desmognathus ochrophaeus (pers. observ.), Ensatina eschscholtzi (pers. observ.), Plethodon glutinosus (Arnold 1972), P. jordani (Arnold 1976; Organ 1958), P. ouachitae (Arnold 1972), and P. yonahlossee (Organ, pers. comm.). Aggressive interactions between courting males have been observed in Eurycea bislineata (Arnold 1972), Pseudotriton ruber (Organ and Arnold pers. observ.), Desmognathus (Organ 1961a; pers. observ.), Plethodon glutinosus (Organ 1960a), P. caddoensis (Arnold 1972), P. ouachitae (Arnold 1972), P. jordani (Arnold 1976; Organ 1958), and P. welleri (Organ 1960b).

James Organ was the first to observe these two kinds of interactions in plethodontids. He proposed that aggressive interactions between males serve the function of sex recognition, but attached no adaptive significance to the phenomenon of homosexual courtship (Organ 1961a; Organ and Organ 1968). Arnold (1976) suggested that homosexual courtship is an adaptive mode of sexual interference in plethodontids rather than an incidental malfunction indicating poor ability for sex recognition. Aggressive interactions between males certainly indicate a capacity for sex identification (since aggression between the sexes is much rarer), but the main function may be to disperse rivals and so guard against homosexual courtship and other kinds of sexual interference (Arnold 1976).

Thurow (1976) reports aggressive interactions between the sexes of various Plethodon species as well as between males and between females. Thurow suggests that aggressive interactions in Plethodon may be important in the allocation of space as well as in the allocation of sexual resources. In contrast to Thurow, I never observed aggressive interactions between females; however, my observations were made only during the courtship season. We need well designed laboratory studies, as well as field experiments, to determine whether aggressive interactions represent contests for retreats and food as well as for mates. It is clear, however, that aggressive behavior is a real part of the biology of many plethodontids.

The Evolution of Courtship Variables and the Impact of Male-Male Interactions

The male's courtship investment per spermatophore varies among salamander taxa. This courtship variable is probably molded by two conflicting selective pressures: sexual success with particular females and sexual competition among males. Each taxon has responded to these two pressures with a characteristic compromise. Lengthy courtship before spermatophore deposition promotes the success of each spermatophore. Selection may favor rapid courtship, however, if sexual rivals are likely to intrude, and if the male has no means of monopolizing the female. This seems to be the most plausible explanation for the very rapid courtship of various Ambystoma species such as A. annulatum, A. maculatum, and A. opacum, which characteristically court in polyandrous aggregations and have no means of monopolizing females. Plethodontids and some Ambystoma species (A. laterale and A. tigrinum) may be able to exploit the advantages of lengthy, elaborate courtship because males are able to defend the female against intruding rivals.

In addition to courtship investment per spermatophore, the number of spermatophores per courtship seems to evolve in response to at least two conflicting selective pressures. No salamander species has a fool-proof system for sperm transfer, and the more spermatophores the male deposits, the greater the likelihood of sperm transfer. The male's spermatophore production is limited, however. If sexually receptive females are available for a long time, selection may favor males that deposit relatively few spermatophores per courtship and can thereby inseminate females encountered later in the season. This principle may explain the salamandrid and plethodontid patterns of low, sustained spermatophore production and the explosive productivity of Ambystoma. The courtship season of Plethodon is much longer than that of Ambystoma (10 weeks vs. 1-2 weeks; Arnold 1976). Unfortunately there are no published data on the length of salamandrid courtship seasons. Thus the length of time during which receptive females are available may determine the male's pattern for spermatophore allocation.

CONCLUSIONS

Salamander courtship is notable for complexity at two levels. Many species have elaborate courtship with complex temporal structure, and there is considerable courtship diversity in the order Urodela. The elaborateness of salamander courtship is largely a response to the difficulties of indirect sperm transfer. Many arthropods, which, like salamanders, attach spermatophores to the substrate, have intricate sexual behavior (Schaller 1971). In contrast, salamanders that practice external fertilization (cryptobranchids and hynobiids) have relatively simple sexual behavior, and the male's activities are directed towards eggs rather than females (Smith 1907; Kerbert 1904; Sasaki 1924; Thorn 1962, 1963, 1966; Rehberg 1962). Taxonomic diversity in sperm transfer behavior can be attributed to the long phylogenetic history of salamanders.

Salamanders have organized their sexual behavior around spermatophores for millions of years. The behavioral radiation we see today is the reflection of an ancient commitment to an intrinsically difficult mode of insemination.

Many archaic courtship activities probably represent important solutions to adaptive hurdles. The plethodontid tail-straddling walk and the 90° turn of salamandrids are major innovations in the process of sperm transfer. Both behaviors lessen the chance that receptive females will fail to find the spermatophore. The plethodontid trait of injecting mental gland secretions with the premaxillary teeth is probably an ancient solution to the problem of female non-receptivity.

Diversification in techniques for female persuasion is one of the striking features in the behavioral evolution of salamanders. This radiation can be understood in a general way as a consequence of rapid coevolution between male ability for persuasion and female responsiveness. Studies of behavioral isolation between populations of Cynops suggest that these two behavioral traits can coevolve rapidly (Kawamura and Sawada 1959; Sawada 1963). The precise dynamics of the evolutionary process are not clear, however. We do not know how female responsiveness affects selection for male behavior within populations. The selective advantage of coyness in females, a second major problem, has only recently been approached with experimentation (Halliday, this volume). Thus, while we can appreciate an extensive radiation in techniques for persuasion, we are still speculating about its causes.

The main thesis of this review is that the sexual dynamics of extant populations will illuminate the macroevolution of salamander courtship. Courtship activities seem to serve several functions: identification of mating partners, female persuasion, coordination during sperm transfer, sexual defense of the female and sexual interference. Thus courtship behavior has probably been molded by separate, and sometimes conflicting, selective pressures exerted by sexual partners and sexual rivals.

I thank Timothy Halliday, Lynne Houck, David Sever and David Wake for helpful discussions and comments on the manuscript and Hymen Marx for making specimens in the Field Museum of Natural History (FMNH) available for study. Edmund Brodie, Jr., Gunther Freytag, Ronald Heyer, Lynne Houck, Jens Knudsen, James Lynch, James Organ, Ronald Nussbaum, Robert Thorn, Stephen Tilley, David Wake, Thomas Wake and Henry Wilbur generously helped in the procurement of specimens. George Nace and Christina Richards kindly allowed me to study several species maintained at the University of Michigan Amphibian Facility. I am particularly indebted to James Organ and Robert Stebbins for many stimulating conversations. They were the first actually to observe sperm transfer in plethodontid salamanders, and their success encouraged me to work with this group. Portions of this work were submitted in partial fulfillment of the degree of Doctor of Philosophy at the University of Michigan. Some of the observations reported here were made possible by a Postdoctoral Fellowship from the Miller Institute for Basic Research in Science of the University of California, Berkeley; a Grant-in-Aid of Research from the Society of Sigma Xi and financial support from the Louis H. Block Fund of the University of Chicago.

LITERATURE CITED

Anderson, J. D. 1961. The courtship behavior of Ambystoma macrodactylum croceum. Copeia 1961(2):132-139.

Anderson, J. D. 1970. Description of the spermatophore of Ambystoma tigrinum. Herpetologica 26(3):304-308.

Arnold, S. J. 1972. The evolution of courtship behavior in salamanders. Ph. D. dissertation, University of Michigan, 570 p.

Arnold, S. J. 1976. Sexual behavior, sexual interference and sexual defense in the salamanders Ambystoma maculatum, Ambystoma tigrinum and Plethodon jordani. Z. Tierpsychol. 42:247-300.

Baker, C. L., L. C. Baker and M. F. Caldwell. 1947. Observation of copulation in Amphiuma tridactylum. J. Tenn. Acad. Sci. 22: 77-87.

Bedriaga, J. 1882a. Über die Begattung bei einigen geschwänzten Amphibien. Zool. Anz. 5:265-268.

Bedriaga, J. 1882b. Über die Begattung bei einigen geschwänzten Amphibien. Zool. Anz. 5:357-359.

Bishop, S. C. 1926. Notes on the habits and development of the mudpuppy, Necturus maculosus (Rafinesque). N. Y. State Mus. Bull. No. 268: 60 pp.

Blanchard, F. N. 1930. The stimulus to the breeding migration of the spotted salamander, Ambystoma maculatum (Shaw). Am. Nat. 64:154-167.

Brame, A. H., Jr. 1968. Systematics and evolution of the Mesoamerican salamander genus Oedipina. J. Herpetol. 2(1-2):1-64.

Brandon, R. A. 1970. Courtship, spermatophores and eggs of the Mexican Achoque, Ambystoma (Bathysiredon) dumerili (Dugès) Zool J. Linn. Soc. 49:247-254.

Breder, R. C. 1927. Courtship of the spotted salamander. Bull. N. Y. Zool. Soc. 30:3.

Briegleb, W. 1962. Zur Biologie und Ökologie des Grottenolms (Proteus anguineus Laur. 1768). Zeit. Morph. Ökol. Tiere 51:271-334.

Cedrini, L. and A. Fasolo. 1971. Olfactory attractants in sex recognition of the crested newt. An electrophysiological research. Monit. Zool. Ital. (N. S.) 5:223-229.

Cope, E. D. 1889. The batrachia of North America. U. S. Nat. Mus. (34):5-525.

Dähne, C. 1926. Die Copula bei Euproctus asper Dug. Blätter Aquar.-Terrar. Kde. 37:198-201.

Davis, W. C. and V. C. Twitty. 1964. Courtship behavior and reproductive isolation in the species of Taricha (Amphibia, Caudata). Copeia 1964(4):601-610.

Despax, R. 1923. Contribution à l'étude anatominique et biologique batraciens urodèles du groupe des Euproctus, et specialement de l'euprocte des Pyrenees. Thèse Facult. Sci. Paris, 258 pp.

Dunn, E. R. 1923. The breeding habits of salamanders and their bearing on phylogeny. Copeia 1923(115):25-27.

Dunn, E. R. 1926. The salamanders of the family Plethodontidae. Smith College, Mass., 441 pp.

Gallien, L. 1953. Elevage et comportment du Pleurodele au laboratoire. Bull. Soc. Zool. France 77:456-461.

Garton, J. S. 1972. Courtship of the small mouthed salamander Ambystoma texanum, in southern Illinois. Herpetologica 28:41-45.

Gasco, F. 1881. Les amours des axolotls. Bull. Soc. Zool. France 6:151-164.

Häfeli, H. P. 1971. Zur Fortpflanzungsbiologie des Alpensalamanders (Salamandra atra Laur.). Rev. Suisse de Zoologie 78(2): 235-293.

Halliday, T. R. 1972. A Study of Behaviour of Newts. D. Phil. Thesis, Oxford Univ.

Halliday, T. R. 1974. Sexual behaviour of the smooth newt, Triturus vulgaris (Urodela, Salamandridae). J. Herpetol. 8(4):277-292.

Halliday, T. R. 1975. An observational and experimental study of sexual behaviour in the smooth newt, Triturus vulgaris (Amphibia: Salamandridae). Anim. Behav. 23(2):291-322.

Halliday, T. R. 1976. The libidinous newt. An analysis of variations in the sexual behaviour of the male smooth newt, Triturus vulgaris. Anim. Behav. 24(2):398-414.

Halliday, T. R. 1977. The effects of experimental manipulation of breathing behaviour on the sexual behaviour of the smooth newt, Triturus vulgaris. Anim. Behav. 25:39-45.

Halliday, T. R. and H. P. A. Sweatman. 1976. To breathe or not to breathe; the newt's problem. Anim. Behav. 24:551-561.

Highton, R. 1962. Revision of North American salamanders of the genus Plethodon. Bull. Florida State Museum 6(3):235-367.

Highton, R. 1970. Evolutionary interactions between species of North American salamanders of the genus Plethodon. Part 1. Genetic and ecological relationships of Plethodon jordani and P. glutinosus in the southern Appalachian Mountains. Evol. Biol. 4:211-241. Appleton-Century-Crofts, New York.

Hilton, W. A. 1902. A structure connected with the mating of Diemyctylus viridescens. Am. Nat. 36:643-651.

Humphries, A. A., Jr. 1955. Observations on the mating behavior of normal and pituitary-implanted Triturus viridescens. Physiol. Zool. 28(1):73-79.

Hutchison, V. C. 1959. Aggressive behavior in Plethodon jordani. Copeia 1959(1):72-73.

Joly, J. 1966. Sur l'ethologie sexuelle de Salamandra salamandra L. Zeit. Tierpsychol. 23(1):8-27.

Jordan, E. O. 1891. The spermatophores of Diemyctylus. Jour. Morph. 5:263-270.

Kawamura, T. and S. Sawada. 1959. On the sexual isolation among different species and local races of Japanese newts. J. Sci. Hiroshima Univ. ser B. div. 1 18:17-31.

Kerbert, C. 1904. Zur Fortpflanzung von Megalobatrachus maximus Schlegel. Zool. Anz. 27(10):305-320.

Knudsen, J. W. 1960. The courtship and egg mass of Ambystoma gracile and Ambystoma macrodactylum. Copeia 1960(1):44-46.

Kumpf, K. F. 1934. The courtship of Ambystoma tigrinum. Copeia 1934(1):7-10.

Kumpf, K. F. and S. C. Yeaton. 1932. Observations on the courtship behavior of Ambystoma jeffersonianum. Am. Mus. Novit. 546:1-7.

Lanza, B. 1959. Il corpo ghiandolare mentoniero dei Plethodontidae (Amphibia, Caudata). Monit. Zool. Ital. 67:15-53.

Licht, L. E. 1969. Observations on the courtship behavior of Ambystoma gracile. Herpetologica 25(1):49-52.

Maynard Smith, J. 1974. The theory of games and the evolution of animal conflicts. J. Theor. Biol. 47:209-221.

Mohr, C. E. 1930. The ambystomid salamanders of Pennsylvania. Proc. Penn. Acad. Sci. 4:50-55.

Noble, G. K. 1927a. The plethodontid salamanders; some aspects of their evolution. Am. Mus. Novit. 249:1-26.

Noble, G. K. 1927b. The value of life history data in the study of the evolution of the amphibia. Ann. N. Y. Acad. Sci. 30:31-128.

Noble, G. K. 1929. The relations of courtship to secondary sex characters of the two-line salamander, Eurycea bislineata (Green). Am. Mus. Novit. 362:1-5.

Noble, G. K. and M. K. Brady. 1930. The courtship of the plethodontid salamanders. Copeia 1930(2):52-54.

Noble, G. K. and M. K. Brady. 1933. Observations on the life history of the marbled salamander, Ambystoma opacum Gravenhorst. Zoologica (N. Y.) 11(8):89-132.

Noble, G. K. and S. H. Pope. 1929. The modification of the cloaca and teeth of the adult salamander, Desmognathus, by testicular transplants and by castration. Brit. J. Exper. Biol. 6(4):399-411.

Organ, J. A. 1958. Courtship and spermatophore of Plethodon jordani metcalfi. Copeia 1958(4):251-259.

Organ, J. A. 1960a. The courtship and spermatophore of the salamander Plethodon glutinosus. Copeia 1960(1):34-40.

Organ, J. A. 1960b. Studies on the life history of the salamander Plethodon welleri. Copeia 1960(4):287-297.

Organ, J. A. 1961a. Studies on the local distribution, life history, and population dynamics of the salamander genus *Desmognathus* in Virginia. *Ecol. Monog.* *31*:189-200.

Organ, J. A. 1961b. Life history of the pygmy salamander, *Desmognathus wrighti*, in Virginia. *Am. Midl. Nat.* *66*:384-390.

Organ, J. A. and D. J. Organ. 1968. Courtship behavior of the red salamander *Pseudotriton ruber*. *Copeia* *1968*(2):217-223.

Rehberg, F. 1962. Beobachtungen über die Fortpflanzung des *Hynobius nebulosus*. *Mitteilungsblatt Salamander Ges. für Terrarinfr. Buchloe.* *9*:62-63.

Ritter, W. E. 1897. *Diemyctylus torosus* Esch. The life history and habits of the Pacific Coast newt. *Proc. Calif. Acad. Sci. (III) Zool.* *1*(2):73-114.

Rogoff, J. L. 1927. The hedonic glands of *Triturus viridescens*; a structural and morphological study. *Anat. Rec.* *34*:132-133.

Salthe, S. N. 1967. Courtship patterns and phylogeny of the urodeles. *Copeia* *1967*(1):100-117.

Salthe, S. N. and J. S. Mecham. 1974. Reproductive and courtship patterns. IN: B. Lofts (ed.), *The Physiology of Amphibia, Vol. II*, pp. 309-521. Academic Press.

Salthe, S. N. and B. M. Salthe. 1964. Induced courtship in the salamander *Pseudoeurycea belli*. *Copeia* *1964*(3):574-576.

Sasaki, M. 1924. On a Japanese salamander in Lake Kuttarush which propogates like the axolotl. *J. Coll. Agric. Hokkaido Imp. Univ.* *15*:12-23.

Sawada, S. 1963. Studies on the local races of the Japanese newt, *Triturus pyrrhogaster* Boie, II. Secual isolation mechanisms. *J. Sci. Hiroshima Univ. ser. B. div. 1.* *21*:167-180.

Schaller, F. 1971. Indirect sperm transfer by soil arthropods. *Ann. Rev. Entomol.* *16*:407-446.

Sever, D. M. 1975. Morphological and seasonal variation of the mental hedonic glands of the dwarf salamander, *Eurycea quadridigitata* (Holbrook). *Herpetologica* *31*(3):241-251.

Sever, D. M. 1976. Morphology of the mental hedonic gland clusters of plethodontid salamanders (Amphibia, Urodela, Plethodontidae). *J. Herpetol.* *10*(3):227-239.

Shoop, C. R. 1960. The breeding habits of the mole salamander, Ambystoma talpoideum (Holbrook), in southeastern Louisiana. Tulane Stud. Zool. 8:65-82.

Smith, B. G. 1907. The life history and habits of Cryptobranchus allegheniensis. Biol. Bull. 13:5-39.

Smith, B. G. 1910. The structure of the spermatophores of Ambystoma punctatum. Biol. Bull. 18(4):204-211.

Smith, R. E. 1941. Mating behavior in Triturus torosus and related newts. Copeia 1941(4):255-262.

Spotila, J. R. and R. J. Beumer. 1970. The breeding habits of the ringed salamander, Ambystoma annulatum (Cope), in northeastern Arkansas. Am. Midl. Nat. 84:77-89.

Stebbins, R. C. 1949. Courtship of the plethodontid salamander Ensatina eschscholtzii. Copeia 1949(4):247-281.

Stewart, M. M. 1958. Seasonal variation in the teeth of the two-lined salamander. Copeia 1958(3):190-196.

Storez, R. A. 1969. Observations on the courtship of Ambystoma laterale. J. Herpetol. 3(1-2):87-95.

Thorn, R. 1962. Protection of the brood by a male salamander, Hynobius nebulosus. Copeia 1962(3):638-640.

Thorn, R. 1963. Contribution à l'étude d'une salamandre japonaise, l'Hynobius nebulosus (Schlegel). Comportement et reproduction en captivité. Inst. Gr.-Duc. Luxembourg, Sect. Sci. nat. phys. math., Archives (N. S.) 29:201-215.

Thorn, R. 1966. Nouvelles observations sur l'éthologie sexuelle de l'Hynobius nebulosus (temminck et Schlegel) (Caudata, Hynobiidae). Inst. Gr.-Duc. Luxembourg, Sect. Sci. nat. phys. math., Archives (N. S.) 32:267-271.

Thurow, G. 1976. Aggression and competition in Eastern Plethodon (Amphibia, Urodela, Plethodontidae). J. Herpetol. 10(4):277-291.

Tinbergen, N. 1953. Social Behaviour in Animals. Methuen. London.

Tsutsui, Y. 1931. Notes on the behavior of the common Japanese newt Diemyctylus pyrrhogaster Boie. I. Breeding habit. Mem. Coll. Sci. Kyoto Imp. Univ., ser. B. 7(4):159-178.

Twitty, V. C. 1955. Field experiments on the biology and genetic relationships of the California species of Triturus. J. Exp. Zool. 129(1):129-148.

Twitty, V. C. 1961a. Experiments on homing behavior and speciation in Taricha. IN: W. F. Blair (ed.), Vertebrate Speciation, pp. 415-459. Univ. Texas Press, Austin.

Twitty, V. C. 1961b. Second-generation hybrids of the species of Taricha. Proc. Nat. Acad. Sci. 47(9):1461-1486.

Uzzell, T. M. 1964. Relations of the diploid and triploid species of the Ambystoma jeffersonianum complex (Amphibia, Caudata). Copeia 1964(2):257-300.

Uzzell, T. M. 1969. Notes on spermatophore production by salamanders of the Ambystoma jeffersonianum complex (Amphibia, Caudata). Copeia 1969(3):602-612.

Wake, D. B. 1966. Comparative osteology and evolution of the lungless salamanders, family Plethodontidae. Mem. S. Calif. Acad. Sci. 4:1-111.

Wake, D. B. and N. Özeti. 1969. Evolutionary relationships in the family Salamandridae. Copeia 1969(1):124-137.

Wyman, R. L. 1971. The courtship behavior of the small-mouthed salamander, Ambystoma texanum. Herpetologica 27(4):491-498.

Zeller, E. 1890. Über die Befruchtung bei den Urodelen. Zeit. f. wissen. Zool. 49:583-601.

Zeller, E. 1905. Untersuchungen über die Samenträger und den Kloakenwulst der Tritonen. Zeit. f. wissen. Zool. 79:171-221.

THE COURTSHIP OF EUROPEAN NEWTS: AN EVOLUTIONARY PERSPECTIVE

T. R. Halliday

Animal Behaviour Research Group, Department of Zoology, South Parks Road, Oxford OX1 3PS, England

INTRODUCTION

A full understanding of a biological system requires parallel consideration of both the form and the function of that system. This basic approach, the essence of post-Darwinian biology, has been widely applied to the morphology and physiology of animals and in recent years to some aspects of their behavior, notably by Niko Tinbergen and his students (Baerends *et al*. 1976). Rather little use of this dual approach has been made in the analysis of the causal mechanisms underlying behavior, however, although McFarland (1976) has shown how such an approach can lead to a better understanding of the mechanisms determining the precise order in which behavior patterns occur. It is the purpose of this paper to show how an understanding of the causation of courtship behavior in the smooth newt, *Triturus vulgaris*, which has been the subject of extensive experimental analysis, can be enhanced by a consideration of various functional aspects of the behavior.

The precise form that a man-made machine like a car or a washing machine takes is the reult of constraints imposed by the properties of basic physical components like metal, gasoline and electricity and of design criteria imposed by the user of the machine, who requires certain levels of performance under certain conditions. Likewise, the behavioral mechanisms of an animal are constrained on the one hand by the properties of tissues and physiological processes and on the other by the environment which, through natural selection, imposes demands on the form of the animal's behavior. In this paper I attempt to identify the design constraints which may have operated in the evolution of newt courtship.

THE EUROPEAN NEWTS

In common with the majority of urodeles, fertilization in the Palearctic newts, genus Triturus, is internal and sperm exchange is effected by means of a spermatophore. While there are striking differences between the various species in the details of the courtship behavior, all conform to the general five-stage pattern described for urodeles by Salthe (1967). The male becomes aware of the female and makes initial overtures to her (Stage A); he then displays to her (Stage B). Following display he moves away from her (Stage C), deposits a spermatophore (Stage D) and finally, as he moves away from it, the female moves over the spermatophore and picks it up (Stage E). Triturus is remarkable among urodeles in two respects. First, the male does not capture or grasp the female during courtship; indeed, contact between the two is minimal (Salthe and Mecham 1974). Second, there is a higher degree of sexual dimorphism in species of Triturus than in any other genus.

In most species of Triturus most of the year is spent in a terrestrial phase with a two to four month aquatic breeding season. Courtship takes place on the bottom of the breeding pond and the eggs are laid under water, each one individually wrapped in a leaf by the female. During aquatic life newts are able to take in dissolved oxygen through their skins but they are dependent upon gaseous air obtained during occasional breathing ascents to the water surface.

COURTSHIP IN THE SMOOTH NEWT, TRITURUS VULGARIS

Of the nine species of European newts, T. vulgaris, with its nine subspecies, has the largest range (Fig. 1). The sexes are markedly dimorphic (Fig. 2). Each breeding season the male develops a deep crenulated crest and tail, skin flaps on each of the hind toes, a conspicuous coloration with dark spots on back and flanks, a black-spotted orange belly and red and blue stripes at the base of the tail behind a greatly swollen, darkly colored cloaca.

Throughout courtship the male is the more active partner and usually initiates sexual encounters (Halliday 1974) (Fig. 3). He swims or walks around the pond or observation tank and when he meets a female he sniffs her intently, especially around her cloaca, and attempts to take up a position in front of her with his body across her line of vision. A female's first reaction is almost invariably to swim or walk away. A sexually active male will follow her and will again attempt to take up his position ahead of her. There is often a prolonged orientation phase in which this sequence of interaction is repeated many times. After a time the interval between successive movements away from the male by the female becomes longer

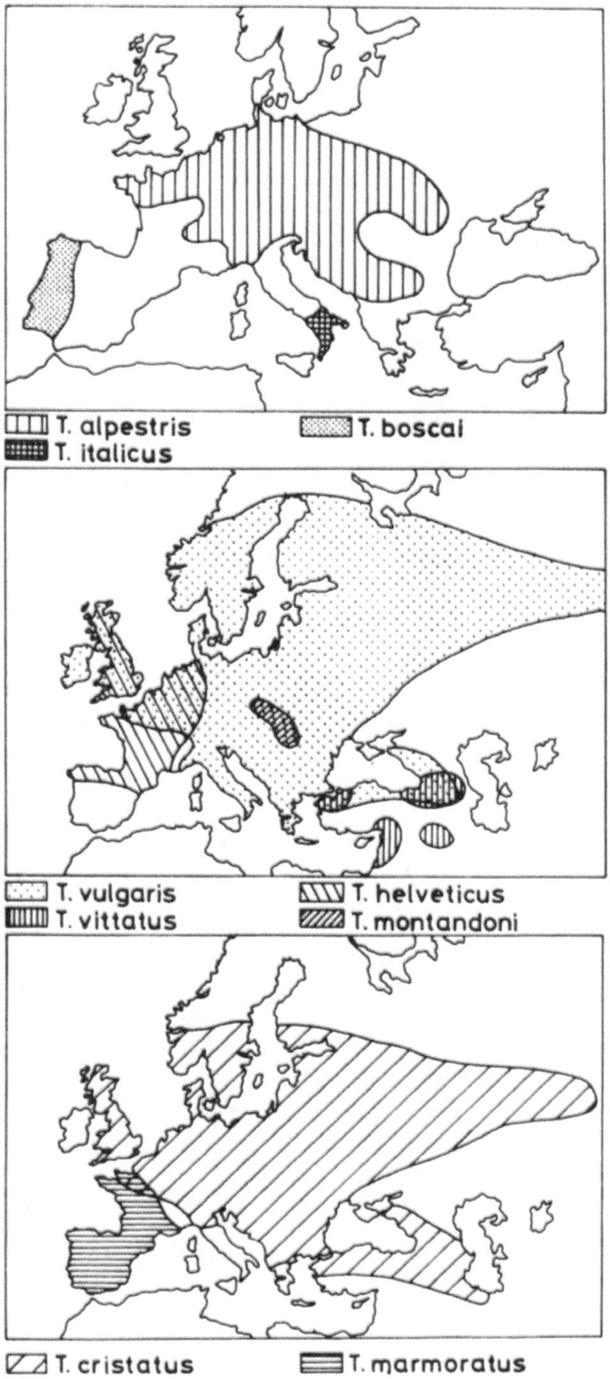

Fig. 1. Distribution maps of nine *Triturus* species.

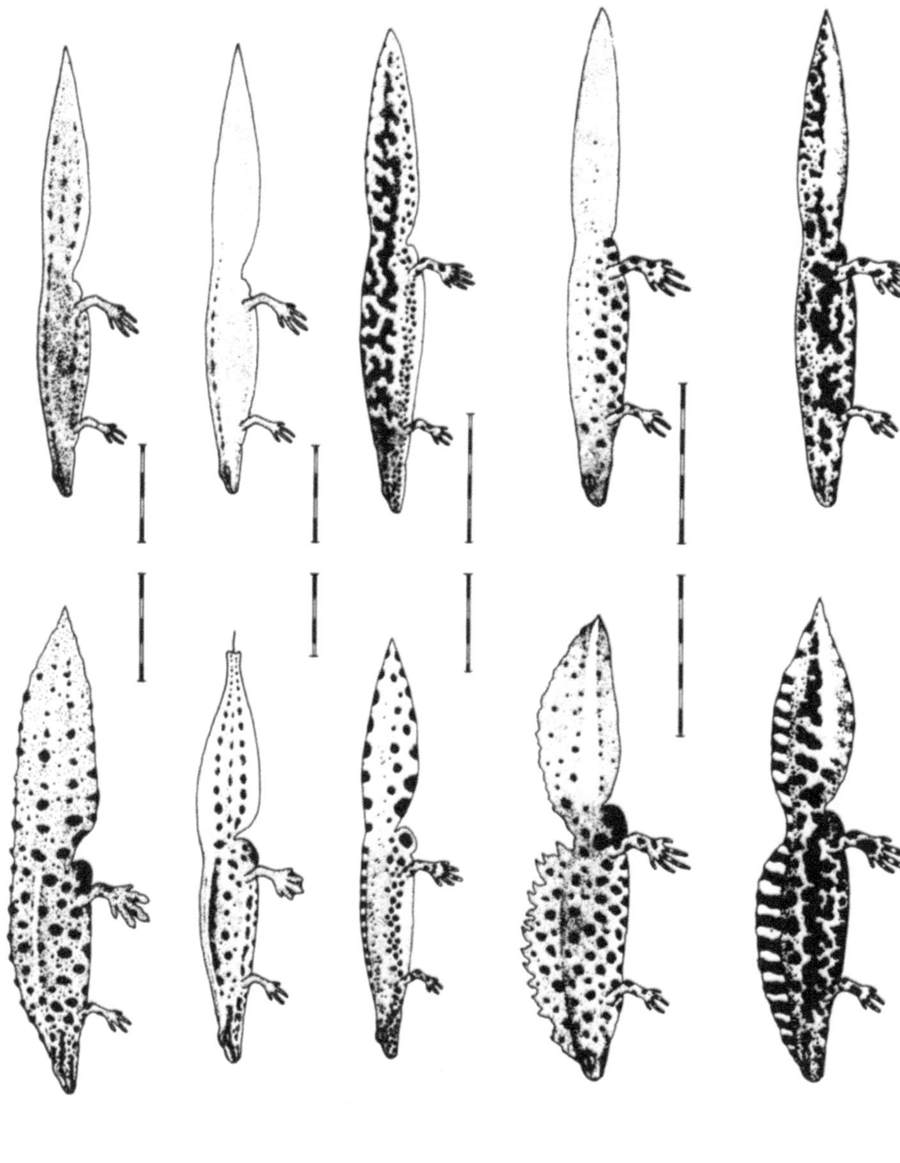

Fig. 2. Males (left) and females of five *Triturus* species, *T. vulgaris* (top), *T. helveticus*, *T. alpestris*, *T. cristatus* and *T. marmoratus*. Each division on the scale bars represents 2 cm.

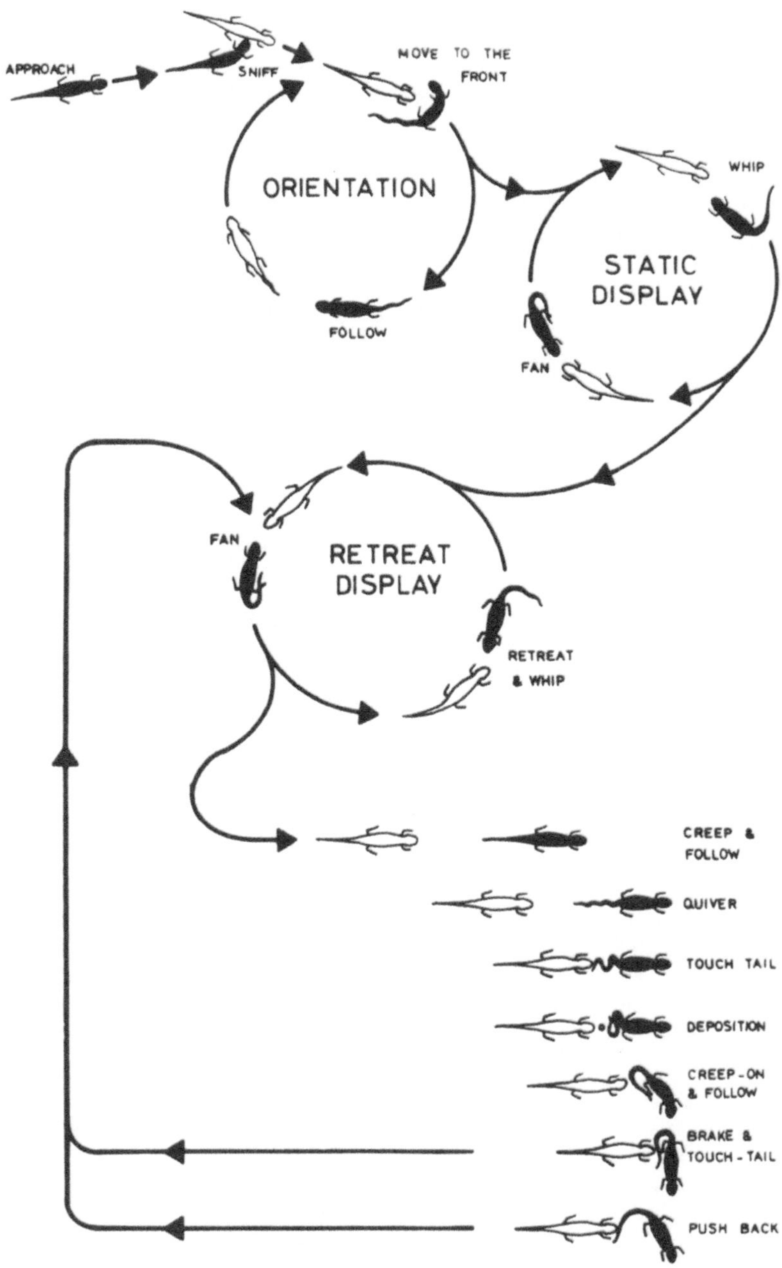

Fig. 3. The courtship sequence of the smooth newt, *T. vulgaris*. The male is in black.

and while she is stationary the male displays to her from his position in front of her. This static display consists of three distinct movements, the wave, the whip and the fan, all of which involve movements of the male's tail (Fig. 4). This heterogeneous display probably stimulates the female visually, tactually and olfactorily.

In the wave, the male holds his tail for up to a second in a position such that he presents to the female a full lateral view of his body and tail. The whip is a sudden and violent movement in which the tail, from the wave position, is lashed against the male's flank, creating a powerful blast of water that may be strong enough to knock the female backwards. By contrast, the fan is a delicate and sustained movement; the tail is curved around against the flank

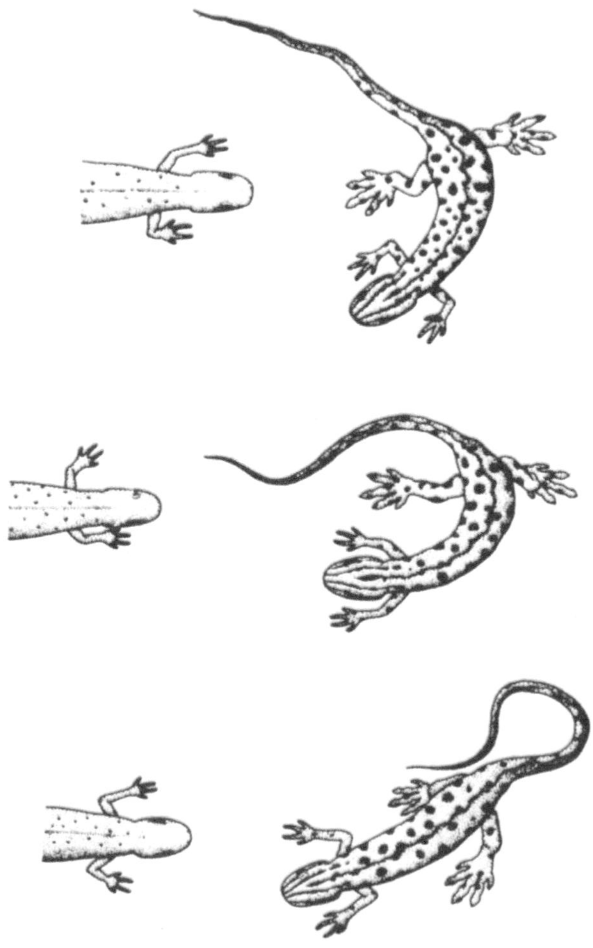

Fig. 4. The three display actions of *T. vulgaris*, the wave (top), the whip and the fan. The male is on the right.

nearest the female and its distal portion is vibrated at approximately six beats per second, generating a water current that, because of the male's orientation, is directed towards the female's nose. This current is thought to carry odor from glands in the male's skin and cloaca to the female (Tinbergen and Ter Pelkwijk 1938; Halliday 1975b). Fanning is performed in bouts lasting up to 32 seconds (mean duration 4.6 seconds).

As the female's negative responses become less frequent, bouts of static display become longer until the female starts to approach the male. He then retreats before her, maintaining both his orientation to her and his display, although this now consists mostly of whips with very few waves or fanning bouts. After this retreat display has been maintained for 5 to 60 seconds, the male ceases to display, turns and moves away from the female (creep); she follows him. After creeping for 5 to 10 cm he stops and moves his tail slowly from side to side (quiver). The female, continuing her movement, touches his tail with her snout; he then folds his tail, raises it and deposits a spermatophore onto the substrate. Immediately after spermatophore deposition the male creeps away, and in doing so moves to one side in a characteristic movement that ends with his body perpendicular to his previous direction of movement with his tail folded against the flank nearest the female (braking). She moves towards him, again touching his tail, and presses against him. He resists her pressure and may actually push her away from him by flexing his tail and body (push-back). This post-deposition maneuver is such that when the male brings the female to a halt her cloaca is above the point on the substrate where the spermatophore was deposited. If her cloaca, which is extended at this time, touches the spermatophore, the latter becomes attached and begins to be taken up into the body. Only the sperm mass on top of the spermatophore is taken up; some time after courtship has ended the female can be observed rubbing her cloaca on the substrate to remove the base.

From his braking position, and often after one or more push-backs, the male turns back towards the female and resumes his display. She approaches him and a new courtship sequence is initiated beginning at the retreat display phase. During a sexual encounter the courtship sequence is usually performed two or three times or, rather rarely, more often (Halliday 1974).

ANALYSIS OF THE CAUSATION OF THE MALE'S BEHAVIOR

By comparison with the complex and varied repertoire of the male, the female's role in courtship is very simple and can be classified into three basic states: moving away from the male, standing still and approaching the male. It is thus a simple matter to mimic her behavior using a simple model or to manipulate it using a harness or "strait-jacket" (Halliday 1975a). By manipulating the female it is possible to investigate her role and that of other factors in the causation of the male's behavior. In a series of experiments three principle causal factors have been identified and their effects and interactions analyzed. These factors are the female's behavior, the male's capacity to produce spermatophores and his need to ascend to the surface to take in a fresh supply of air.

The Role of the Female

A spermatophore can be successfully transferred to the female only if the various components of the courtship sequence are performed in the correct order. There are a number of points in the sequence at which the male will not proceed from one activity to the next unless he receives appropriate feedback from the female (Halliday 1975a). He will not proceed from orientation behavior to static display unless the female becomes stationary, he will not start retreat display unless she starts to approach him and he will not usually deposit a spermatophore unless she touches his tail. If the tail-touch is withheld, he usually turns back after a few seconds and resumes his display. While the female thus clearly plays a major role in determining the progress of a courtship sequence, her behavior is not the principal causation in certain transitions, notably that from retreat display to creep. Although the female must show a persistent positive response if the male is to make this transition, its precise timing is variable and is apparently controlled by factors internal to the male.

There is apparently no feedback from the female to the male when she successfully picks up a spermatophore. He is as likely to proceed to further sequences after a successful sequence as he is after an unsuccessful one (Halliday 1975a).

The Role of Spermatophore Production

In a study designed to investigate variation in male behavior due to internal factors, five male newts were tested over 40 consecutive days in a standard test condition in which the female's behavior was carefully controlled (Halliday 1976). This revealed a number of variables in male behavior that correlate with one another and with spermatophore production. The most important of these variables concern the duration of various parts of the sequence, the nature of the transition from one phase to the next and the rate at which the male performs his display actions. In a trial in which three sequences are completed there are striking variations in the temporal and sequential patterning of those sequences (Fig. 5).

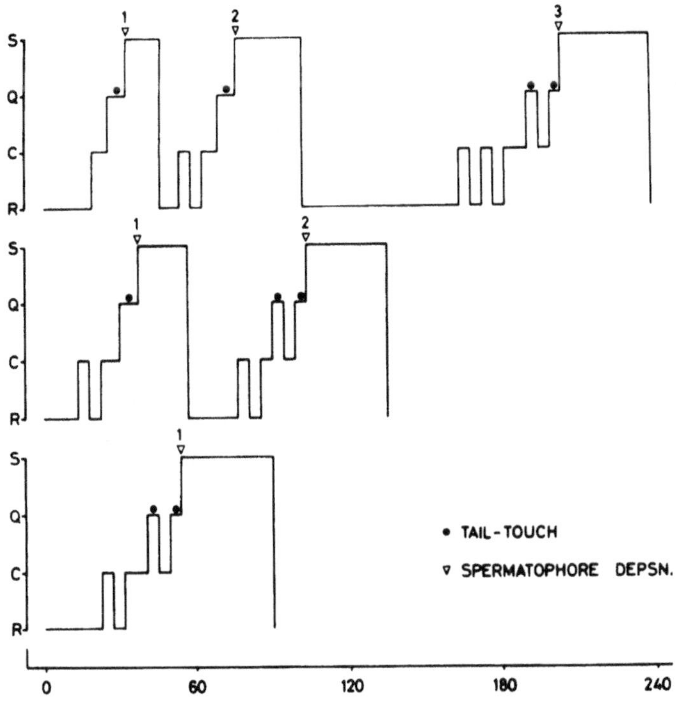

Fig. 5. Diagrammatic representation of the temporal patterning of courtship in trials incorporating three (top), two (middle) and one (bottom) complete sequences. The abscissa represents time from the beginning of the first period of retreat display, the ordinate successive stages in the sequence. R: retreat display, C: creep, Q: quiver, S: spermatophore deposition. (From Halliday, 1976.)

During the first of three sequences, the male typically displays at a high rate, retreat display is of short duration, he makes a simple, quick transition from retreat display to creep and he usually completes the spermatophore transfer phase quickly, requiring only one tail-touch before depositing the spermatophore. In contrast, during the third sequence he displays at a slow rate, retreat display is long, he vacillates between retreat display and creep before proceeding with a lengthy spermatophore transfer phase during which he requires an average of two tail-touches before he will deposit the spermatophore. In a two-sequence trial the patterning of sequences is very similar to that of the second and third sequences in the three-sequence trial. Similarly, the only sequence in a one-sequence trial is similar to the last sequence in a two- or three-sequence case. There is thus a clear relationship between the temporal patterning of a sequence and the number of sequences that will follow it in a trial, suggesting that the timing of the male's behavior is influenced by his capacity to produce spermatophores. This hypothesis assumes that during a trial a male uses up some finite number of spermatophores that are immediately available for deposition and that following a trial he is unable for a time to deposit another. There is some circumstantial evidence for this assumption (Halliday 1976) but until appropriate physiological investigations of spermatophore production can be made it cannot be directly tested and it is not known whether the limiting factor is the supply of spermatozoa or of the materials required for making the spermatophore base, if either.

The Role of Breathing

The frequency with which newts make breathing ascents outside the courtship context is correlated with their level of activity (Halliday and Worsnop 1976). It is not surprising that ascents occur in close association with courtship, which involves a high level of activity (Halliday and Sweatman 1976). On rare occasions ascents occur during a courtship sequence, although they never occur during the spermatophore transfer phase. Most commonly they occur immediately after the completion of a sequence (Halliday and Sweatman 1976). It is probable that, in natural situations, a breathing ascent will often terminate a courtship encounter because it is unlikely that a male will be able to find the same female on his return to the pond floor.

To some extent courtship and breathing may be regarded as competing tendencies; the probability that breathing will occur during display increases with the intensity of the display. During spermatophore transfer, breathing is apparently totally suppressed, although it may be disinhibited on completion of the sequence (Halliday and Sweatman 1976). Breathing and courtship, however, cannot be regarded as being causally independent of one another because the male's need to breathe seems to affect the temporal patterning of his courtship. By comparison with normal conditions, males will complete courtship sequences more quickly under oxygen-deficient conditions and more slowly under oxygen-rich conditions (Halliday 1977).

Synthesis

The central point of interest in this analysis of newt courtship is the temporal patterning of male behavior. The sequence is relatively predictable in the sense that the various phases follow one another in a stereotyped order but there is considerable variation in the timing of transitions from one activity to the next. The female plays a crucial role in controlling this timing because of the male's dependence on feedback from her. If, for example, she fails to respond positively the male is "locked" in the static display phase until he terminates courtship by breathing (Halliday and Sweatman 1976). Within the constraints imposed by the female, the male can speed up or slow down his behavior depending on the number of spermatophores he has deposited or is going to deposit. He can also speed it up or slow it down according to his oxygen requirement. The magnitude of the effect of oxygen supply on the temporal patterning of the behavior is small compared to that of spermatophore limitation.

These three causal factors have been incorporated in a deterministic model that produces an accurate simulation of male behavior (Houston et al. 1977). It thus appears that we have successfully identified the important causal factors and the way in which they interact with each other.

FUNCTIONAL CONSIDERATIONS OF NEWT COURTSHIP

Spermatophore Transfer Success

Before considering the functional and evolutionary aspects of newt courtship it is necessary to examine its effectiveness in terms of how often spermatophores are successfully transferred from male to female. In *T. vulgaris* less than 50 percent of all spermatophores deposited by the male are picked up by the female, but the probability of successful pick-up is not the same for all spermatophores. Spermatophores deposited in early sequences are less likely to be picked up than those deposited in later sequences (Table 1). This effect occurs not only in *T. vulgaris* but also in at least three other species of *Triturus* as well (Table 1). Why are early sequences so unsuccessful?

I shall argue below that the female's tendency to miss early spermatophores can be explained functionally as a mechanism of sexual selection, increasing her chances of mating with genetically fitter males, but first I shall consider what the immediate cause of the variation in pick-up success might be. Spermatophores are usually missed because the female passes to one side of them; only occasionally do they attach to an inappropriate part of her body such as a leg. It might be supposed that because early sequences are completed so quickly the female is hurried into following an incorrect path. This hypothesis is not supported, however, by analysis of pick-up success in one-sequence trials which, despite the fact that they are completed slowly, are very unsuccessful (Halliday 1974). It may be that multi-sequence trials give the female time to perfect her line of movement by a learning process but as yet it is not possible to identify any variable in the courtship behavior of either animal that constitutes an immediate cause of the failure or the success of the pick-up process.

Table 1. Success of spermatophore transfer in successive sequences in four Triturus species. Bracketed figures indicate percentages based on less than 10 observations.

Order of Sequence	1st	2nd	3rd	4th	Total
T. vulgaris					
Spermatophores deposited	45	38	21	2	106
Spermatophores picked up	13	19	13	1	46
Percentage picked up	29	50	62	(50)	43
T. helveticus					
Spermatophores deposited	22	16	7	2	47
Spermatophores picked up	12	11	5	0	28
Percentage picked up	54	69	(71)	(0)	60
T. alpestris					
Spermatophores deposited	20	13	5	1	39
Spermatophores picked up	6	5	0	1	12
Percentage picked up	30	38	(0)	(100)	31
*T. cristatus**					
Percentage picked up	15	30	36	67	

*Data from Nelson (1959).

The Functions of Courtship

The ultimate function, or survival value, of all courtship behavior is to maximize the representation of an individual's genes in succeeding generations. The proximate function of newt courtship is to ensure the successful transfer of sperm from a male to a female of the same species, but the long-term interests of both partners are best served if fertilization is achieved with a partner of the highest genetic quality. This discussion examines how newt courtship may have evolved so as to fulfill both the short- and the long-term functions.

Ethologists have made functional analyses of courtship from a number of points of view. Tinbergen (1953) has summarized these as orientation, persuasion, synchronization and reproductive isolation.

Orientation is not an important aspect of newt courtship, the first meeting of male and female usually being by chance; neither sex shows any long-distance attraction behavior. Gauss (1961) presented evidence that male crested newt T. cristatus are dispersed in the field and described scent-marking, which enables wandering females to detect the presence of a nearby male. My observations of T. vulgaris in the wild suggest that females are relatively sedentary and that the males wander widely throughout the pond. Persuasion is the primary function of the male's display, which apparently serves to suppress her tendency to swim away from the male and to stimulate her to approach him. The synchronization function is nicely shown by the causal relationships between male and female actions. The fact that the male does not proceed from one stage in the sequence to the next until the female has shown an appropriate response means that the two animals cannot get out of step with one another. Most importantly, the male does not take the crucial step of depositing a spermatophore until the female has first shown the persistent positive response necessary to initiate his creep and until she has finally given him the tail-touch that actually elicits spermatophore deposition.

Trivers (1972) criticized what he considered to be the rather limited functional analyses of courtship made by ethologists and suggested that more attention should be paid to the role that courtship may play in the assessment and selection of prospective mates. The actual form of mate assessment in a given species depends upon many aspects of its reproductive biology. For example, in those species in which there is a high level of parental care of the young, courtship may enable individuals to assess the capacity of their potential mates to provide food or protection for the young (Trivers 1972). In newts, parental care is minimal, consisting simply of the wrapping of each fertilized egg in a leaf by the female. The assessment aspect of newt courtship can therefore be expected to be confined to the assessment of a potential mate's genetic fitness, not his or her capacity to protect or provide for the young.

Bateman (1948) showed for Drosophila melanogaster that the female, having the more limited reproductive potential, was more discriminating in her choice of mates than the reproductively more prolific male. This dichotomy of mating strategies between the sexes is characteristic of many species (Trivers 1972). Each year the female newt produces 100 to 300 large eggs, with an increase in number annually (Bell 1973). She lays each egg individually, a process that takes several days for a full clutch. By contrast males produce large numbers of spermatophores during the breeding season, each of which probably contains sufficient sperm to fertilize a full clutch of eggs (Smith 1969). It is thus not surprising that males take the early initiative in courtship and are rather indiscriminate in their choice of objects to display to whereas females are initially slow to respond and frequently fail to respond to a male at all. That females make a much higher parental investment than males is born out by Bell's (1973) study of the life history of T. vulgaris. His analysis suggests that oogenesis is metabolically very expensive so that by reproducing, females incur a cost in terms of potential growth, and thereby in potential fecundity since larger females are capable of producing more eggs. Males incur costs too; during the aquatic phase they suffer a mortality rate 10 percent higher than that of females, probably due to their conspicuous coloration and movements.

Mate Assessment by the Female

The benefit that a female derives from a successful sexual encounter is a quantity of sperm with which to fertilize her eggs. Her offspring presumably will be more fit if she somehow obtains her sperm from the most genetically fit males available to her. Probably single clutches of eggs are fertilized by sperm from a number of males since inseminated females remain responsive to males for some days. However, the possibility that some form of sperm competition occurs in newts should not be discounted; such an effect has been found in some insects (Parker 1970). The choice of one mate from among several is the essence of sexual selection and there are a number of features of male newts, notably their highly developed secondary sexual characters and their complex display, that suggest that sexual selection has operated in their evolution.

A male is more likely to fertilize a female successfully if he can complete several courtship sequences with her, because he has more opportunities to deposit spermatophores, the later ones of which are more likely to be picked up. Thus the female's failure to pick up early spermatophores favors males with the capacity to produce several spermatophores in a single encounter. The ability of a male to deposit several spermatophores is limited by two factors, the number he can produce and the amount of time during which he can sustain continuous sexual behavior. The latter is in turn limited by his ability to maintain activity without taking in a fresh supply of air. The female, by her behavior, forces the male towards the limits of his spermatophore and oxygen supplies.

The crucial question is, are the abilities to deposit several spermatophores and to remain submerged for long periods indicators of genetic fitness? On these points I can only present speculative arguments in the absence of hard evidence. Under laboratory conditions males are capable of producing spermatophores daily for a period of several days, although there is a slow decline in the number produced each day (Halliday 1976). In a prolonged experiment males can be stimulated to a point of being apparently incapable of producing more spermatophores but they are often able to produce just as many spermatophores in their next test 24 hours later. These observations suggest that males are able to produce large numbers of spermatophores over the course of a season but rather few during a single sexual encounter, perhaps because some component of the spermatophore, either spermatozoa or the base, is available in only limited amounts in the short term. The level of spermatophore production over a season may be an indicator of male fitness, highly productive males demonstrating a higher capacity to invest nutritive resources in excess of those required for the maintenance of daily life. A male that deposits several spermatophores in a single encounter is not necessarily one who is able to do so over several days, however; he may have accumulated them be abstaining for several days or by a failure to attract females. Thus a male's spermatophore production during an encounter is not a wholly reliable indicator of his fitness unless it is allied to some other factor.

Reproductive males make sexual investment not only in spermatophores but also in secondary sexual characters, notably the crest. Is the crest a good indicator of a male's fitness? Zahavi (1975) has argued that sexually selected characters constitute handicaps to an individual's survival. Such handicaps are a useful cue, according to Zahavi, to the selecting sex since their degree of development serves as a mark of quality; thus sexual selection favors individuals who can carry the biggest handicap. This theory has been criticized on theoretical grounds by Davis and O'Donald (1976) and Maynard-Smith (1976), although the latter has argued that this "handicap principle" may be true for characters that are not wholly genetically determined but depend for their development on an animal's ability to obtain the energetic and nutritive resources required. The male newt's crest appears to be such a character. When newts arrive in their breeding ponds their crests are partially developed and they grow during the succeeding days. If captive males are kept on a limited food supply their crests rapidly regress, suggesting that only if they can keep up a high rate of food intake can they maintain their crests. The size of a male's crest may, therefore, be a good indicator of his capacity to maintain a level of metabolism above that required for the maintenance of basic functions. If this capacity has a genetic basis, crest size may indicate genetic fitness. It is possible that the crest confers a more direct advantage on the male, perhaps by increasing his surface area and thus his ability to absorb oxygen from the water (Foxon 1964).

Spermatophore production seems also to be related to the male's size; on a number of occasions I have observed males near the upper limit of the size distribution depositing five, six or seven spermatophores in a single trial. Since larger males have larger testes and since size is a function of age (Bell 1973), older males may be expected to maintain a higher level of spermatophore production than younger ones. On the argument that longevity is an indicator of fitness, female selection of older, larger males would be adaptive. This theory predicts, therefore, that females will show a preference for large males with big crests. Such a preference, combined with their immediate selection of more productive males, would tend to ensure that individual females mate with fitter males.

In their breeding ponds newts are generally difficult to see, being cryptically colored on the back and tending to remain concealed beneath the vegetation. They become conspicuous, however, when they swim to the surface to breathe. Although their skin produces distasteful secretions, they are eaten by such predators as herons (Witherby et al. 1938; Owen 1955) which are probably most likely to detect them during a breathing ascent. Animals able to minimize the frequency of their breathing ascents would be at a selective advantage and if females, by forcing males to maintain courtship for prolonged periods, favor those least dependent on atmospheric air, they will contribute to the survival of their offspring.

Mate Assessment by the Male

Since males apparently have many more opportunities to mate than females, they need not be as discriminating about their sexual partners as females are. The catholic variety of objects that males will approach and court suggests that indeed they are not very selective. Nonetheless, natural selection might be expected to favor males that fertilize those females with the larger clutches of eggs. Certain features of the male's behavior may be mechanisms ensuring that they are more likely to mate with more responsive females. If one assumes a positive relationship between a female's fecundity and her level of responsiveness, such mechanisms would be adaptive.

In the initial orientation phase the male often nudges the female along the length of her body with his snout, apparently sniffing her. Sniffing may provide him with olfactory cues indicative of her reproductive state and tactile feedback may enable him to assess the distension of her belly due to the oocytes within her. I believe, however, that a much more important mechanism for testing the female's condition is to be found in the variations in the temporal patterning of courtship sequences, which, as described above, appear to be related to spermatophore production.

One can regard the courtship sequence as being "driven" through its various phases by a combination of three causal factors, the female, male sperm supply and male oxygen debt. The slowing-down that occurs from one sequence to the next is apparently due to a change in the male's internal state. The corollary of this effect is that more "responsibility" for the completion of the sequence is shifted to the female. This effect is well illustrated by the results of a recent experiment in which the female's tail-touch was withheld immediately prior to spermatophore deposition in every sequence in a trial in which the male was tested to the limit of his immediate spermatophore production. In early sequences he was as likely to deposit the spermatophore as he was to turn back and resume his display, regardless of the lack of the tail-touch stimulus. In later sequences, however, he was very unlikely to deposit and almost invariably turned back (Fig. 6). In later sequences the male may require as many as four tail-touches before he will deposit (Halliday 1976). Thus during a courtship encounter the male raises his criterion of female responsiveness required for spermatophore deposition from one sequence to the next. Likewise, the increase of the duration of retreat display in successive sequences by the male requires that females must show persistent positive response for a longer period before he will creep.

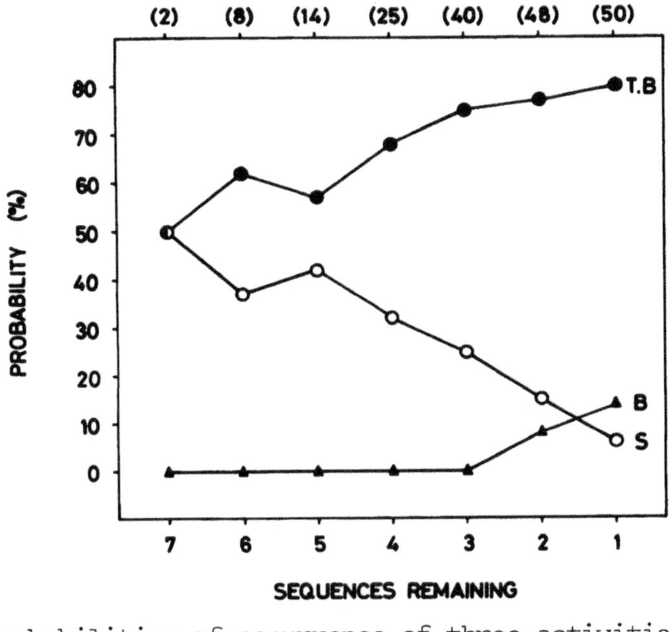

Fig. 6. Probabilities of occurrence of three activities following withholding of the female's tail-touch plotted against the number of sequences subsequently completed. T. B.: turn-back, S: spermatophore deposition, B: breathe.

These effects can be interpreted as adaptive responses to the female's tendency to miss early spermatophores. By failing to pick them up she reduces their value to the male and, correspondingly, the amount of time and energy it is worth his while to invest in them. By applying a lower criterion of feedback from the female he can complete early sequences quickly, thus leaving more of his limited time budget to later sequences. In the later sequences his behavior is such that only if females show persistent and strong positive response throughout the sequence will he deposit sperm. From the strength of her response he might assess the likelihood that she will pick up a spermatophore; this in turn might influence his decision to deposit it. The argument is that by taking more time over later sequences the male reduces sperm wastage by ensuring that the female is fully responsive.

Reproductive Isolation

In the western part of its range T. vulgaris is extensively sympatric with the very closely related palmate newt, T. helveticus (Fig. 1). The various mechanisms that prevent or reduce interbreeding between species were classified by Mayr (1963) under two headings, pre-mating and post-mating mechanisms. Liley (1966) listed as premating factors: geographical barriers, habitat and seasonal differences, mechanical barriers and behavioral mechanisms. Not only do the geographical ranges of T. vulgaris and T. helveticus overlap considerably, but overlap exists also at the local level (Taylor 1948; Smith 1969; Steward 1969). There is no precise information regarding the habitat preferences of the two species, but in general T. vulgaris is found in lowland areas and on alkaline soils whereas T. helveticus is found in upland areas and on acid soils, although the two species frequently inhabit the same ponds (Smith 1969; Steward 1969; Bell 1970). The breeding seasons of the two species are similar and pairs of both species have been observed courting in the same pond on the same day (Smith 1969). Since spermatophores are used by both species to transfer sperm and since the two species are similar in size and morphology, mechanical barriers can be discounted. This assumption is confirmed by the fact that under certain conditions to be described below T. helveticus females will pick up spermatophores deposited by T. vulgaris males.

Hybrids between these two species have reduced viability and fertility (Spurway and Callan 1960). The absence of any reports of naturally occurring hybrids (Lantz 1947; Spurway and Callan 1960) suggests the existence of a highly effective premating mechanism which, by elimination, is probably behavioral.

I have carried out some experiments to investigate the possible role of sexual behavior as an isolating mechanism between individuals of the two species. In all these experiments a female of one of the species was put into a tank and paired successively with conspecific and heterospecific males. In the first four experiments a female was paired first with a heterospecific, then with a conspecific male, the pairings being within an hour of each other; two females from each species were tested in this way. In all four tests males pursued and displayed to heterospecific and conspecific females with apparently equal intensity. In contrast, females appeared oblivious to heterospecific males. When paired with males of their own species, two females, one of each species, responded positively after failing to do so to heterospecific males; the other two females showed no positive response to either male.

In a second set of four experiments two females of each species were paired first with conspecific, then with heterospecific males only if they had shown a positive response in the first part of the test. Then, if they were unresponsive to the heterospecific male, they were again paired with a conspecific male. As in the previous experiments, all males displayed to conspecific and heterospecific females with equal vigor, but neither *T. vulgaris* female, though responsive to a conspecific male both before and after, showed any positive or passive behavior when displayed to by a *T. helveticus* male. In contrast, both *T. helveticus* females did respond to the display of *T. vulgaris* males, and spermatophores were deposited and some picked up in each experiment. The behavior of these females, however, was markedly different from the normal behavior of responsive females. Instead of a steady movement towards the male, they moved in sudden jerks. The heterospecific pairing in these experiments was conducted within minutes of the first conspecific pairing, and it seems likely that the positive response shown by these females to heterospecific males was due to their first being aroused by conspecific males.

Males in these experiments also lacked selectiveness. One cannot discount, however, the possibility that in natural situations males exercise some degree of discrimination, possibly on the basis of the female odor perceived during the sniff behavior. It is possible that males kept in the laboratory apart from females become less selective.

Some of the results suggest that females will respond only to the display of conspecific males. In the two instances when females responded to heterospecific males they had immediately before become responsive to conspecific males. There is no evidence that males can elicit positive behavior in initially unresponsive heterospecific females.

A behavioral isolating mechanism based on female discrimination has been demonstrated for three- and ten-spined sticklebacks (Wilz 1970); males of both species are unselective. Liley (1966) showed that female selectivity is an important isolating mechanism in guppies, although males are selective to a certain extent, depending on their previous experience.

The hypothesis that reproductive isolation between *T. vulgaris* and *T. helveticus* is the result of female selectivity is consistent with other observations:

1. There are marked differences between the epigamic characters of the males of the two species, whereas the females are similar in appearance.

2. The sexual behavior of the two species, though similar in overall plan, differs in many details. These differences are in the display behavior (see below), the later parts of the courtship sequence being similar. It has been found for a number of species that, where behavior is important in species isolation, it is behavior that occurs early in the sexual process that differs most markedly between sympatric species. Examples include pair-formation in cardueline finches (Hinde 1959), orientation songs of grasshoppers (Perdeck 1958), and the mating calls of several frogs (Blair 1964; Ball and Jameson 1966).
3. Certain secondary sexual characters of the males of these two species exhibit the phenomenon of character displacement (Grant 1972; Halliday 1975b). In the area of sympatry males of the two species are morphologically very distinct, but in *T. vulgaris* subspecies from the southern and eastern limits of the species range, where *T. helveticus* is absent, males show a number of characters in common with *T. helveticus* males. These morphological differences may have evolved in the region of sympatry in association with differences in courtship through selection against hybrids (Halliday 1975b). Such selection, called reinforcement by Dobzhansky (1951), has been suggested as the basis of mating call differences in sympatric populations of hylid frogs (Blair 1964; Littlejohn in this volume).

The experiments described above are too few other than to suggest that reproductive isolation between *T. vulgaris* and *T. helveticus* is brought about by selective responsiveness of females to conspecific and heterospecific male display. If such a mechanism is involved in newt courtship, however, it will have constituted a considerable selective influence on the male's display. I suggest that it has influenced the evolution of newt courtship in two ways. First, the need for females to make an accurate discrimination will be an additional factor tending to favor those females who delay their first positive response to the male. Secondly, female discrimination may have been the selective pressure that led to the development of three distinct movements in the male's display. The three movements seem to be adapted to stimulate the female through three different modalities: visual, olfactory and tactile. If males of different species are distinctive in all three modalities, the female will receive three sources of information on which to base her discrimination, which will be more reliable than a discrimination based on one or two cues. T. vulgaris and T. helveticus males are highly distinctive in appearance, perform the fanning display, which is probably a vibrational stimulus, at markedly different speeds (Halliday 1975b), and it is likely that they also have distinctive odors.

Synthesis

In this section I have suggested that much of the variation in newt courtship is due to selective pressures exerted by each sex on the other that have arisen as the result of the disparity between the sexes in the allocation of their reproductive investment. Females, which produce relatively few "expensive" eggs, maximize their reproduction by selecting the fittest males. Males, which produce large numbers of "cheap" sperm, maximize their reproduction by mating with as many females as possible, the quality of those females being a lesser consideration.

Female mate selection may operate in three ways, by her initial slowness to respond, by her tendency to miss early spermatophores and possibly by preference for large males and for males with large crests. These pressures will tend to favor males that can maintain sexual activity for long periods without taking in air, that can produce many spermatophores and that can maintain a large crest. In their turn males exercise a degree of selection over females by raising the criteria of female responsiveness that will elicit spermatophore deposition in later sequences.

Females probably also select males on the basis of their species, apparently differentiating between conspecific and heterospecific males on the basis of their appearance, odor and display. This has led to the evolution in males of species-specificity in secondary sexual characters and in a number of aspects of display behavior.

COMPARATIVE ASPECTS OF TRITURUS COURTSHIP

In the previous section I discussed the courtship of T. vulgaris from a functional point of view in an attempt to identify some of the factors that have been important in its evolution. The evolution of behavior patterns can also be deduced by a comparative examination of homologous activities in closely related species

(Tinbergen 1959; Cullen 1972). In this section I shall widen the discussion of newt courtship behavior to include other Triturus species in an attempt to construct the evolutionary history of courtship in the genus.

The Palmate Newt, Triturus helveticus

In many respects this species is similar to the smooth newt and was not recognized as a separate species in Britain until 1843 (Baker 1848), though it had earlier been described in Europe (Razoumowsky 1788). The females of the two species are easily confused but the males are quite distinct (Fig. 2). The male palmate newt has a very low, smooth-edged crest and is generally less dramatically colored and patterned. The ground color is variable, ranging from greenish to brown, and the spots, smaller than in T. vulgaris, are black, brown or reddish. The belly is pale cream in color. The distinctive features of the palmate male are a dark, webbed hind foot and the shape of the tail with its smooth outline, truncated tip and characteristic terminal filament. There is a prominent dorso-lateral ridge running the length of the body (Halliday 1975b). Female T. helveticus differ from female T. vulgaris in that the former usually have very few dark spots and frequently none at all.

Descriptions in the literature of courtship in this species are fragmentary. A rapid fanning movement was described by Rusconi (1821) and Cummings (1911); otherwise sexual behavior is described as being identical to that of T. vulgaris (Smith 1969; Steward 1969). I have observed a total of 52 sexual encounters, involving 20 males and 19 females and providing records of 47 sequences. In all general aspects the courtship sequence in this species if similar to that of T. vulgaris, with the same overall pattern of four phases. There are a number of differences between the displays of the two species; their spermatophore transfer phases are identical.

The male's display incorporates the same three tail movements as in T. vulgaris, but they differ in their form and in the relative frequencies with which they are performed. Waves are less conspicuous, the "showing-off" movement in which the tail is held still for a moment being rarely seen. The whip is a less violent movement; whereas T. vulgaris uses the full length of its tail to generate a violent blast of water, T. helveticus moves only the distal portion of his tail, creating much less turbulence. There are distinct differences between the two species in the manner in which they fan. Although the fanning posture is identical (Fig. 7), in T. helveticus the tail movement is twice as rapid, about 12 beats per second, and the bouts tend to be much longer, lasting up to 230 seconds (mean 17.1 seconds; Fig. 8).

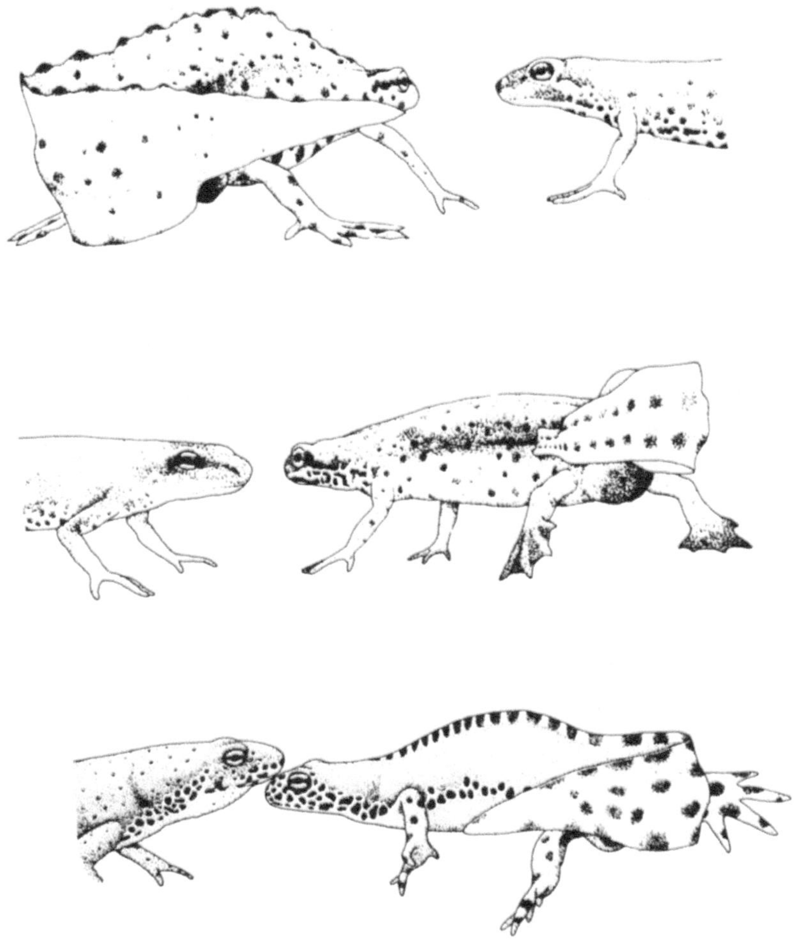

Fig. 7. The male's fanning display in three species, T. vulgaris (top), T. helveticus and T. alpestris.

There are significant differences in the relative frequencies of the male's three display acts in the two species (Fig. 9). Fan bouts are more common in T. helveticus, this effect being particularly marked in the retreat display phase. In T. vulgaris retreat display consists almost entirely of whips with only occasional fanning bouts. In T. helveticus whips and fans are almost equally common and more or less alternate with one another.

It is in the fanning of T. helveticus that the probable function of the dorso-lateral ridge is apparent (Halliday 1975b). The effect of this ridge is to create below it a broad, shallow groove bounded ventrally by the swell of the belly. During fanning the male holds his body in such a way that this groove becomes deeper on the side of the body on which he is displaying. This effect

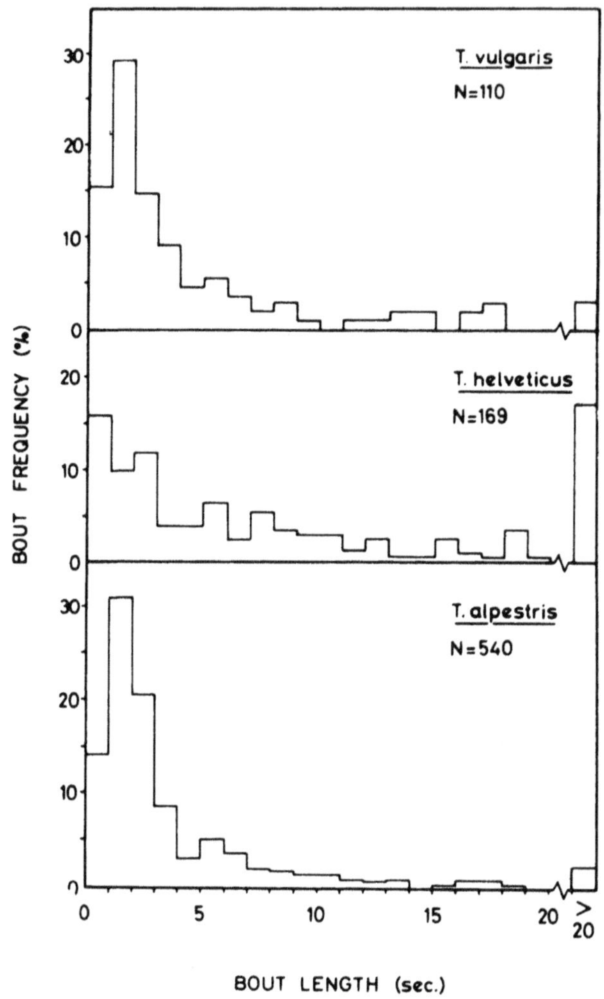

Fig. 8. Frequency histograms of fan bout durations in three species.

has also been described by Cummings (1911) and Arnold (1972). The vibrating tail tip is positioned in the posterior end of the groove; this apparently has the effect of channeling the water current generated by the tail along the male's flank and towards the female. A similar, although less well-developed, ridge and groove are present in the male T. vulgaris, in which they probably serve a similar function (Halliday 1975b).

The Alpine Newt, Triturus alpestris

Sexual dimorphism in this species is less marked than in other Triturus species. Both sexes are blue, mottled with black on the

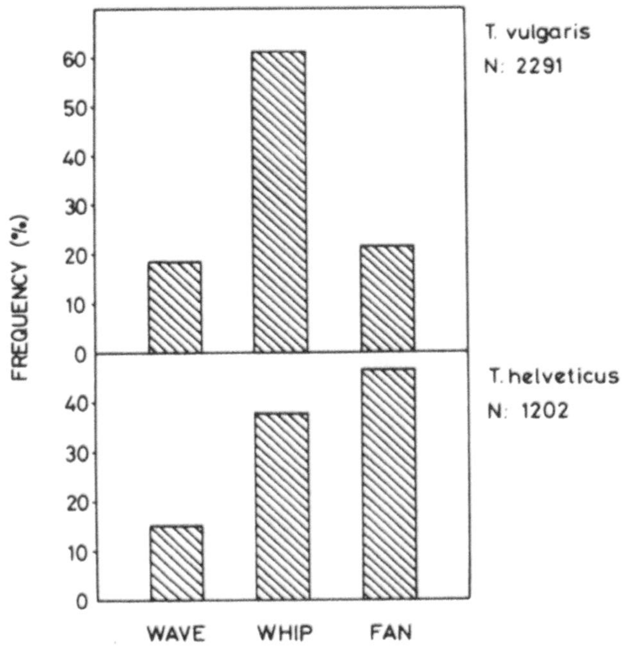

Fig. 9. Frequency histograms of display acts in T. vulgaris and T. helveticus.

upper surface and flanks, and brilliant orange on the belly. Both sexes have a lateral band of black spots, but in the male these are larger and ringed in white. The male has a low crest of alternating black and creamy-white stripes and large irregular spots on the tail. There is no sexual dimorphism of the hind-limb and the male does not possess dorso-lateral ridges. The female is considerably larger than the male (Fig. 2).

Some elements of the sexual behavior of this species have been described by Finkler (1923) and Prechtl (1951). T. alpestris is more sluggish than either T. vulgaris or T. helveticus, both in its general habits and during courtship. I have observed a total of 36 encounters, involving 13 males and 12 females, and incorporating 44 courtship sequences.

In its overall pattern the courtship sequence of T. alpestris differs markedly from that of the previous two species. Orientation behavior is similar and is followed by static display, but the latter is much more prolonged than in the other species, and there is no retreat display (Fig. 10). Instead, the male creeps from his static display position, simply stopping his display and moving away from the female. Usually she follows him and spermatophore transfer is essentially the same as in the other species. The male's quiver movement is slightly different in this species; instead

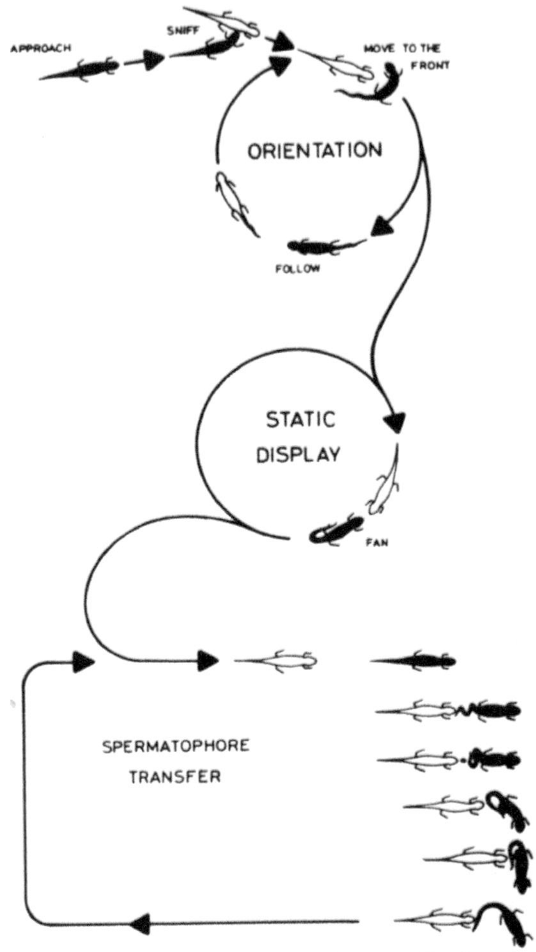

Fig. 10. The courtship sequence of the Alpine newt, *T. alpestris*. The male is in black.

of slow, lateral waves the male makes spasmodic vertical jerks with his tail.

The male's display differs from that of *T. vulgaris* and *T. helveticus* in that only one tail movement, the fan, is used (Fig. 7). Display consists of a series of fanning bouts, most of them of short duration (Fig. 8), with a brief pause of half a second to a second between bouts. When the male has been fanning for some time a new behavior is combined with his display, the lean-in. In between fanning bouts he raises his body by extending his legs and leans over towards the female so that his flank or tail touches her snout. The incorporation of this behavior into his display, which does not invariably occur, is indicative that the male is about to turn away from the female and start to creep.

The fanning movement of T. alpestris differs in form from the similar movements in the two previous species. The tail is stiffer and the individual tail beats, approximately three to four per second, are at a slower frequency. The orientation of the fanning male relative to the female is the same as in the other species (Fig. 6), but there is no dorso-lateral ridge or lateral groove.

The name fan has been used to describe a display action in three different species and attention has been drawn to the differences between species in the form of this movement. I argue below that the fanning movements of these species are homologous in the evolutionary sense, but the use of the same term for three different actions should not be taken at present to imply any relationship other than similarity of form.

A striking difference between the courtship of T. alpestris and that of other species is that, following completion of the first sequence, i.e. at the brake or push-back stage, the male does not generally revert to display behavior but turns away to creep, thus starting the next sequence at the beginning of the spermatophore transfer phase (Fig. 10). Up to three sequences were observed to follow one another in this way. If the female turns or moves away after a sequence, the male follows her, resumes his display position and reverts to display behavior.

The Crested Newt, Triturus cristatus

This, the largest of the three species found in Britain, shows spectacular sexual dimorphism (Fig. 2). The male has a high, deeply crenulated crest that is not continuous with the very deep tail. In the breeding season the background skin color, which is black in the terrestrial phase, pales to a brownish or bluish gray, bringing into prominence large, irregular black patches. The belly is vivid orange, patterned with black markings. The lower flanks are spotted with white dots that in this area coincide with the raised "warts" that give this species one of its common names. On the tail there is a bluish-white band that tapers toward the tail tip. The female lacks a crest and the white markings and is less heavily spotted.

Sexual behavior in the crested newt has been described in varying degrees of detail by Rusconi (1821), Prechtl (1951), Nelson (1959, 1960) and Arnold (1972). I have made only a few observations, but these agree very closely with Nelson's and Arnold's accounts on which the description that follows is based.

Orientation behavior is essentially the same as that of the other species. The male is the more active partner, approaching the female, sniffing her and making "eager rushes" round to the front of her head. His display differs in many respects from that of the other species described, in particular in the male's distinctive display posture (Fig. 11). He lifts his hind legs and body off the substrate and stands on his forelimbs; his throat is parallel to the ground and his body is sharply kinked in the middle so that his thorax is vertical while the rest of his body and tail are parallel to the substrate, and angled over slightly towards the female. The male's display consists of slow beats of the tail. In each beat the tail is flexed towards the female's head, and then back to a more-or-less straight position, each beat incorporating a slight flick of the tail tip at its furthest extension towards the female. At first these tail beats are irregular in amplitude and in frequency but they quickly settle into a steady rhythm of uniform movements, between one and two beats per second. They are performed in bouts of between three and eight beats per bout with a brief pause between bouts. The tail beats are so powerful that the whole rear part of the male's body is swung from side to side and the dorsal crest is waved in a conspicuous manner. As he displays the male makes small body movements that apparently offset the tendency of his tail beats to topple him over. As display progresses the male begins to incorporate a lean-in movement similar to that seen in T. alpestris. This appears first in the pauses between bouts of tail beats but becomes more frequent until it tends to alternate with individual tail beats. The lean-in movement becomes progressively more pronounced until the male's tail begins to strike the female's head. At this stage the tail beats are called lashes, and the incorporation

Fig. 11. The display posture of the male crested newt, T. cristatus: the female's view.

of lashing into his display signifies that the male is about to creep. At no time during display does the female move; as in T. alpestris there is no retreat display.

The spermatophore transfer phase is much like that of the other species described. Nelson (1959, 1960) and Arnold (1972) described a tail-touch action by the female but this does not seem to be as invariable a precursor to spermatophore deposition as it is in the other species. Sometimes the female does not start to follow the male until after he has turned side-on to her at the end of his creep-on movement. The male's movement after deposition is the same as in the other species except that he does not stop rigidly in the brake position but moves sideways away from the female in a crab-like fashion. There is no mention in any account of a push-back movement. Following the completion of the courtship sequence, the male reverts to his display and the sequence is repeated. Nelson reported encounters incorporating up to five complete sequences.

Nelson (1959, 1960) classified female behavior into three states: uncooperative, receptive and intense. Uncooperative females move away from the male, receptive females remain passive, and intense females approach the male. A very responsive female may move towards a displaying male so vigorously that she knocks him over and prevents him from displaying to her. Females may also show pseudo-male behavior, namely taking up the display posture, regular tail beats and lashes. Similar behavior has been observed in female T. vulgaris (Halliday 1974).

Other Triturus species

There is little information in the literature concerning sexual behavior in other Triturus species. Steward (1969) gave a brief account, which was intended to cover all species.

Steward (1969) stated that the courtship of the marbled newt, T. marmoratus, is similar to that of T. cristatus, and from my limited observations this seems to be true. The display posture is similar with the sharply kinked body and hand stand position, and I have observed similar regular tail beats.

The male of the smallest species, T. boscai, has no crest but shows sexual dimorphism in other respects (Steward 1969). Prechtl (1951) gave a brief description of sexual behavior in this species. The male performs a "tail-wagging" display similar to that of other species, but in a different display orientation. Male and female are aligned side by side and head to tail in such a way that the male wags his tail just in front of the female's nose. Bouts of tail movement are said to be very long, lasting about a minute, and display continues for several hours before a spermatophore is deposited.

The male banded newt, T. vittatus, has the most highly developed crest of all the species and has a conspicuous white stripe along the lower flank. Display in this species is said to incorporate vertical head-bobbing and a movement in which the male slaps his body against the female's flank (Oeser 1951).

Of the courtship of the Carpathian newt, T. montandoni, which is closely related to T. vulgaris and T. helveticus (Lantz 1947; Steward 1969), nothing seems to be known. Males have virtually no crest, but do possess dorso-lateral ridges and a terminal tail filament, both reminiscent of T. helveticus.

Interspecific Differences in Courtship

In its general pattern courtship is similar in all the Triturus species described. The male first orients himself in front of the female; there follows a period of display involving tail movements, leading to spermatophore transfer. Within this general pattern the principle variations between species concern male display behavior. I will discuss two aspects of display, variations in the form and number of the tail movements and whether there are one or two separate display phases.

There is considerable variety in the form of the tail movements performed during display, notably in their speed, amplitude and frequency. There are slow, large-amplitude movements of the whole tail: the wave of T. vulgaris and the tail beats of T. cristatus. There are high-frequency, small-amplitude movements of the distal portion of the tail: fanning in T. vulgaris, T. helveticus and T. alpestris. Finally, there are sudden, violent movements of all or part of the tail: the whip of T. vulgaris and T. helveticus and the lash of T. cristatus.

There is also interspecific variation in the number of different display movements incorporated in the display. In T. alpestris there is only one movement, fanning, whereas in T. vulgaris and T. helveticus there are three, wave, whip and fan. A quite different type of display is seen in T. cristatus. Different tail movements have been described, namely irregular and regular tail beats and lashes, but these are not discrete actions but form a gradation of movements which Nelson (1959) interpreted as being indicative of successively higher states of arousal. This is quite unlike the display of T. vulgaris and T. helveticus, which have distinct tail movements that do not intergrade and do not form a rigid succession.

The three display movements of these two species differ in the length of tail involved and in both the amplitude and the frequency of the movements, whereas the movements of T. cristatus all involve the whole tail, have similar frequency and vary only in amplitude. On the basis of limited observations it appears that T. marmoratus has a similar display.

The display repertoires of all Triturus species described have three features in common. All are repetitive, all include at least one rhythmic component and, with the possible exception of T. boscai, all are performed in a position just in front of the female. Also common to the displays of all these species is the lean-in movement. In T. alpestris this punctuates fanning towards the end of the static display, in T. cristatus it appears to be an incipient form of the tail-lash and in T. vulgaris and T. helveticus it is sometimes apparent in the wave display. It is likely that the common ancestor of living Triturus species showed these four features. Evolution seems to have taken place along at least three lines from some ancestral display.

1. Modification into a single, stereotyped movement: fanning in T. alpestris.

2. Modification into a complex, graded movement with constant frequency, but variable amplitude: the tail beats of T. cristatus and T. marmoratus.

3. Differentiation into a number of discrete movements: wave, whip and fan in T. vulgaris and T. helveticus.

An important factor to be considered is the nature of the stimuli presented to the female by the various male display movements. Newts detect food items by a combination of vision, olfaction and sensitivity to water currents through lateral line organs (Tinbergen and Ter Pelkwijk 1938; Himstedt 1967, 1968, 1969; Cooke and Fulford 1971; Matthes 1924a, 1924b, 1927), so all these senses could be used during courtship. The fanning of T. vulgaris, T. helveticus and T. alpestris presents little visual stimulation to the female and is generally assumed to carry odor to the female from the male's cloacal and skin glands (Tinbergen and Ter Pelkwijk 1938; Thorpe 1972). Newts possess lateral line organs that are particularly dense over the head, and fanning may also be perceived by the female as a vibrational stimulus. That these three species each has a characteristic fanning speed suggests that this may be an important variable, perhaps playing a part in species recognition by the female (Halliday 1975b).

In contrast to fanning, the wave of T. vulgaris and the tail beats of T. cristatus and T. marmoratus expose the full length of the male's tail to the female's view, and these are also the slowest of all the display movements described. It is likely that these are visual displays and it is in these species that conspicuous tail patterns are most highly developed in the male, spots and red and blue stripes in *T. vulgaris* and the broad white stripe in *T. cristatus* and *T. marmoratus*.

It is improbable that the violent whip movement of *T. vulgaris* and *T. helveticus* presents anything but tactile stimuli to the female; the movement is too sudden to be an effective visual one and odor would be widely dispersed by it. In the lash movement of *T. cristatus* the tail actually makes contact with the female; this clearly seems to be a tactile stimulus.

Triturus displays thus show many of the features commonly shown by communicative actions that have undergone ritualization (Huxley 1966; Cullen 1972). Most are stereotyped and repetitive, many are rhythmic and some are associated with specialized structures (crests, tails and dorso-lateral ridges) and conspicuous color patterns. An important trend in the evolution of displays that play a role in species recognition is the reduction of ambiguity between the displays of sympatric forms (Cullen 1966). This effect is well shown by the striking differences between male visual characters and in the different fanning speeds shown by *T. vulgaris*, *T. helveticus* and *T. alpestris* (Halliday 1975b).

A major variation in the display behavior of *Triturus* species is that in two species, *T. vulgaris* and *T. helveticus*, there are two distinct phases, static and retreat display, whereas in the other species described there is only static display. As static display is common to all species, it is probable that it is a primitive feature and that retreat display is a more recent evolutionary innovation.

Retreat display comes about because the male does not creep away from the female as soon as she starts to approach him; instead he continues to display while retreating before her. Alternatively, retreat display can be regarded as a function of the fact that the female starts to approach the male before he has begun to creep. Experiments on display behavior in *T. vulgaris* have shown that the male does not creep unless the female has shown persistent approach for some seconds (Halliday 1975a). Both *T. vulgaris* and *T. helveticus* are very active species and the female shows frequent switches between moving away from the male, approaching him and remaining motionless. Thus her first movement towards the male might not be a sustained positive response and for him to respond to it by creeping immediately would be inappropriate. Insteady, by maintaining his display through a retreat phase, the male ensures that he creeps only in response to a persistently responsive female (Halliday 1974).

In T. alpestris and T. cristatus there is no retreat display phase; the male stops displaying, turns away to creep and the female usually follows him. The observer cannot predict whether or not the female will follow, and there is no apparent way by which the male could predict this either. It is possible that the male responds to cues not apparent to the human observer but in many T. alpestris sequences I have observed the male creep, receive no tail-touch because the female had not followed him, and then turn back and resume his display. A similar effect was observed in T. cristatus by Nelson (1959) and by Arnold (1972). The retreat display phase eliminates this element of uncertainty and seems to be a more sophisticated system than that involving only static display. This variation in courtship may be related to ecological factors; possibly T. vulgaris and T. helveticus live in habitats in which space is less restricted and it is possible for the pair to move about over a wide area during courtship. Alternatively, it may be correlated with the fact that T. vulgaris and T. helveticus females are more active during courtship than are females of other species and are therefore more likely to make a "false positive" move towards the male.

Retreat display not only follows static display and precedes spermatophore deposition but also occurs between successive spermatophore transfer phases; herein is another variation in the overall pattern of courtship between species. In T. alpestris successive spermatophore transfer phases generally follow one after another whereas in the other species they are separated by a period of display, static display in T. cristatus and retreat display in T. vulgaris and T. helveticus. The adaptive value of these intervening display periods may be that they provide an opportunity for the male to assess or to increase the female's responsiveness before the deposition of each spermatophore.

On the basis of these comparisons I suggest that the species for which courtship has been described fall into three groups:

1. T. vulgaris and T. helveticus. Both show two display phases, the same three distinct display movements and a period of retreat display between successive spermatophore transfer phases.

2. T. cristatus and T. marmoratus. Both show a distinctive display posture, a single display phase, a single, variable-amplitude display movement and a period of static display between successive spermatophore transfer phases.

3. T. alpestris. This species shows a single display phase, a single, stereotyped display movement and there is frequently no display between successive spermatophore transfer phases.

Taxonomic Affinities of *Triturus* Species

It is an important adjunct of a comparative approach to behavior patterns that there should be other criteria by which taxonomic relationships can be independently deduced (Hinde and Tinbergen 1958). On the basis of morphological characters the nine *Triturus* species can be classified into three groups (Mertens and Muller 1940; Thorn 1968). The largest group consists of six species, *T. vulgaris*, *T. helveticus*, *T. boscai*, *T. montandoni*, *T. vittatus* and *T. italicus*; these are all relatively small forms, generally under 10 cm in length. Authorities vary as to whether the Italian newt is a true species, *T. italicus* (Lantz 1947; Spurway 1953; Mancino 1961), or a subspecies, *T. vulgaris italicus* (Steward 1969). The results of Mancino's (1961) hybridization experiments between this form and *T. vulgaris* provide strong grounds for regarding it as a full species. Two larger species, *T. cristatus* and *T. marmoratus*, both about 15 cm in length, form a second group, leaving one species, *T. alpestris*, which is of intermediate size, in a category of its own. Spurway (1953) groups *T. boscai* with *T. alpestris* because these two species readily court each other in captivity; but in view of the undiscriminating behavior shown by males of other species this seems a poor criterion for grouping species together.

There is a considerable body of research into the genetic affinities of the several species and subspecies within the genus, and a number of experimental crosses have been made between newts from different subspecies of the same species, between species within the same species group, and between species in different groups. Crosses at the subspecific level have been made between *T. cristatus* subspecies (Spurway and Callan 1950; Callan and Spurway 1951; Spurway 1953). *T. cristatus* and its close relative, *T. marmoratus*, have overlapping ranges and a naturally occuring hybrid (*T. blasii*; de l'Isle 1862, 1872; Lantz 1947) has been described. Experimental hybrids between these species have been produced by a number of workers (Wolterstorff 1908; Pariser 1936; White 1946; Lantz 1947; Spurway 1953; Lantz and Callan 1954). Within the *T. vulgaris* group hybrids have been obtained artifically between *T. vulgaris* and *T. helveticus* (Wolterstorff 1925; Pariser 1935, 1936; Freytag 1950; Mancino 1959; Spurway and Callan 1960), between *T. vulgaris* and *T. italicus* (Mancino 1961) and between *T. boscai* and *T. helveticus* (Pariser 1936). No natural hybrids between any of these species have been described, although their ranges overlap extensively (Fig. 1). Across the species groups hybrids have been obtained from the following artificial crosses: *T. vulgaris* x *T. cristatus* (Poll 1909; Pariser 1932, 1935, 1936; Hamburger 1935, 1936), *T. vulgaris* x *T. marmoratus* (Pariser 1936), *T. vulgaris* x *T. alpestris* (Pariser 1935, 1936), *T. helveticus* x *T. cristatus* (Poll 1909; Bataillon 1927; Pariser 1932, 1935, 1936; Hamburger 1935, 1936; Baltzer 1936), *T. helveticus* x *T. marmoratus* (Bataillon

1927), T. helveticus x T. alpestris (Bataillon 1927; Pariser 1935, 1936), T. alpestris x T. cristatus (Pariser 1935, 1936), T. alpestris x T. marmoratus (Bataillon 1927), and T. boscai x T. cristatus and T. marmoratus (Pariser 1936).

As a result of such studies much is known about the degree of genetic breakdown in hybrids between newts at all three taxonomic levels. Genetic breakdown in the form of hybrid sterility and inviability is quantitatively greater in the results of crosses made at the species level than that found in crosses between subspecies. Similarly such effects tend to be less extreme in hybrids between species from the same species group than in those made across the species groups. The relationships between the various forms shown by their genetic affinities are thus consistent with the relationships established on the basis of morphological similarities. Furthermore, the species groupings I have derived above on the basis of courtship behavior are likewise consistent with morphological and genetic affinities.

Newts have particularly large and readily observed chromosomes, which at meiosis reveal a number of reciprocal translocations, each species and subspecies studied showing a characteristic set of these translocations. There are thus gross differences between the species in the form of their chromosomes and hybrid sterility seems to be largely the mechanical result of the lack of homology between the chromosomes contributed by the two parents (Spurway 1953). Wright (1941) argued that reciprocal translocations will only become established in a sexually reproducing population if that population consists of "numerous isolated populations that pass through phases of extreme reduction in numbers." Spurway (1953) argued that newts live under just these conditions, particularly at the edges of their range, and that during colonization of new areas populations may have become established that consisted of the offspring of a single fertilized female. Such conditions are favorable for a translocation to become established in a population which would in consequence be to a large degree genetically isolated from other populations (Spurway 1953). Spurway also argued that selection would tend to reinforce genetic isolation of this sort by favoring secondary isolating mechanisms that reduce the frequency of wasteful hybrid matings. As I have argued above, this isolating mechanism appears to be based on female discrimination of male courtship display.

Courtship in Some Other Urodeles

The courtship behavior of urodeles has been reviewed by Salthe (1967) and by Salthe and Mecham (1974), who discussed it in relation to ecological factors and to the taxonomic affinities of the various genera. In an unpublished study Arnold (1972) made original observations of courtship in a large number of species.

The genera most closely related taxonomically to Triturus are Taricha and Notophthalmus from western and eastern North America, respectively, and Cynops from Japan (Wake and Ozeti 1969). The later parts of sexual behavior are essentially the same in all four genera; the male deposits a spermatophore, creeps away, and the female follows and picks it up. There are, however, marked differences between these genera in the earlier parts of the courtship sequence (Fig. 12). In Taricha the male mounts the female's back and clasps her pectoral region with his fore and hindlimbs, which in the breeding season are equipped with horny pads (Twitty 1966; Davis and Twitty 1964; Arnold 1972). He applies his submandibular gland to the female's snout and at the same time strokes her cloaca with his hind feet. After a time he dismounts and deposits a spermatophore.

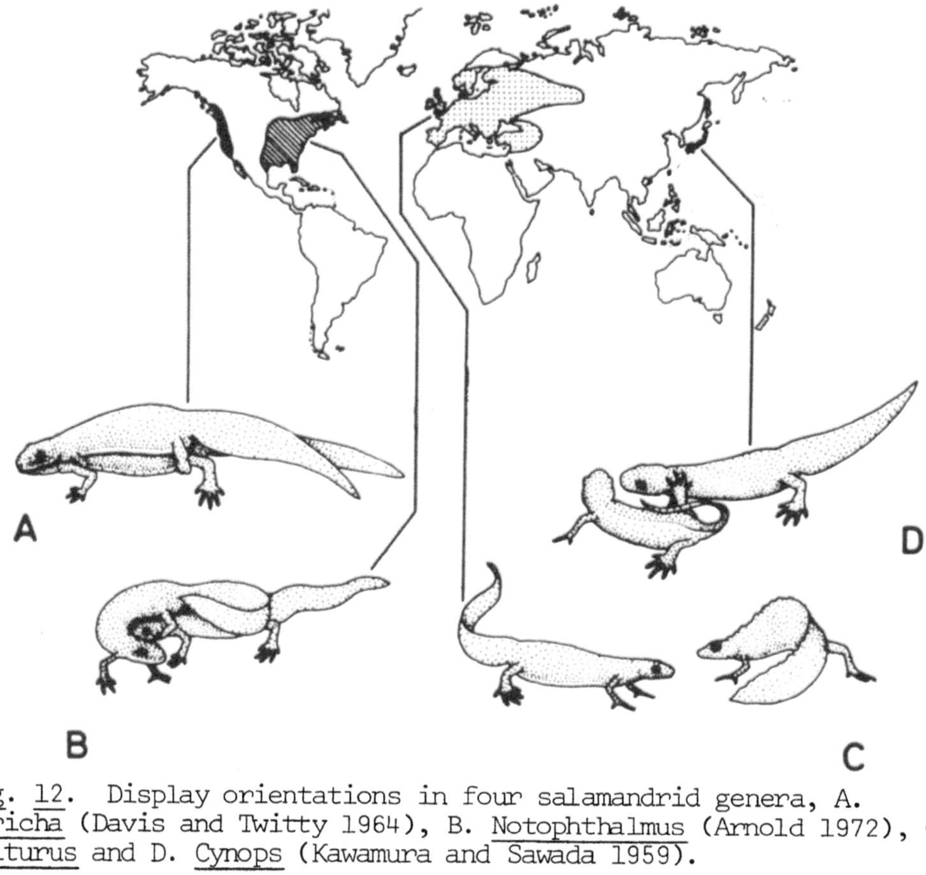

Fig. 12. Display orientations in four salamandrid genera, A. Taricha (Davis and Twitty 1964), B. Notophthalmus (Arnold 1972), C. Triturus and D. Cynops (Kawamura and Sawada 1959).

In Notophthalmus the male similarly mounts the female dorsally, but grasps her with his hindlimbs which, like the forelimbs of Taricha, are equipped with horny pads (Jordan 1893; Hilton 1902; Arnold 1972). He rubs the female's head with his snout and simultaneously performs a "fanning" movement with his tail that creates water currents around her head (Fig. 12). After a time he dismounts and spermatophore deposition follows after a tail-touch. On some occasions when the female is very responsive amplexus does not occur; instead, after a brief visual interaction between male and female, the former proceeds directly to spermatophore transfer (Arnold 1972).

In Cynops the male does not fully grasp the female at any stage of the courtship sequence (Prechtl 1951; Arnold 1972). He nudges her along the length of her body and then swims round to block her path in the manner shown by Triturus. In his position in front of her the male rubs the female's head with his snout and performs a "fanning" movement with his tail (Fig. 12). His behavior is thus intermediate between that of Triturus and Notophthalmus, differing from the latter in that he does not grasp the female, though in some species he restrains her by putting one or two legs on her neck. Freytag (1970) described how the male Cynops pyrrhogaster restrains the female with his hindleg around her neck and points out that this represents an intermediate condition between amplexus and complete female freedom. He also quoted earlier accounts by Klinge (1915), who described male restraint by the front and hind limbs of one side and of Tsutsui (1931), who described male restraint by means of one forelimb locked with one of the female's forelimbs. From the account of Kawamura and Sawada (1959) it is clear that there is considerable variation between local races of this species in the form of restraint exercised by the male.

The fact that the later parts of the courtship sequence are essentially similar in these four genera suggests that this method of spermatophore transfer, in which there is minimal contact between the sexes, evolved before these genera became distinct. The success of this method depends on the female following the male closely; that is, she must be responsive to him. The variation between Triturus, Taricha and Notophthalmus in the earlier phases of courtship suggests that each genus has evolved its own method of making the female responsive. In three of them the male uses a rhythmic tail movement, and in three genera the male rubs his hedonic glands against the female's snout. The males of two genera use amplexus to hold the female, one genus employing the forelimbs, the other the hindlimbs. Triturus differs from the other three genera in lacking both amplexus and hedonic rubbing but instead shows elaboration and multiplication of the tail movements that are of crude form in Cynops and Notophthalmus. The fanning display of Triturus may be comparable to the rubbing of the other genera in that it is an alternative method by which male odor is conveyed to the female.

An interesting feature of this comparison is that these four genera differ in the extent to which the female is free to move during courtship. In Triturus she is not restrained in any way by the male, but can move away from him at any time; in the other genera she is, to a greater or lesser extent, restrained by the male during the initial stages. The Triturus pattern allows greater opportunity for the female to exercise discrimination between males, the situation in which sexual selection of male characters might be expected to occur. It is significant in this context that Triturus species generally show more marked sexual dimorphism than any of the closely related genera.

In their analysis of the evolutionary relations of salamandrids Wake and Ozeti (1969) concluded that the four genera discussed above have close affinities with Euproctus, Neurergus, Paramesotriton, Hypselotriton and Pachytriton. There is very little information about courtship behavior in these genera. Euproctus sexual behavior is quite unlike that of any other urodele; there is a curious amplexus in which the male wraps his tail around the female's sacral region in such a way that their cloacas are opposed. This posture is maintained throughout the sexual encounter, which may last for several hours (de Bedriaga 1895; Ahrenfeldt 1960). This behavior may be an adaptation to the fast-flowing mountain stream habitat of this genus (Salthe 1967), although descriptions of the courtship suggest that they generally mate in quiet pools. An incomplete account of Neurergus courtship is given by Schmidtler and Schmidtler (1970), which described a display in which the male beats his tail rapidly in a position in front of the female's head; they did not observe spermatophore transfer. The courtship of Paramesotriton is much like that of Cynops; there is a rapid tail-beating display from a position in front of the female and the spermatophore deposition procedure is essentially similar (Arnold 1972). No information is available concerning courtship in Hypselotriton or Pachytriton.

In his scheme for the evolution of urodele courtship Salthe (1967) suggested that from a primitive mountain-dwelling ancestor there have arisen three taxonomic groups that invaded lowland habitats, the ambystomatids, the plethodontids and a group of salamandrids that includes Triturus. In all three lines there is a general trend towards decreased contact between male and female. The ancester of this lineage may have maintained amplexus throughout courtship, as does Euproctus. In all three groups amplexus has been lost during spermatophore transfer but has been retained during the early stages of courtship except in Triturus, Cynops, Neurergus and Paramesotriton.

Salthe suggested that the spermatophore evolved as an adaptation to life in fast-flowing streams in which sperm in milt form would be quickly dispersed. Likewise, amplexus during spermatophore transfer would reduce the chance of the spermatophore being swept away. Those forms that invaded still, lowland waters were able to dispense with amplexus. The spermatophore method of sperm transfer has other advantages; it makes internal fertilization possible, reducing the occurrence of unfertilized eggs, and it enables the female to exercise choice over where and when to lay her eggs so as to maximize their chances of survival. These advantages would be as important in still water as they are in flowing, and the possibility that the spermatophore evolved in lowland habitats, perhaps independently of montane forms, should not be discounted.

Synthesis

Reconstructing the phylogeny of behavior patterns is a highly speculative exercise, partly because of the lack of fossil evidence, partly because behavior appears to be evolutionarily very plastic and to undergo dramatic changes in relatively short periods of time. The phylogeny of *Triturus* courtship that I have proposed above concords well, however, with affinities established on morphological and genetic evidence. This phylogeny suggests that in one group of species, represented here by *T. vulgaris* and *T. helveticus*, a courtship pattern has evolved in which the female plays a more active and mobile part than in the other species. There is considerable diversity in the form and extent of restraint exercised by the male on the female during courtship in the ambystomatids and plethodontids (Salthe 1967), although whether, as Salthe suggested, there is a trend towards greater female freedom in these groups is open to question.

CONCLUSIONS

In this paper I have tried to identify the various factors that may have been significant in the evolution of *Triturus* courtship. I have derived relevant information from three sources, the results of an experimental analysis of the causal mechanisms underlying the behavior, a largely theoretical consideration of the biological functions of courtship and some comparative data on the courtship behavior of a number of *Triturus* species and of some closely related genera.

The total absence of amplexus during any stage of courtship is shared by all *Triturus* species and by some related genera, suggesting that it was lost before *Triturus* became differentiated from an ancestral stock. The greater freedom of movement that this loss of amplexus allows the female appears to have had a profound effect on the evolution of Triturus courtship, principally because it enables her to exercise a choice from among a number of potential mates. There is limited experimental evidence, and some circumstantial evidence, that reproductive isolation is effected by specificity in the female's response to male display. It is argued that the tendency of the female to be slow in responding to the male, and her relative failure to pick up his early spermatophores, are adaptive mechanisms by which she selects fitter males. In effect her behavior favors males that achieve a high spermatophore output in a time period constrained by their oxygen requirements. In other words, she "tests" males by forcing them towards the limits of two of the physiological constraints acting upon their behavior. Certain features in the sequential and temporal patterning of the male's behavior appear to be an adaptive response to the selective pressures imposed by the female, in that they may ensure that males mate with more responsive females.

The secondary sexual characters that are so distinctive a feature of the males of Triturus species seem also to have evolved as a consequence of female liberation. These characters are highly species-specific and it is probable that they serve as cues that enable females to recognise conspecific males. They may also have evolved through sexual selection, although it has yet to be demonstrated that females show any preference for one male over another.

In a comparative study of courtship in three American salamanders, Ambystoma maculatum, A. tigrinum and Plethodon jordani, Arnold (1976) has shown that the length of the breeding season and social context in which mating occurs are important factors that have influenced the evolution of courtship in these three species. Males not only behave so as to promote their own chances of transferring spermatophores to females, they also show sexual interference towards other males, which tends to suppress the reproductive success of rivals. While this behavior does not appear to have any parallels in Triturus, Arnold's study does illustrate one of the evolutionary consequences that arise from the indirect transfer of sperm via a spermatophore.

There is considerable scope for further studies of the courtship behavior of urodeles. Further information about other Triturus species is required before a complete picture of the evolution of courtship in the genus can be obtained. In addition detailed studies of other genera provide vital insights into questions of wider interest, for example, is amplexus a primitive feature of courtship that has been reduced or lost in various taxa and, if so, what selection pressures favored this trend? Studies of courtship should go beyond mere description. What is required is a detailed analysis, through both observational and experimental techniques, of the physiological, social and environmental constraints that have been selective forces in the evolution of urodele courtship.

ACKNOWLEDGMENTS

I am deeply indebted to Richard Sibly and Steve Arnold for lengthy discussions of the substance of this paper. Steve Arnold and David Wake provided valuable comments on an earlier manuscript. The work was financed by a Science Research Council grant to David McFarland to whom I am grateful for the provision of facilities and support. Finally, my thanks to Brian Nelson, Graham Bell and Steve Arnold for allowing me to quote extensively from their unpublished theses and papers.

REFERENCES

Ahrenfeldt, R. H. 1960. Mating behaviour of Euproctus asper in captivity. Brit. J. Herpetol. 2:194-197.

Arnold, S. J. 1972. The evolution of courtship behavior in salamanders. Ph. D. thesis, Univ. of Michigan, Ann Arbor.

Arnold, S. J. 1976. Sexual behavior, sexual interference and sexual defense in the salamanders Ambystoma maculatum, Ambystoma tigrinum and Plethodon jordani. Z. Tierpsychol. 42:247-300.

Baker, W. 1848. Description of a species of newt. The Zoologist 6:2198.

Ball, R. W., and D. L. Jameson. 1966. Premating isolating mechanisms in sympatric and allopatric Hyla regilla and Hyla californiae. Evolution 20:533-551.

Baerends, G., C. Beer and A. Manning (eds.) 1976. Function and evolution in behaviour. Clarendon Press, Oxford.

Baltzer, F. 1938. Die Fruhentwicklung des Bastards Triton cristatus ♀ x Triton palmatus ♂. Rev. suisse Zool. 45:391-394.

Bataillon, E. 1927. Les croisements chez les Urodeles et l'androgenèse hybride. C. r. Seanc. Soc. Biol. Paris 97:1715-1717.

Bateman, A. J. 1948. Intra-sexual selection in Drosophila. Heredity 2:349-368.

de Bedriaga, J. 1895. On the Pyrenean Newt, Molge aspera Duges. Proc. zool. Soc. Lond. (1895):150-161.

Bell, G. A. C. 1970. The distribution of amphibians in Leicestershire. *Trans. Leics. Lit. and Phil. Soc.* *64*:122-143.

Bell, G. A. C. 1973. The life of the Smooth Newt, *Triturus vulgaris* Linn. D. Phil. thesis, Oxford.

Blair, W. F. 1964. Isolating mechanisms and inter-species interactions in anuran amphibians. *Q. Rev. Biol.* *39*:334-344.

Callan, H. G., and H. Spurway. 1951. A study of meiosis in interracial hybrids of the newt, *Triturus cristatus*. *J. Genet.* *50*: 235-249.

Cooke, A. S., and W. G. Fulford. 1971. Observations on the feeding behaviour of a blind Warty newt (*Triturus cristatus*). *Brit. J. Herpetol.* *4*:216-217.

Cullen, J. M. 1966. Ritualisation of animal activities in relation to phylogeny, speciation and ecology. Reduction of ambiguity through ritualisation. *Phil. Trans. Roy. Soc. Ser. B* *251*:363-374.

Cullen, J. M. 1972. Some principles of animal communication. IN R. A. Hinde (ed.), *Non-verbal communication*. Cambridge Univ. Press.

Cummings, B. F. 1911. Some features of behavior in the courtship display of the Palmate newt (*Molge palmata* Schneid.). *J. Anim. Behav.* *1*:305-306.

Davis, J. W. F., and P. O'Donald. 1976. Sexual selection for a handicap: a critical analysis of Zahavi's model. *J. Theor. Biol.* *57*:345-354.

Davis, W. C., and V . C. Twitty. 1964. Courtship behavior and reproductive isolation in the species of *Taricha* (Amphibia, Caudata). *Copeia* (*1964*):601-610.

Dobzhansky, T. 1951. *Genetics and the origin of species*. Columbia Univ. Press, New York. 3rd ed.

Finkler, W. 1923. Analytical studies on the factors causing the sexual display in the Mountain newt (*Triton alpestris*). *Proc. Roy. Soc. Ser. B* *95*:356-364.

Foxon, G. E. H. 1964. Blood and respiration. IN: J. A. Moore (ed.), *Physiology of the amphibia*. Academic Press, New York.

Freytag, G. E. 1950. Über F_1 Bastarde zwischen Fadenmolch (*Triturus helveticus helveticus*) und Teichmolch (*Triturus vulgaris*) insbesondere über Lantz' Bastarde mit T. v. meridionalis. Abh. Ber. Mus. Magdeburg 8:2.

Freytag, G. E. 1970. Beobachtungen zum Paarungsverhalten von Cynops pyrrhogaster (Amphibia, Salamandridae). Salamandra 6: 15-17.

Gauss, G. H. 1961. Ein Beitrag zur Kenntnis des Balzverhaltens einheimischer Mölche. Z. Tierpsychol. 18:60-66.

Grant, P. R. 1972. Convergent and divergent character displacement. Biol. J. Linn. Soc. 4:39-68.

Halliday. T. R. 1974. Sexual behaviour of the Smooth newt, Triturus vulgaris (Urodela, Salamandridae). J. Herpetol. 8:277-292.

Halliday. T. R. 1975a. An observational and experimental study of sexual behaviour in the Smooth newt, Triturus vulgaris (Amphibia: Salamandridae). Anim. Behav. 23:291-322.

Halliday, T. R. 1975b. On the biological significance of certain morphological characters in males of the Smooth Newt Triturus vulgaris and of the Palmate Newt Triturus helveticus (Urodela: Salamandridae). Zool. J. Linn. Soc. 56:291-300.

Halliday, T. R. 1976. The libidinous newt. An analysis of variations in the sexual behaviour of the male Smooth newt, Triturus vulgaris. Anim. Behav. 24:398-414.

Halliday, T. R. 1977. The effect of experimental manipulation of breathing behaviour on the sexual behaviour of the Smooth newt, Triturus vulgaris. Anim. Behav. 25:39-45.

Halliday, T. R., and H. P. A. Sweatman. 1976. To breathe or not to breathe; the newt's problem. Anim. Behav. 24:551-561.

Halliday, T. R., and A. Worsnop. 1977. Correlation between activity and breathing rate in the Smooth newt, Triturus vulgaris. J. Herpetol. 11:244-246.

Hamburger, V. 1935. Malformations of hindlimbs in species hybrids of Triton taeniatus, Leyd. (and Triton palmatus, Duges) x Triton cristatus, Laur. J. Exp. Zool. 73:319-373.

Hamburger, V. 1936. The larval development of reciprocal species hybrids of Triton taeniatus, Leyd. (and Triton palmatus, Duges) x Triton cristatus, Laur. J. Exp. Zool. 73:319-373.

Hilton, W. A. 1902. A structural feature connected with the mating of Diemyctylus viridescens. Am. Nat. 34:643-651.

Himstedt, W. 1967. Beobachtungen zum Paarungsverhalten des Feuersalamanders. Zool. Anz. 175:295-300.

Himstedt, W. 1968. Die Reaktionen des Teichmolchs (Triturus vulgaris) auf intermittierende Lichtreize. Verh. der Deut. Zool. Ges. Innsbruck 1968:357-361.

Himstedt, W. 1969. Zur Funktion eines Reizfiltermechanismus im visullen System von Urodelen. Z. vergl. Physiol. 62:197-204.

Hinde, R. A. 1959. Behaviour and speciation in birds and lower vertebrates. Biol. Rev. 34:85-128.

Hinde, R. A., and N. Tinbergen. 1958. The comparative study of species-specific behaviour. IN: A. Roe and G. G. Simpson (eds.), Behaviour and evolution. Yale Univ. Press, New Haven.

Houston, A. I., T. R. Halliday and D. J. McFarland. 1977. Towards a model of the courtship of the Smooth newt, Triturus vulgaris, with special emphasis on problems of observability in the simulation of behaviour. Med. Biol. Eng. 15:49-61.

Huxley, J. 1966. A discussion on ritualization of behaviour in animals and man. Phil. Trans. Roy. Soc. Ser. B 251:249-271.

de l'Isle, A. 1862. Notice zoologique sur un nouveau Batracien Urodele de France (Triton blasii). Ann. Sci. Nat. 17:363-371 (4th ser., zool.).

de l'Isle, A. 1872. De l'hybridation chez les amphibies. Ann. Sci. Nat. 17 (5th ser., paleontol.).

Jordan, E. O. 1893. The habits and development of the newt (Diemyctylus viridescens). J. Morphol. 8:269-366.

Kawamura, T., and S. Sawada. 1959. On the sexual isolation among different species and local races of Japanese newts. J. Sci. Hiroshima Univ. Ser. B, Div. 1 18:17-30.

Klinge, W. 1915. Triton pyrrhogaster. Wschr. Aquar. Terrar. Knde. 12:427-431.

Lantz , L. A. 1947. Hybrids between Triturus cristatus Laur. and Triturus marmoratus Latr. Proc. zool. Soc. Lond. 117:247-258.

Lantz, L. A., and H. G. Callan. 1954. Phenotypes and spermatogenesis of interspecific hybrids between *Triturus cristatus* and *T. marmoratus*. *J. Genet.* *52*:165-185.

Liley, N. R. 1966. Ethological isolating mechanisms in four sympatric species of Poeciliid fishes. *Behaviour*, Supplement 13.

Mancino, G. 1959. La struttura dell'ovario nell'ibrido *Triturus helveticus* ♀ x *Triturus vulgaris* ♂. *Boll. Zool.* *26*:317-328.

Mancino, G. 1961. Anomalie meiotiche nella spermatogenesi dell' ibrido *Triturus italicus* ♀ x *Triturus vulgaris* ♂. *Boll. Zool.* *28*:691-701.

Matthes, E. 1924a. Das Geruchsvermogen von Triton beim Aufenthalt unter Wasser. *Z. vergl. Physiol.* *1*:57-83.

Matthes, E. 1924b. Das Geruchsvermogen von Triton beim Aufenthalt an Land. *Z. vergl. Physiol.* *1*:590-606.

Matthes, E. 1927. Der Einfluss des Mediumwechsels auf das Geruchsvermogen von Triton. *Z. vergl. Physiol.* *5*:83-166.

Maynard-Smith, J. 1976. Sexual selection and the handicap principle. *J. Theor. Biol.* *57*:239-242.

Mayr, E. 1963. *Animal species and evolution*. Oxford University Press.

McFarland, D. J. 1976. Form and function in the analysis of behaviour sequences. IN: P. P. G. Bateson and R. A. Hinde (eds.), *Growing points in ethology*, Cambridge Univ. Press, Chap. 2.

Mertens, R., and L. Muller. 1940. Die Emphibien und Reptilien Europas. *Abh. senchekburg naturf. Ges.* *451*:1-56.

Nelson, J. B. 1959. The courtship behaviour of *Triturus cristatus*. Honours thesis, St. Andrews Univ.

Nelson, J. B. 1960. Courtship display in the Italian Crested newt (*Triturus cristatus carnifex*). *Biol. J.* (St. Andrews Univ.) *1*: 6-9.

Oeser, R. 1951. Haltung and Zucht des *Triturus vittatus* Gray. *Aquar. Terrar. Zeit.* *4*:103-106.

Owen, D. F. 1955. The food of the heron *Ardea cinerea* in the breeding season. *Ibis* *97*:276-295.

Pariser, K. 1932. Verschiebung des Geschlechtsverhältnisses bei kunstlich del genero Triton bastarden. Biol. Zentralbl. 52: 654-659.

Pariser, K. 1935. Deformidades y otras anomalias en hibridos interspecíficos del género Triton (Anfibios). Revta. esp. Biol. 4:5-12.

Pariser, K. 1936. El desarrollo y la relación numérica entre los sexos en los hibridos interspecíficos obtenidos por fecundation artificial en el género Triton (Molge). Revta. esp. Biol. 5: 11-93.

Parker, G. A. 1970. Sperm competition and its evolutionary consequences in the insects. Biol. Rev. 45:525-567.

Perdeck, A. C. 1958. The isolating value of specific song patterns in two sibling species of grasshoppers (Chorthippus brunneus and C. biguttulus). Behaviour 12:1-75.

Poll, H. 1909. Mischlinge von Triton cristatus Laur. und Triton vulgaris L. Biol. Zentralbl. 29:30-31.

Prechtl, H. F. R. 1951. Zur Paarungsbiologie einiger Molcharten. Z. Tierpsychol. 8:337-348.

Razoumowsky. 1788. Histoire naturele du Jurat 1. Lausanne.

Rusconi, M. 1821. Amours des salamandres aquatiques. Milan.

Salthe, S. N. 1967. Courtship patterns and the phylogeny of the urodeles. Copeia (1967):100-117.

Salthe, S. N., and J. S. Mecham. 1974. Reproductive and courtship patterns. IN: B. Lofts (ed.), Physiology of the amphibia, Vol. 2, Academic Press, New York.

Schmidtler, J. J., and F. F. Schmidtler. 1970. Morphologie, Biologie und Verwandtschaftsbeziehungen von Neurergus strauchii aus der Turkei (Amphibia: Salamandridae). Senckenberg. Biol. 51:41-53.

Smith, M. 1969. The British amphibians and reptiles. Collins, London. 4th edition.

Spurway, H. 1953. Genetics of specific and subspecific differences in European newts. Symp. Soc. Exptl. Biol. 7:200-237.

Spurway, H., and H. G. Callan. 1950. Hybrids between some members of the rassenkreis Triturus cristatus. Experientia 6:95.

Spurway, H., and H. G. Callan. 1960. The vigour and male sterility of hybrids between the species Triturus vulgaris and Triturus helveticus. J. Genet. 57:84-116.

Steward, J. W. 1969. The tailed amphibians of Europe. David and Charles, Newton Abbot.

Taylor, R. H. R. 1948. The distribution of reptiles and amphibia in the British Isles, with notes on species recently introduced. Brit. J. Herpetol. 1:1-38.

Thorn, R. 1968. Les salamandres d'Europe, d'Asie, et d'Afrique du Nord. Encyclopedie pratique du naturaliste 35, Paris.

Thorpe, W. H. 1972. The lower vertebrates and the invertebrates. IN: R. A. Hinde (ed.), Non-verbal communication, Cambridge Univ. Press.

Tinbergen, N. 1953. Social behaviour in animals. Methuen, London.

Tinbergen, N. 1959. Comparative studies of the behaviour of gulls (Laridae): a progress report. Behaviour 15:1-70.

Tinbergen, N., and J. J. Ter Pelkwijk. 1938. De Kleine Watersalamander. Levende Nat. 43:232-237.

Trivers, R. L. 1972. Parental investment and sexual selection. IN: B. Campbell (ed.), Sexual selection and the descent of man, Aldine Atherton, Chicago.

Tsutsui, Y. 1931. Notes on the behaviour of the common Japanese newt, Diemictylus pyrrhogaster Boie. I. Breeding habit. Mem. Coll. Sci., Kyoto Imp. Univ. Ser. B 7:159-179.

Twitty, V. C. 1966. Of scientists and salamanders. Freeman, San Francisco.

Wake, D. B., and N. Ozeti. 1969. Evolutionary relationships in the family Salamandridae. Copeia (1969):124-137.

White, M. J. D. 1946. The spermatogenesis of hybrids between Triturus cristatus and T. marmoratus (Urodela). J. Exp. Zool. 102:179-208.

Wilz, K. J. 1970. Reproductive isolation in two species of Stickleback (Gasterosteidae). Copeia (1970):587-590.

Witherby, H. F., F. C. R. Jourdain, N. F. Ticehurst and B. W. Tucker. 1938. The handbook of British birds. Vols. 1-5, Witherby, London.

Wolterstorff, W. 1908. Nochmals Kummers Bastarde. _Wschr_. _Aquar_. -_u_. _Terrarienk_ _5_, Supplement "Lacerta" _2_:8.

Wolterstorff, W. 1925. Katalog der Amphibiensammlung im Mus. Nat. -u. Heimatk. Magdeburg. _Abh_. _Mus_. _Nat_. -_u_. _Heimatk_. _Magdeburg_ _4_:259-289.

Wright, S. 1941. On the probability of fixation of reciprocal translocations. _Am_. _Nat_. _75_:513-522.

Zahavi, A. 1975. Mate selection--a selection for a handicap. _J_. _Theor_. _Biol_. _53_:205-214.

THE COURTSHIP OF FROGS

Kentwood D. Wells

Biological Sciences Group, University of Connecticut
Storrs, Connecticut 06268

Frogs were once thought to have relatively uninteresting social behavior, but recent studies on tropical and temperate zone species have revealed a surprising diversity of social systems. There is a growing literature on anuran mating systems, aggression, spacing behavior, and territoriality (Wells 1977b). Unfortunately, many aspects of courtship have been neglected. I use the term "courtship" to refer to interactions between males and females leading to pair formation and mating, and I include the use of both long-range and short-range signals by courting males. The purpose of this paper is to bring together the largely anecdotal literature on anuran courtship and to outline a number of problems that require more detailed investigation.

Some frogs have virtually no courtship in the usual sense of the word. Males assemble at breeding areas and call as a group to attract females. Males locate individual females by active searching and trial-and-error clasping, producing a scene of chaotic activity at the breeding site. Such behavior is largely confined to explosive breeders in which reproductive activities are strongly clumped in both space and time (see Wells 1977b for further discussion).

In species with prolonged breeding periods, females usually approach individual males; they can choose among a number of potential mates. In some species, courtship is very simple. A male calls until a female contacts him; he then clasps the female (Noble and Noble 1923; Eibl-Eibesfeldt 1952; Littlejohn 1958; Mortof and Thompson 1958; Wells 1977b).

Nevertheless, there is increasing evidence of more complex interactions in the courtship of some species. These include changes in long-range vocal signals in response to the calls of other males, as well as changes in calling behavior in response to the approach of other individuals. In some species, particularly among the dendrobatids, there are elaborate male-female interactions involving visual, tactile, and vocal signals.

In the first part of this paper, I will discuss various aspects of long-range and short-range vocal courtship, with reference primarily to males in large choruses. Then I will discuss more elaborate forms of courtship, especially in dendrobatids. Finally, I will discuss possible functions of courtship in frogs as a group and try to provide a framework for understanding the evolution of their courtship behavior. Because of the anecdotal nature of the available literature, my interpretations should be regarded as hypotheses for future testing rather than definitive explanations.

VOCAL COURTSHIP

Most male frogs have vocalizations with which they advertise their species identity, sex, reproductive state, and spatial location to other frogs. These signals are broadcast at large and are not directed at particular individuals. There is no doubt that these vocalizations attract females (Martof and Thompson 1958; Littlejohn and Michaud 1959; Blair and Littlejohn 1960; Littlejohn et al. 1960; Littlejohn 1961; Awbrey 1965; Littlejohn and Loftus-Hills 1968; Schmidt 1969, 1971; Gerhardt 1973; Littlejohn and Watson 1974). They also may serve other functions such as territorial advertisement (Bogert 1960; Emlen 1968; Wiewandt 1969; Wells 1976, 1977b).

Bogert (1960) and subsequent workers have used the term "mating call" for these long-range vocalizations. I prefer the term "advertisement call" because this does not specify whether the intended recipients are males or females (Wells 1976, 1977b). The term "mating call" also can be confusing if males use both long-range and short-range calls to attract females, as discussed below.

A frog calling in a large chorus has at least three major problems: 1) he must reduce acoustic interference from other males to maximize the distinctness of his own calls, 2) he must attempt to make his calls more attractive to females than those of his competitors, and 3) he may have to produce a signal which simultaneously attracts females and conveys an agonistic message to other males. Several possible strategies for dealing with these problems are discussed below.

Call Alternation

The problem of acoustic interference in large choruses has been solved in similar ways by frogs and many insects. In both groups, males tend to space their signals so that those of one individual do not overlap those of his near neighbors (Alexander 1975; Wells 1977b). In some species, call alternation is very precise and may lead to the formation of duets, trios, or other call groups within a large chorus. In others, such as Rana clamitans, calls are spaced, but the interval between calls varies so that no consistent pattern of duets or trios results (Wells 1976). Call alternation has been reported in most of the major anuran families (Goin 1949; Eibl-Eibesfeldt 1952; Anderson 1954; Jameson 1954; Martof and Thompson 1958; Hardy 1959; Bogert 1960; Snyder and Jameson 1965; Capranica 1965; Duellman 1967; Foster 1967; Schneider 1967, 1968; Heinzmann 1970; Lemon 1971; Loftus-Hills 1971, 1974; Rosen and Lemon 1974; Whitney and Krebs 1975a; Wells 1977b).

While call alternation would allow a male to broadcast his calls more effectively, this alone would provide little competitive advantage over other males in attracting females. Presumably most males in a chorus would be alternating calls with their near neighbors in a similar fashion. It has been suggested that males that consistently begin calling bouts ("chorus leaders") might attract more females (Duellman 1967; Brattstrom and Yarnell 1968; Whitney and Krebs 1975a; Wilson 1975). Whitney and Krebs (1975a) showed that a loudspeaker that initiated an experimental chorus attracted more female Hyla regilla than other loudspeakers, but no other data are available to test this hypothesis (see Wells 1977b for further discussion).

It is also conceivable that females might choose males that produced the largest number of signals per unit time, regardless of whether this were achieved by initiating call bouts, ending call bouts, or calling more frequently in each bout. One would expect males to try to outsignal their competitors by producing louder calls, producing more calls per unit time, or producing more distinctive calls (Alexander 1975).

Changes in Call Intensity

There is evidence that certain individuals in frog choruses call louder than others (Whitney and Krebs 1975a; Gerhardt 1975). In some cases, this may be a function of body size (Gerhardt 1975). Rand (in ms.) found, however, that male Physalaemus pustulosus increase calling intensity in response to the calls of other males or tape recorded calls. Presumably some other species do this as well. The usefulness of this behavior as a competitive courtship strategy is probably limited, since the relative intensity of calls perceived by incoming females would depend on their positions relative to different males. Female Hyla versicolor, however, can apparently discriminate intensity differences of as little as 2 db, and will orient toward a louder call in an experimental situation (Fellers 1976). The chief advantage of a loud call may be in inhibiting other males from calling nearby (Allan 1973; Rosen and Lemon 1974; Whitney and Krebs 1975b; Fellers 1976) and in enabling a male to maintain a large individual distance around his calling site (Fellers 1976; Wells 1977b).

Changes in Call Duration and Structure

There are several ways that a frog might increase the information content of its calls relative to those of other males. The strategy adopted would depend on the basic structure of the call (Alexander 1975). When the call is a trill or buzz, a male might lengthen the call in response to the calls of other males (Fig. 1a). When the call is a series of rapidly repeated notes, the note repetition rate of the call might be increased (Fig. 1b). When the call is a single note or group of notes repeated with relatively long intervals between calls, a male might increase the number of repeated elements in the call (Fig. 1c) or add new elements to the original call (Fig. 1d). These alterations in call duration and structure would be useful only if this did not affect the encoding of species identity in the call.

I know of no clearly documented case of frogs with trilled calls increasing trill duration in response to the calls of other individuals, but such interactions occur in a number of insects (Otte 1972; Alexander 1975). Increases in note repetition rate (Fig. 1b) in response to the calls of other frogs are also difficult to document from existing literature. Several species of Puerto Rican Eleutherodactylus, including E. coqui, E. portoricensis, E. antillensis, and E. brittoni call in bouts of a dozen or more notes. The note repetition rate often increases progressively within each bout (Drewry 1970). Whether such behavior occurs only in interactions between calling males or also when males call in isolation is not known.

Individuals of several species are known to call more frequently as chorus size increases. These include Hyla crucifer (Rosen and Lemon 1974), H. cinerea (Garton and Brandon 1975), H. regilla (Foster 1967), Rana catesbeiana (Emlen 1968), and R. clamitans (Wells 1976). It often is difficult, however, to tell whether the increased calling rate of each individual is due to interactions with other males or to a common response to some environmental variable such as light or temperature. In R. clamitans, calling rates increase as light levels decrease immediately after sunset (Wells 1976).

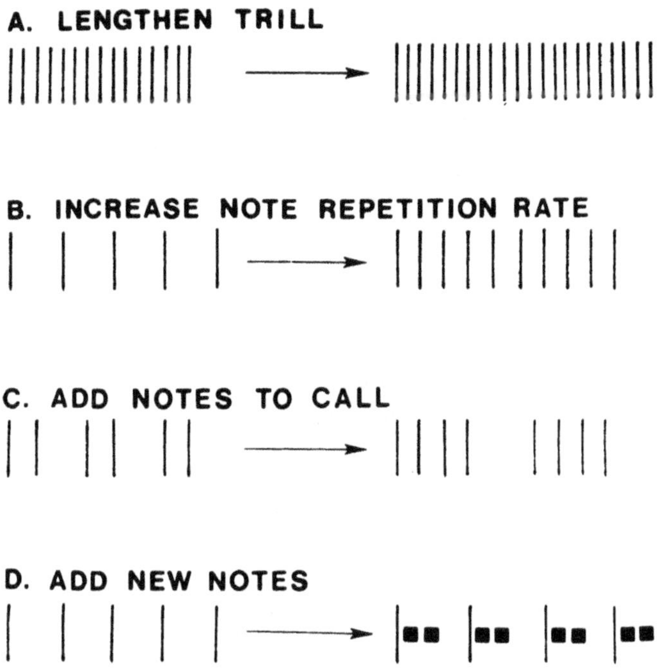

Fig. 1. Four possible methods of increasing the information content of an advertisement call in response to the calls of other males.

In active choruses R. clamitans and R. catesbeiana males may change not only the frequency with which calls are produced, but also the number of notes in each call. R. clamitans males normally give single notes spaced at intervals of several seconds, but may give three or four notes separated by only a few milliseconds during active chorusing (Wells 1976). Similarly, male R. catesbeiana may give calls composed of three or four notes, or up to fifteen notes (Capranica 1968). Is it possible that addition of notes to the call makes an individual's calls more distinct to females and other males, thereby enabling him to attract more mates or defend a territory more effectively? Experimental manipulations using playbacks of calls with varying numbers of notes are needed to test this possibility.

This hypothesis has been tested in experimental work with the leptodactylid frog, Physalaemus pustulosus. Rand (in ms.) found that males produce calls of varying degrees of complexity. The simplest is a "whine" consisting of a single tone-like note with multiple harmonics (Fig. 2). More complex notes are produced by adding one to three "chucks" onto the end of the "whine" note

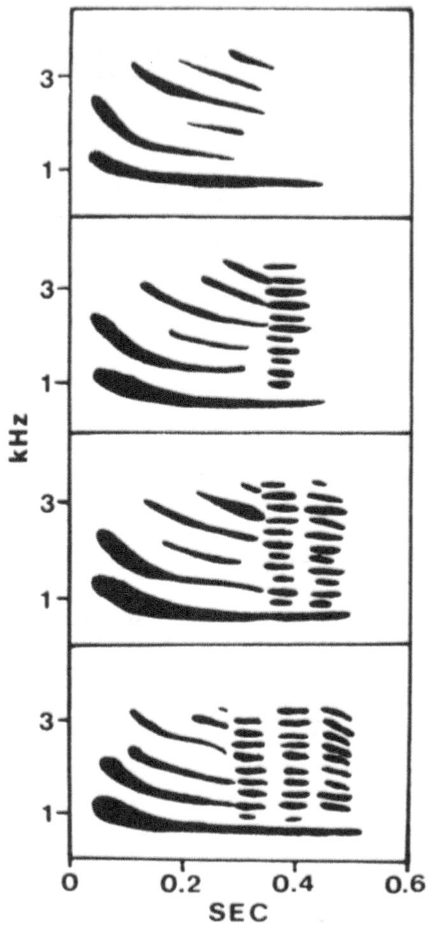

Fig. 2. Tracings of sonograms of Physalaemus pustulosus advertisement calls, showing increasing call complexity (top to bottom). Reproduced courtesy of A. S. Rand.

(Fig. 2). Rand (in ms.) found that social interactions were important in changing the complexity of a male's call.

Through observations of natural choruses and experiments with recorded calls, Rand (in ms.) found that call complexity increased 1) as the number of males in a chorus increased, 2) in response to playbacks of a background chorus, 3) in response to increases in call intensity by nearby males, 4) in response to playbacks of more complex calls, and 5) in response to the approach of other frogs into the visual field of the calling male.

Rand (in ms.) demonstrated that, by adding "chuck" notes to the basic "whine" call, a male Physalaemus provides better directional information for females in a large chorus in which vocal competition is intense. He confirmed this hypothesis by showing that females nearly always chose higher complexity calls in two-choice playback experiments. Working against selection for better directional cues in the call would be the threat of predation. Presumably any call which enabled a female to locate a male more easily also would enable predators to find him (Rand, in ms.). There is evidence that Bufo marinus, a predator on Physalaemus, uses vocalizations as a cue to locate prey (Jaeger 1976).

Changes in Response to Approaching Individuals

In most of the examples cited above, changes in calling behavior occur in response to the calls of other males. In some species, males also change their vocalizations in response to the approach of other frogs. Some examples are given in Table 1. Most of these examples are based on anecdotal accounts and we must be cautious in interpreting the behavior.

There seem to be two general classes of response. In some species, males change the rate or intensity of the advertisement call. For example, male Gastrophryne carolinensis, Chiromantis rufescens, Hyla cinerea, and H. squirella are all reported to increase the call rate as females approach (Table 1). In other species, such as Syrrhophus marnocki, several species of Pseudophryne, and Rana adenopleura, males apparently give calls that are distinct from the normal advertisement call. In Alytes obstetricans, females answer male advertisement calls with calls of their own, and males often give softer "excitement" calls as females approach (Heinzmann 1970). Female calling has also been reported in Tomodactylus angustidigitorum, and males are said to switch from a peeping call to a trill as females approach (Dixon 1957).

It is difficult to assign specific functions to these close-range vocalizations without additional observations and experiments, but it is possible that they serve as "courtship" calls in some species (e.g., Pengilley 1971). I should point out, however, that there is little evidence at present to suggest that males can distinguish the sex of approaching individuals. Presumably they can do so when females also call, and they may be able to do so visually in some species.

Close-range courtship signals occur in many insects and may provide additional directional cues to females at close range or signal a male's lack of aggressiveness and readiness to mate (Alexander 1975). The latter would be particularly important for territorial frogs in which females must enter a male's territory to

Table 1. Examples of changes in calling behavior in response to approaching individuals.

Species	Behavior	Sources
Microhylidae		
Gastrophryne carolinensis	Males increase call rate when other frogs approach	Anderson 1954
Hylidae		
Hyla cinerea	Males turn to face approaching females; increase call rate and pitch	Garton and Brandon 1975; Fellers 1976
Hyla squirella	Males increase call rate as females approach	Fellers 1976
Hyla crucifer	Males increase note duration and interval between calls when females approach	Rosen and Lemon 1974; Fellers 1976
Hyla versicolor	Males produce longer calls when females approach	Fellers 1976
Hyla regilla	Males switch from diphasic to monophasic calls when other frogs are nearby (may not be courtship)	Snyder and Jameson 1965; Allan 1973
Rhacophoridae		
Chiromantis rufescens	Males call in groups; all males increase call rate when females approach	Coe 1967, 1974
Ranidae		
Rana adenopleura	Males add a "series of low unmusical notes" to advertisement calls when females approach	Liu 1950
Leptodactylidae		
Physalaemus pustulosus	Males increase call rate and call complexity when other frogs approach	Rand, in ms.

Table 1 (cont.)

Species	Behavior	Sources
Syrrhophus marnocki	Males have a rapid call that is given only when females are present	Jameson 1954, 1955a, 1955b
Pseudophryne corroboree dendyi bibroni	Males have distinct "courtship" calls given when females are in close proximity in burrows	Pengilley 1971; Woodruff 1976
Tomodactylus angustidigitorum	Females answer male advertisement calls with calls of their own; males switch from peeps to trill as females approach	Dixon 1957
Discoglossidae		
Alytes obstetricans	Females answer male advertisement calls with calls of their own; males give softer calls when females approach; these calls are sometimes interspersed with advertisement calls	Heinzmann 1970

mate. Males are known to be aggressive or territorial in many of the species listed in Table 1 (Pengilley 1971; Rosen and Lemon 1974; Garton and Brandon 1975; Fellers 1975; Whitney and Krebs 1975b; Wells 1977b).

Although males of some species of frogs are reported to increase call intensity when females approach (Coe 1974), others are reported to give calls which are softer than the normal advertisement call (Heinzmann 1970). Such behavior also occurs in many insects (Alexander 1975) and may be advantageous in allowing a male to attract a nearby female without signaling to competing males that a female is in the vicinity. One might expect low intensity courtship calls to evolve in species in which male-male competition is intense, particularly if some males adopt a "satellite" or "parasitic" role and attempt to intercept females attracted to other calling males (Alexander 1975; Wells 1977b).

Experiments using playbacks of natural and synthetic calls would be useful in determining the function of close-range vocalizations in some of these species. It would be particularly interesting to know whether females are attracted to "courtship" calls in preference to normal advertisement calls. Garton and Brandon (1975) reported that male H. cinerea increase the pitch of their calls in addition to the call rate when females approach, although they did not present data from recordings. Gerhardt (1976) reported that the low frequency components of this species' call serve to attract females from a distance, while high frequency components function primarily at close range. It may be that males increase the intensity of the high frequency components relative to that of the low frequencies when females are detected nearby, and that females are better able to localize males giving high frequency calls.

Compound Advertisement Calls

Despite the complexities apparent in the vocal behavior of some frogs, the structure of an individual call often is very simple. In most cases, the advertisement call consists of a simple buzz or trill, a single note given at various intervals, or a rapidly repeated series of notes. In some species, however, the advertisement call consists of two or more distinctly different notes delivered as a single vocalization. I have already mentioned one example of this in the discussion of Physalaemus pustulosus. Additional examples occur in species from several different families.

Several Eleutherodactylus, including E. coqui, E. portoricensis, E. antillensis, and E. martinicensis, have calls consisting of one or more tone-like notes followed by one or more frequency-modulated whistled notes (Drewry 1970; Lemon 1971). Other species, such as E. eneidae and E. locustus, have tone-like notes followed by long series of clicking or tinkling notes (Drewry 1970). Frogs in the Hyla microcephala group from Central and South America have compound calls consisting of a "primary" note followed by a series of shorter "secondary" notes (Duellman 1970). Similarly, the hylids Smilisca sila and S. puma have calls consisting of a low-pitched "squawk" followed by one or more "rattling secondary notes" (Duellman 1970). Compound calls with two or more distinct notes are also found in African rhacophorids of the genus Afrixalus (Schiøtz 1967) and in the Australian genus Pseudophryne (Pengilley 1971). Finally, compound advertisement calls have been reported in some pipids. Pipa carvalhoi, for example, gives a series of clicks followed by a buzz (Weygoldt 1976).

What is the function of these compound advertisement calls? At present, we have direct experimental evidence bearing on this question for only one species, E. coqui from Puerto Rico. This species has an advertisement call composed of an introductory tone-like "co" note and a frequency-modulated "qui" note. Narins and Capranica (1976) showed that females will approach speakers playing the normal call or the "qui" note alone, but are not attracted to the "co" note. Males, on the other hand, will alternate calls with a tape recording of "co" notes but seem to ignore the "qui" note (Narins 1976). Males give only the "co" note when challenging other calling males (Drewry 1970; Narins 1976). This suggests that the "co" note conveys an agonistic message to males, while the "qui" note serves to attract females (Narins 1976; Narins and Capranica 1976). Thus the call has been partitioned to carry separate messages to males and females.

The compound calls of other species may serve similar functions. Indirect evidence for this comes from comparisons of agonistic calls, close-range courtship calls, and elements of the advertisement call. Weygoldt (1976) recorded an agonistic buzz call in Pipa carvalhoi similar in structure to the buzz portion of the advertisement call. Schiøtz (1967) recorded apparent agonistic calls from male Afrixalus which were similar in structure to one note of the advertisement call. In Pseudophryne corroboree, the structural components of the advertisement call appear to be related to several agonistic calls and a "courtship" call (Fig. 3) (Pengilley 1971). In Hyla minuta, fighting males give a call which resembles the introductory note of the advertisement call (personal observations).

In all of these species, compound advertisement calls may have evolved from simple advertisement calls as in Fig. 4. According to this scheme, separate agonistic and courtship calls might have originated as modifications of the advertisement call. While retaining their separate functions, elements of these calls might have been incorporated into a new compound advertisement call which conveys several messages simultaneously. A similar hypothesis has been advanced by Alexander (1975) to explain multipart songs of some insects.

COMPLEX COURTSHIP

In my discussion so far, I have focused on strategies that a male frog might use to attract a female to his calling site. In many species, courtship effectively ends at this point. Once the female arrives, she is quickly clasped without preliminary close

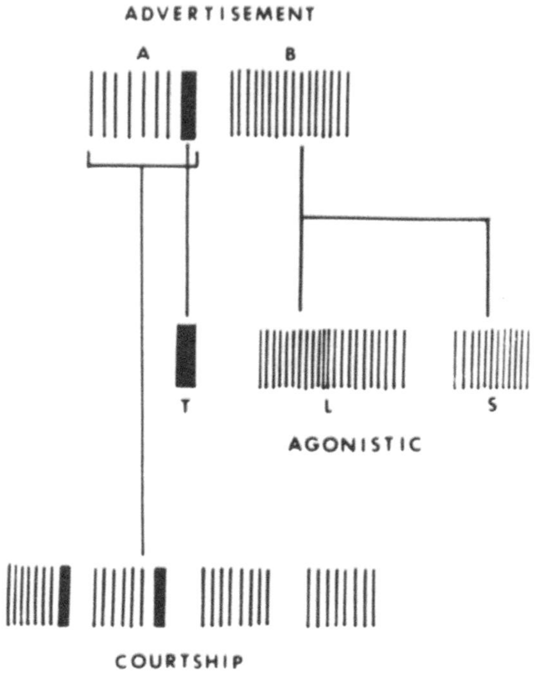

Fig. 3. Diagrammatic representation of structural relationships between advertisement calls, agonistic calls, and presumed courtship calls of *Pseudophryne corroboree*. A = two syllable component; B = one syllable component; T = "tock" call; L = long aggressive call; S = short aggressive call. Modified from Pengilley (1971).

range courtship. In some species, however, there are relatively prolonged associations between individual males and females, and there may be complex courtship interactions involving visual and tactile cues as well as vocalizations. Most examples of complex courtship reported to date come from studies of dendrobatid frogs, perhaps because these conspicuous diurnal species are more easily observed than most nocturnal frogs. However, there are a few examples of complex courtship in other groups which I will mention first.

Prolonged associations between individual males and females have been reported in a number of hole or crevice nesting species, including *Hylactophryne augusti* (Jameson 1954), *Platyhyla grandis* (Blommers-Schlösser 1975), several species of *Pseudophryne* (Pengilley 1971; Woodruff 1976), *Syrrhophus marnocki* (Jameson 1954, 1955b), and *Eleutherodactylus coqui* (Drewry 1974). In some cases, a female apparently stays with a male inside his retreat for many

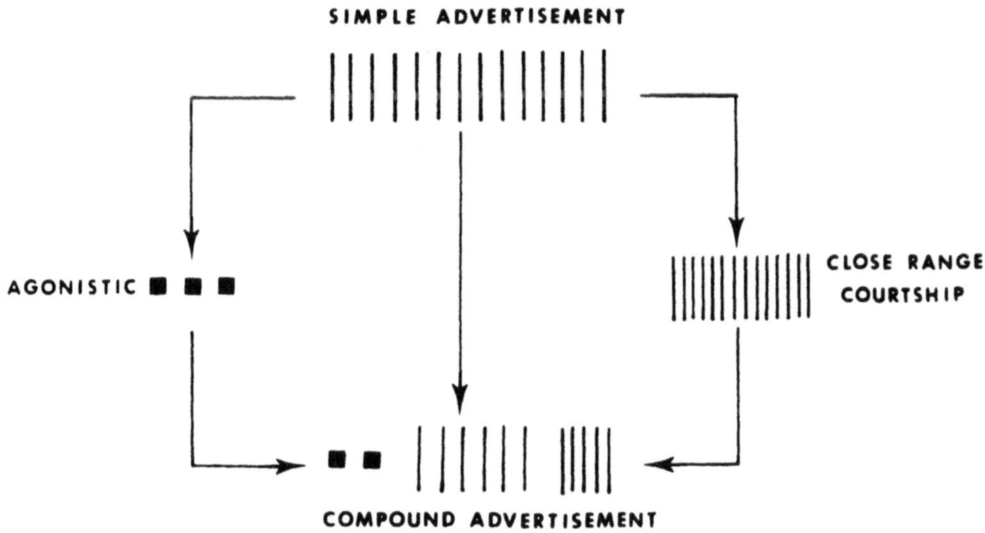

Fig. 4. Hypothetical steps in the evolution of a compound advertisement call from a simple advertisement call. Modifications of the original advertisement call produce distinct agonistic and courtship calls, which recombine with the original call to form a compound advertisement call.

hours, or even days. *Pseudophryne* males are reported to give courtship calls when females are present, as are *Syrrhophus* males. The latter also stroke females with their hind legs before clasping (Jameson 1954, 1955a, 1955b).

Stroking has been reported in several aquatic pipid frogs. In the African species, *Hymenochirus boettgeri*, males seize any females which enter their territories. A male that has clasped a female reaches up and strokes the female's head with his back feet (Österdahl and Olsson 1963; Rabb and Rabb 1963). Stroking occurs in other pipids, including *Pipa*, *Xenopus*, and *Pseudohymenochirus* (Rabb and Rabb 1960, 1963; Weygoldt 1976). In some cases, the behavior has been reported only during oviposition, but it may be part of the preliminary courtship in some species.

The courtship of dendrobatids is more elaborate than that reported for any other group. Although the details of courtship vary from one species to another, the basic pattern is similar in all dendrobatids studied so far. Summaries of the social systems and courtship behavior of several *Dendrobates* and *Colostethus* are given in Tables 2 and 3.

When males are territorial, as in *D. pumilio*, *D. granuliferus*, *C. inguinalis*, and *C. palmatus*, courtship usually begins with the

Table 2. Summary of social systems and courtship behavior in several species of *Dendrobates*.

Species	Social System	Courtship Behavior
D. granuliferus Goodman 1971; Crump 1972; Silverstone 1975	Males defend territories on ground or low vegetation. Females reported to carry tadpoles; not known whether males carry tadpoles	Female approaches calling male in territory; male leads female to oviposition site in litter; elaborate tactile courtship precedes oviposition; amplexus absent; male and female place vents together as eggs are laid
D. pumilio Duellman 1966; Bunnell 1973; Rand, pers. comm.	Males defend territories on ground or low vegetation; in captivity, females carry tadpoles most frequently, but males care for eggs and may carry tadpoles	Female approaches calling male in territory; male leads female to oviposition site on leaf or low plant; pair may visit several plants before mating; amplexus absent; male wets leaf and may deposit sperm before female lays eggs
D. histrionicus Silverstone 1973	Males defend territories near ground; parental care unknown	Initial approach of female to male not reported; male pursues female and attempts to court as female apparently searches for oviposition site; elaborate tactile interactions; oviposition and mating not observed
D. auratus Senfft 1936; Dunn 1941; Wells, pers. obs.	Males do not defend territories and do not call continuously; calling males may attack other males calling nearby; male performs all parental	Female approaches male calling in hole or under log; male leads female through leaf litter until oviposition site is found; female actively courts

	care, including carrying tadpoles; can care for more than one clutch at a time	male by jumping on his back, prodding with her forelegs, etc. May be some female-female aggression when more than one female courts a single male
D. azureus Polder 1974	Not observed in field; males show no signs of territoriality in captivity; female-female aggression observed, possibly in connection with courtship of males; males perform most parental care, but females may also carry tadpoles	Female seems to initiate courtship and actively stimulates male by rubbing, jumping on his back, etc. Very similar to D. auratus

Table 3. Summary of social systems and courtship behavior in several species of Colostethus.

Species	Social System	Courtship Behavior
C. inguinalis Duellman 1966; Wells, pers. obs.	Males defend permanent territories on rocks in or near streams; females defend temporary territories but move around extensively; females carry tadpoles for up to eight days	Female enters male territory in low posture; male responds with aggressive encounter call, but gives advertisement calls in courtship; male calls while touching female, sometimes circling around her or mounting; male leads female to oviposition site in crevice or leaf
C. palmatus Lüddecke 1974, 1976	Males and females both territorial; degree of permanence unknown; males care for eggs and carry tadpoles up to seven days	Female enters male territory in low posture; male assumes dark color during courtship; male changes orientation to female while calling and may circle; physical contact infrequent; male leads female to oviposition site in crevice; elaborate tactile courtship precedes oviposition; male rests on female's back during egg deposition, but does not clasp
C. trinitatis Test 1954; Sexton 1960; Kenny 1969; Wells, pers. obs.	Females very aggressive; defend territories on rocks in streams; males more mobile but may defend territories; courting males aggressive to one another; males carry tadpoles	Male calls from position near several female territories; possibly moves into female territory and attempts to court resident; male turns jet black and jumps up and down in front of female while calling; male leads female to oviposition site in crevice

C. *collaris* Dole and Durant 1974; Durant and Dole 1975	Females very aggressive; defend territories on rocks in streams; males apparently more mobile; males carry tadpoles	Males may enter female territories to court residents, but this is not certain; calling males perform "toe dance" in front of females; male then leads female to oviposition site; oviposition not observed

female entering a male's territory, approaching the resident, and indicating her willingness to court. In some cases, the female contacts the male or crouches in front of him (Lüddecke 1974; personal observations on C. inguinalis). In others, the male leaves his calling site and approaches the female (Crump 1972; Silverstone 1973). Exceptions to this general pattern occur in C. trinitatis and C. collaris. In these species, females are highly territorial, and males may enter female territories to court (Test 1954; Sexton 1960; Dole and Durant 1974; Durant and Dole 1975). Males have elaborate visual courtship displays. In C. trinitatis, males turn jet black and jump up and down in front of females (Kenny 1969; personal observations). In C. collaris males "dance" on their hind legs (Dole and Durant 1974).

In most species, the male leads the female to the oviposition site (Dunn 1941; Crump 1972; Dole and Durant 1974; Lüddecke 1974; Silverstone 1976), although females are reported to lead males in D. histrionicus (Silverstone 1973). This phase of courtship lasts anywhere from a few minutes to many hours. The pair may inspect several potential oviposition sites before mating. In some species, males vocalize less frequently at this stage, but may stop and call at females that lag behind (Crump 1972).

The movement of the pair to the oviposition site may be interrupted by bouts of tactile courtship. In D. histrionicus the male leaps on the female's back while "pursuing" her, but the female often shakes him off. Sometimes a pair pauses on a dead leaf and engages in mutual courtship involving various postural displays, circling, rubbing, etc. (Silverstone 1973). Very similar tactile interactions have been observed in D. granuliferus (Crump 1972), D. pumilio (A. S. Rand, pers. comm.), and Phyllobates vittatus (D. Travis, cited in Silverstone 1976).

In D. auratus, the female takes the more active role, jumping on the male's back and prodding him with her front feet (Senfft 1936; Dunn 1941; personal observations). A female may also sit on a male's back, with her hind legs on his shoulders and her body facing away from the male (personal observations). Similar behavior has been reported in D. histrionicus, but in that species the male normally sits on the female (Silverstone 1973). Females may take the more active role in courtship in D. azureus as well, although this species has been studied only in captivity (Polder 1974). This apparent sex role reversal in courtship roles may be related to parental investment by males exceeding that of females (Trivers 1972). More detailed discussions of this problem are given elsewhere (Wells 1977b, in preparation).

Mating and oviposition have been observed in only a few species. Male and female D. granuliferus place their vents together while facing away from each other; presumably sperm and eggs are released simultaneously (Crump 1972). In D. pumilio, the male wets a leaf, possibly depositing sperm, and moves his legs in a sort of wiping motion across the leaf. The female then deposits eggs on the wet leaf (A. S. Rand, pers. comm.). Male D. auratus and D. azureus are believed to fertilize eggs by sitting on them after they are laid (Senfft 1936; Polder 1974). Similar behavior apparently occurs in Phyllobates vittatus (D. Travis, cited in Silverstone 1976). In C. palmatus, the male rests loosely on the female's back during oviposition, but does not clasp her (Lüddecke 1974).

THE EVOLUTION OF COURTSHIP BEHAVIOR

Why is there such wide variation in duration and complexity of anuran courtship? To answer this question, we need to consider the function of courtship from both the male's and female's point of view. Courtship usually is said to have several independent functions, including 1) serving as a premating isolating mechanism between species, 2) advertising a male's readiness to mate, 3) overcoming agonistic tendencies of the partner, 4) bringing the female into physiological condition to mate, and 5) synchronizing mating activities of males and females (Morris 1970).

When courtship is viewed in this way, it is difficult to account for variations in complexity. Why should dendrobatid females require more stimulation than other frogs? Why should the vocal repertoire of one species be more complex than that of another? Perhaps it would be more useful to view courtship as an exchange of communication signals between males and females and to try to understand the types of information being transmitted. Courtship by a male can be considered an attempt to "persuade" a female to mate with him instead of a competing male. At the same time, courtship provides females with a potential source of information about the relative qualities of different males. Vocal courtship is really a contest among males for the attention of females (Trivers 1972; Ghiselin 1974; Alexander 1975; Wilson 1975). The form of the contest will depend on ecological factors such as the temporal pattern of breeding and the spatial distribution of oviposition sites (Wells 1977b).

To illustrate this point, I will discuss a number of specific examples. First, let us consider species where males gather at aquatic breeding sites but do not call from water. Such behavior is characteristic of many North American hylids and other treefrogs in all parts of the world. Typically, males call from bushes or other vegetation near water. Amplexus occurs at the male's calling site and the female carries the male to water (Fellers 1975,

1976; Garton and Brandon 1975; Whitney and Krebs 1975b; Wells 1977b).

In many species, males defend calling sites against other males, but do not defend individual oviposition sites (Fellers 1975, 1976; Garton and Brandon 1975; Whitney and Krebs 1975b; Wells 1977b). Thus females are unable to use the quality of a male's territory as a source of information about his desirability as a mate. Females probably select mates at a distance largely on the basis of vocalizations, and a male's reproductive success may depend on his ability to produce distinctive signals. This could lead to various forms of vocal competition, including the use of special close range signals to provide better directional cues to approaching females. Males might also seek out calling sites unobstructed by vegetation to allow more effective broadcasting of signals (Fellers 1976).

It seems unlikely that there would be prolonged close range associations between individual males and females prior to amplexus in these species. A female would gain little information about a male at close range that she had not already received at a distance. Indeed, there is evidence that some females in these species complete ovulation before approaching individual males (Martof and Thompson 1958; Gosner and Rossman 1959; Oplinger 1966; Crump 1974; Salthe and Mecham 1974; Garton and Brandon 1975).

When males call from territories that include individual oviposition sites, different selective pressures come into play. If territories vary in quality, then females have an important source of information on which to base their choice of mates. Females might be expected to spend some time in close association with individual males to gauge the relative quality of their territories. They also might delay ovulation until a favorable site is found. Females apparently choose mates on the basis of territory quality in *Rana clamitans* (Wells 1976, 1977a) and *R. catesbeiana* (R. D. Howard, pers. comm.). In both of these species, a female may spend many hours sitting near several calling males before choosing a mate (Emlen 1968; Wells 1976, 1977a). There seems to be no evidence, however, of any close-range courtship behavior in these species. The prolonged associations between individual males and females in many hole and crevice nesting species also may be related to the length of time females require to assess territory quality. This certainly seems to be the case in *Eleutherodactylus coqui*. In this species, pairs may stay together for several days and visit a number of potential oviposition sites in tree holes before mating (Drewry 1974).

The behavior of dendrobatids resembles that of *E. coqui*. Males usually call from rocks, logs, or other elevated sites away from oviposition sites. Because these species lay terrestrial eggs, the distribution of suitable oviposition sites probably is highly unpredictable. Males and females may spend hours together inspecting potential sites (Tables 2 and 3). It is probably important for a female to avoid committing herself to mating with a particular male until an oviposition site is located. She might approach a calling male and indicate her willingness to court, but remain "coy" until an oviposition site is chosen. The male would be expected to persist in his courtship only if the female gave some indication of becoming receptive in the near future. This could lead to a complex exchange of signals between male and female. Such behavior would not necessarily indicate that dendrobatid females require more physiological stimulation than other female frogs.

Dendrobatids differ from many frogs in another important way. Because they are terrestrial breeders and do not move to special breeding areas to mate, males and females are found together in the same habitat throughout the year. It is quite likely that males frequently encounter both receptive and nonreceptive females in their territories. If a male had some means of determining a female's receptivity early, he could avoid wasting time attempting to mate with non-gravid females. Furthermore, non-gravid females should signal their lack of receptivity to males to avoid constant harassment. Thus there might be strong selective pressures for two-way communication between males and females, and various vocal, tactile, and visual signals might evolve.

I would suggest that pipid frogs have complex courtship for exactly the same reason. In these permanently aquatic frogs, males would encounter both gravid and non-gravid females. In *Hymenochirus boettgeri*, for example, a male seizes every female that enters his territory. Males should not waste time trying to mate with non-gravid females, and the stroking behavior reported in the species may be a means of determining female receptivity. Nonreceptive females do not respond to this tactile stimulation and usually are released (Rabb and Rabb 1963).

The problem of distinguishing receptive from nonreceptive females is particularly critical when males and females compete for the same resources. In many hole nesting species, for example, the same holes may be used for oviposition sites and for shelter by both males and females (Drewry 1970). If holes are scarce, females might try to take over male shelters. A male should attempt to determine whether an intruding female is a potential mate or a potential competitor. Instead of driving a female away immediately, a male might first try to court her. If she proved to be unreceptive, then he might drive her from the shelter. This could account for the apparent courtship calls of *Pseudophryne* (Pengilley

1971; Woodruff 1976) and the special calls and stroking behavior of Syrrhophus (Jameson 1954, 1955a, 1955b).

A similar line of reasoning would apply to dendrobatids in which both males and females are aggressive. One or both partners must signal to the other that it is more interested in mating than in contesting a territory. I would expect this to favor visual or postural displays that would be effective without actual physical contact. The most elaborate visual displays reported to date are the color changes and "dances" of male Colostethus collaris, C. trinitatis, and C. palmatus (Table 3). Both sexes are aggressive in all three species (Test 1954; Sexton 1960; Lüddecke 1974, 1976; Dole and Durant 1974; Durant and Dole 1975). The crouching behavior of female C. palmatus and C. inguinalis approaching calling males (Table 3) also may signal a lack of aggressive intent.

CONCLUSION

Anuran courtship frequently has been interpreted in a very limited way. In the case of vocal courtship, most studies have focused on the role of vocalizations as species isolating mechanisms (Bogert 1960; Blair 1964, 1968; Salthe and Mecham 1975). With the exception of some work on agonistic encounter calls, there has been surprisingly little research on the functioning of vocalizations as intraspecific communication signals.

Future workers should pay particular attention to individual variations in calling behavior within a chorus. Also needed are more experimental studies to determine how males change their calling behavior in response to other calling males, or in response to the approach of other individuals. The few investigations reported to date have focused on temporal shifts to minimize acoustic interference (Lemon 1971; Loftus-Hills 1974; Rosen and Lemon 1974). The problem of changes in call duration and structure has received little attention.

When males change their vocalizations in response to other approaching frogs, it is important to test the responses of females to the normal and altered calls. This is the only way to determine whether the behavior is a courtship strategy which enhances a male's reproductive success. When males have compound advertisement calls, it would be worthwhile to test the responses of females to different parts of the call. In fact, the whole question of what cues females use to select mates is an open problem which is just beginning to be investigated (Whitney and Krebs 1975a; Fellers 1976; Wells 1976, 1977a).

In interpreting the more complex courtship of pipids, dendrobatids, and a few other species, most workers have suggested that such behavior "stimulates" or "excites" the female and brings her into the proper physiological condition for mating (Jameson 1955a, 1955b; Rabb and Rabb 1963; Pengilley 1971; Rabb 1973; Dole and Durant 1974). There is no doubt that courtship can stimulate females physiologically (e.g., Crews 1974). It seems likely that some action patterns, such as the stroking of pipids, originally served to stimulate the female during amplexus and oviposition (Rabb and Rabb 1963). However, it is equally likely that many of these behaviors have been partially divorced from their original functions and now are used primarily as communication signals in the early stages of courtship.

In focusing on the effects of courtship on females, one should not overlook the possibility that courtship also provides males with information about the motivations and physiological states of females. Additional studies of close range interactions between males and females are needed. Investigators should pay close attention to the relationships of resource competition, aggression in the two sexes, and courtship complexity. Furthermore, observers should study not only the responses of females to courting males, but also the responses of males to nonreceptive females. Complex courtship in dendrobatids, pipids, and other frogs should not be viewed as simple one-way broadcasting of information by males, but as a system of two-way communication between males and females attempting to maximize their own fitness.

ACKNOWLEDGMENTS

I thank A. Stanley Rand for making unpublished material on Physalaemus available to me. Janice Derr, Alan Jaslow, and A. S. Rand read and commented on the manuscript. This paper was prepared while I was supported as a postdoctoral fellow of the Smithsonian Tropical Research Institute, Balboa, Canal Zone.

REFERENCES

Alexander, R. D. 1975. Natural selection and specialized chorusing behavior in acoustical insects. IN: D. Pimentel (ed.), Insects, Science, and Society, pp. 35-77. New York: Academic Press.

Allan, D. M. 1973. Some relationships of vocalizations to behavior in the Pacific treefrog, Hyla regilla. Herpetologica 29:366-371.

Anderson, P. K. 1954. Studies in the ecology of the narrow-mouthed frog, Microhyla carolinensis carolinensis. Tulane Stud. Zool. 2:15-46.

Awbrey, F. T. 1965. An experimental investigation of the effectiveness of anuran mating calls as isolating mechanisms. Ph. D. dissertation, Univ. of Texas.

Blair, W. F. 1964. Isolating mechanisms and interspecies interactions in anuran amphibians. Quart. Rev. Biol. 39:334-344.

Blair, W. F. 1968. Amphibians and reptiles. IN: T. A. Sebeok (ed.), Animal Communication, pp. 289-310. Bloomington: Indiana Univ. Press.

Blair, W. F., and M. J. Littlejohn. 1960. Stage of speciation of two allopatric populations of chorus frogs (Pseudacris). Evolution 14:82-87.

Blommers-Schlösser, R. M. A. 1975. Observations on the larval development of some Malagasy frogs, with notes on their ecology and biology (Anura: Dyscophinae, Scaphiophryninae, and Cophylinae). Beaufortia 24:7-26.

Bogert, C. M. 1960. The influence of sound on amphibians and reptiles. IN: W. E. Lanyon and W. N. Tavolga (eds.), Animal Sounds and Communication, pp. 137-320. Washington: American Institute of Biological Science.

Brattstrom, B. H., and R. M. Yarnell. 1968. Aggressive behavior in two species of laptodactylid frogs. Herpetologica 24:222-228.

Bunnell, P. 1973. Vocalizations in the territorial behavior of the frog Dendrobates pumilio. Copeia 1973:277-284.

Capranica, R. R. 1968. The vocal repertoire of the bullfrog (Rana catesbeiana). Behaviour 31:302-325.

Coe, M. J. 1967. Co-operation of three males in nest construction by Chiromantis rufescens Günther (Amphibia: Rhacophoridae). Nature 214:112-113.

Coe, M. J. 1974. Observations on the ecology and breeding biology of the genus Chiromantis (Amphibia: Rhacophoridae). J. Zool. London 172:13-34.

Crews, D. 1974. Effects of group stability, male-male aggression, and male courtship behavior on environmentally induced ovarian recrudescence in the lizard Anolis carolinensis. J. Zool. London 172:419-441.

Crump, M. L. 1972. Territoriality and mating behavior in Dendrobates granuliferus (Anura: Dendrobatidae). Herpetologica 28: 195-198.

Crump, M. L. 1974. Reproductive strategies in a tropical anuran community. Misc. Publ. Univ. Kansas Mus. Nat. Hist. 61:1-68.

Dixon, J. R. 1957. Geographic variation and distribution of the genus Tomodactylus in Mexico. Texas J. Sci. 9:379-409.

Dole, J. W., and P. Durant. 1974. Courtship behavior in Colostethus collaris (Dendrobatidae). Copeia 1974:988-990.

Drewry, G. 1970. The role of amphibians in the ecology of Puerto Rican rain forest. IN: Puerto Rico Nuclear Center Rain Forest Project Annual Report, pp. 16-85. San Juan: Puerto Rico Nuclear Center.

Drewry, G. 1974. Ecology of Eleutherodactylus coqui Thomas in montane rain forest of eastern Puerto Rico (abstract). 54th Annual Meeting, Am. Soc. of Ichthyol. Herpetol., p. 5.

Duellman, W. E. 1966. Aggressive behavior in dendrobatid frogs. Herpetologica 22:217-221.

Duellman, W. E. 1967. Social organization in the mating calls of some neotropical anurans. Am. Midl. Nat. 77:156-163.

Duellman, W. E. 1970. The Hylid Frogs of Middle America. 2 vols. Lawrence: Univ. of Kansas Press.

Dunn, E. R. 1941. Notes on Dendrobates auratus. Copeia 1941:88-93.

Durant, P., and J. W. Dole. 1975. Aggressive behavior in Colostethus (= Prostherapis) collaris (Anura: Dendrobatidae). Herpetologica 31:23-26.

Eaton, T. H. 1941. Notes on the life history of Dendrobates auratus. Copeia 1941:93-95.

Eibl-Eibesfeldt, I. 1952. Vergleichende Verhaltensstudien an Anuren: I. Zur Paarungsbiologie des Laubfrosches, Hyla arborea L. Z. Tierpsychol. 9:383-395.

Emlen, S. T. 1968. Territoriality in the bullfrog, Rana catesbeiana.

Copeia 1968:240-243.

Fellers, G. M. 1975. Behavioral interactions in North American treefrogs (Hylidae). Chesapeake Sci. 16:218-219.

Fellers, G. M. 1976. Social interactions in North American treefrogs (Hyla). Ph. D. dissertation, Univ. of Maryland.

Foster, W. A. 1967. Chorus structure and vocal response in the Pacific tree frog, Hyla regilla. Herpetologica 23:100-104.

Garton, J. S., and R. A. Brandon 1975. Reproductive ecology of the green treefrog, Hyla cinerea, in southern Illinois (Anura: Hylidae). Herpetologica 31:150-161.

Gerhardt, H. C. 1975. Sound pressure levels and radiation patterns of the vocalizations of some North American frogs and toads. J. Comp. Physiol. 102:1-12.

Gerhardt, H. C. 1976. Significance of two frequency bands in long distance vocal communication in the green treefrog. Nature 261:692-694.

Ghiselin, M. T. 1974. The Economy of Nature and the Evolution of Sex. Berkeley: Univ. of California Press.

Goin, C. J. 1949. The peep order in peepers; a swamp water serenade. Quart. J. Fla. Acad. Sci. 11:59-61.

Goodman, D. E. 1971. Territorial behavior in a neotropical frog, Dendrobates granuliferus. Copeia 1971:365-370.

Gosner, K. L., and D. A. Rossman. 1969. Observations on the reproductive cycle of the swamp chorus frog, Pseudacris nigrita. Copeia 1959:263-266.

Hardy, D. F. 1959. Chorus structure in the striped chorus frog, Pseudacris nigrita. Herpetologica 15:14-16.

Heinzmann, U. 1970. Untersuchungen zur Bio-Akustik und Ökologie der Geburtshelferkröte, Alytes o. obstetricans (Laur.). Oecologia 5:19-55.

Jaeger, R. G. 1976. A possible prey-call window in anuran auditory perception. Copeia 1976:833-834.

Jameson, D. L. 1954. Social patterns in the leptodactylid frogs Syrrhophus and Eleutherodactylus. Copeia 1954:36-38.

Jameson, D. L. 1955a. Evolutionary trends in the courtship and mating behavior of Salientia. Syst. Zool. 4:105-119.

Jameson, D. L. 1955b. The population dynamics of the cliff frog, Syrrhophus marnocki. Am. Midl. Nat. 54:342-381.

Kenny, J. S. 1969. The Amphibia of Trinidad. Studies on the Fauna of Curacao and other Caribbean Islands 108:1-78.

Lemon, R. E. 1971. Vocal communication by the frog, Eleutherodactylus martinicensis. Can. J. Zool. 49:211-217.

Littlejohn, M. J. 1961. Mating call discrimination by females of the spotted chorus frog (Pseudacris clarki). Texas J. Sci. 13: 49-50.

Littlejohn, M. J., M. J. Fouquette, Jr., and C. Johnson. 1960. Call discrimination by female frogs of the Hyla versicolor complex. Copeia 1960:47-49.

Littlejohn, M. J., and J. J. Loftus-Hills. 1968. An experimental evaluation of premating isolation in the Hyla ewingi complex (Anura: Hylidae). Evolution 22:659-663.

Littlejohn, M. J., and T. C. Michaud. 1959. Mating call discrimination by females of Strecker's chorus frog (Pseudacris streckeri). Texas J. Sci. 11:86-92.

Littlejohn, M. J., and G. F. Watson. 1974. Mating call discrimination and phonotaxis by females of the Crinia laevis complex (Anura: Leptodactylidae). Copeia 1974:171-175.

Liu, C. C. 1950. Amphibians of Western China. Fieldiana: Zool. Memoirs 2:1-400.

Loftus-Hills, J. J. 1971. Neural correlates of acoustic behavior in the Australian bullfrog, Limnodynastes dorsalis (Anura: Leptodactylidae). Z. Vgl. Physiol. 74:140-152.

Loftus-Hills, J. J. 1974. Analysis of an acoustic pacemaker in Strecker's chorus frog, Pseudacris streckeri (Anura: Hylidae). J. Comp. Physiol. 90:75-87.

Lüddecke, H. 1974. Ethologische Untersuchungen zur Fortpflanzung von Phyllobates palmatus (Amphibia, Ranidae). Ph. D. dissertation, Johannes Gutenberg-Univ., Mainz.

Lüddecke, H. 1976. Einige Ergebnisse aus Feldbeobachtungen an Phyllobates palmatus (Amphibia, Ranidae) in Kolumbien. Mitt.

Inst. Colombo-Aleman Invest. Cient. 8:157-163.

Martof, B. S., and E. F. Thompson. 1958. Reproductive behavior of the chorus frog, Pseudacris nigrita. Behaviour 13:243-257.

Morris, D. 1970. The function and causation of courtship ceremonies. IN: D. Morris (ed.), Patterns of Reproductive Behavior, pp. 128-152. New York: McGraw-Hill.

Narins, P. M. 1976. Auditory processing of biologically meaningful sounds in the treefrog, Eleutherodactylus coqui. Ph. D. dissertation, Cornell Univ.

Narins, P. M., and R. R. Capranica. 1976. Sexual differences in the auditory system of the tree frog Eleutherodactylus coqui. Science 192:378-380.

Noble, G. K., and R. C. Noble. 1923. The Anderson tree frog (Hyla andersonii Baird). Observations on its habits and life history. Zoologica 2:417-455.

Oeser, R. 1932. Die Zucht des Baumsteigers (Dendrobates tinctorius). Bl. Aquar. Terrarienk. 43:196-200.

Oplinger, C. S. 1966. Sex ratio, reproductive cycles, and time of ovulation in Hyla crucifer crucifer Wied. Herpetologica 22:276-283.

Österdahl, L., and R. Olsson. 1963. The sexual behavior of Hymenochirus boettgeri. Oikos 14:35-43.

Otte, D. 1972. Simple vs. elaborate behavior in grasshoppers: an analysis of communication in the genus Syrbula. Behaviour 42: 291-322.

Pengilley, R. K. 1971. Calling and associated behavior of some species of Pseudophryne (Anura: Leptodactylidae). J. Zool. London 163:73-92.

Polder, W. N. 1974. Pflege und Fortpflanzung von Dendrobates azureus und anderer Dendrobatiden II. Aquar. Terrarien Z. 27: 28-32.

Rabb, G. B. 1973. Evolutionary aspects of the reproductive behavior of frogs. IN: J. L. Vial (ed.), Evolutionary Biology of the Anurans, pp. 213-227. Columbia: Univ. of Missouri Press.

Rabb, G. B., and M. S. Rabb. 1960. On the mating and egg-laying behavior of the Surinam toad, Pipa pipa. Copeia 1960:271-276.

Rabb, G. B., and M. S. Rabb. 1963. On the behavior and breeding biology of the African pipid frog Hymenochirus boettgeri. Z. Tierpsychol. 20:215-241.

Rand, A. S. The displays of Physalaemus pustulosus (Leptodactylidae). Unpublished ms.

Rosen, M., and R. E. Lemon. 1974. The vocal behavior of spring peepers, Hyla crucifer. Copeia 1974:940-950.

Salthe, S. N., and J. S. Mecham. 1974. Reproductive and courtship patterns. IN: B. Lofts (ed.), Physiology of the Amphibia, 2:209-251. New York: Academic Press.

Schiøtz, A. 1967. The treefrogs (Rhacophoridae) of West Africa. Copenhagen: Bianco Lunos Bogtrykkeri.

Schmidt, R. S. 1969. Preoptic activation of mating call orientation in female anurans. Behaviour 35:114-127.

Schmidt, R. S. 1971. Neural mechanisms of mating call orientation in female toads (Bufo americanus). Copeia 1971:545-548.

Schneider, H. 1967. Rufe und Rufverhalten des Laubfrosches, Hyla arborea arborea (L.). Z. Vgl. Physiol. 57:174-189.

Schneider, H. 1968. Bio-akustische Untersuchungen am Mittelmeerlaubfrosch. Z. Vgl. Physiol. 61:369-385.

Senfft, W. 1936. Das Brutgeschäft des Baumsteigerfrosches (Dendrobates auratus Girard) im Gefangenschaft. Zool. Gart. 8: 122-136.

Sexton, O. J. 1960. Some aspects of the behavior and of the territory of a dendrobatid frog, Prostherapis trinitatis. Ecology 41:107-115.

Silverstone, P. A. 1973. Observations on the behavior and ecology of a Columbian poison-arrow frog, the Kokoe-pa (Dendrobates histrionicus Berthold). Herpetologica 29:295-301.

Silverstone, P. A. 1975. A revision of the poison-arrow frogs of the genus Dendrobates Wagler. Nat. Hist. Mus. Los Angeles County Sci. Bull. 21:1-55.

Silverstone, P. A. 1976. A revision of the poison-arrow frogs of the genus Phyllobates Bibron in Sagra (Family Dendrobatidae). Nat. Hist. Mus. Los Angeles County Sci. Bull. 27:1-53.

Snyder, W. F., and D. L. Jameson. 1965. Multivariate geographic variation of mating call in populations of the Pacific tree frog (*Hyla regilla*). *Copeia* *1965*:129-142.

Test, F. H. 1954. Social aggressiveness in an amphibian. *Science* *120*:140-141.

Trivers, R. L. 1972. Parental investment and sexual selection. IN: B. G. Campbell (ed.), *Sexual Selection and the Descent of Man*, pp. 136-179. Chicago: Aldine Press.

Wells, K. D. 1976. Territorial behavior of the green frog (*Rana clamitans*). Ph. D. dissertation, Cornell Univ.

Wells, K. D. 1977a. Territoriality and male mating success in the green frog (*Rana clamitans*). *Ecology*, in press.

Wells, K. D. 1977b. The social behaviour of anuran amphibians. *Anim. Behav.*, in press.

Weygoldt, P. 1976. Beobachtungen zur Biologie und Ethologie von *Pipa* (*Hemipipa*) *carvalhoi* Mir. RiB. 1937. (Anura, Pipidae). *Z. Tierpsychol.* *40*:80-99.

Whitney, C. L., and J. R. Krebs. 1975a. Mate selection in Pacific tree frogs, *Hyla regilla*. *Nature* *255*:325-326.

Whitney, C. L., and J. R. Krebs. 1975b. Spacing and calling in Pacific tree frogs, *Hyla regilla*. *Can. J. Zool.* *53*:1519-1527.

Wiewandt, T. A. 1969. Vocalization, aggressive behavior, and territoriality in the bullfrog, *Rana catesbeiana*. *Copeia* *1969*:276-285.

Wilson, E. O. 1975. *Sociobiology: The New Synthesis*. Cambridge: Harvard Univ. Press.

Woodruff, D. S. 1976. Courtship, reproductive rates, and mating systems in three Australian *Pseudophryne* (Amphibia, Anura, Leptodactylidae). *J. Herpetol.* *10*:313-318.

LONG-RANGE ACOUSTIC COMMUNICATION IN ANURANS: AN INTEGRATED AND EVOLUTIONARY APPROACH

M. J. Littlejohn

Department of Zoology, University of Melbourne
Parkville, Victoria, 3052, Australia

INTRODUCTION

Acoustic communication constitutes an important and conspicuous part of the breeding biology of most anurans: in the establishment and maintenance of territories by males, in facilitating the attraction of conspecific mates to males, in courtship, and in the identification of sex and reproductive state. The aim of this review is to examine the long-range acoustic communication system of anurans through an integrated and evolutionary approach, and to consider the constraints on the system in the context of reproduction. A useful framework on which to build such an approach is provided by the model of a communication system initially developed by Shannon and Weaver (1949), and applied to biological systems by Cherry (1957), Johnston (1976), Moles (1963) and others. This biocommunication system consists of four main elements: the common repertoire (set of signs), the source (emitter or transmitter), the communication channel and associated noise, and the destination or receiver (Fig. 1). Each functional subsystem, emitter complex and receiver complex, is normally contained within the structure of one individual. A message is selected from the common repertoire, encoded and introduced into the channel as a signal. Perturbations of the channel (noise) tend to reduce the intelligence of the signal (information content), which is subsequently accepted by the receiver, decoded and compared with the contents of its repertoire. The result is an output that completes the bio-communicative sequence. The incorporation of redundancy into the message, through synonymy or repetition, helps to overcome the problem of a noisy channel, but at the expense of energy, time and transmission rate.

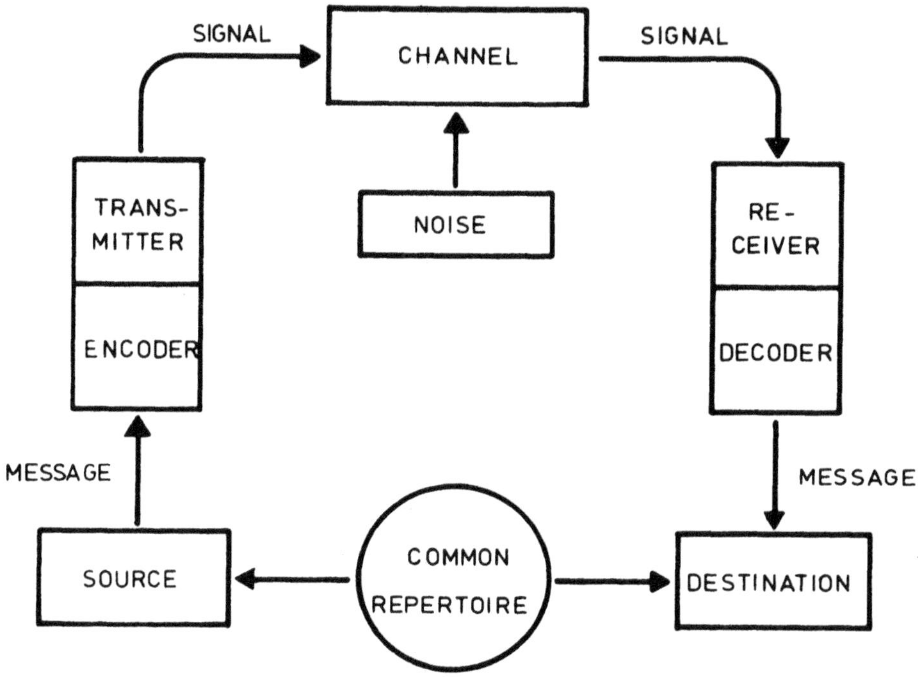

Fig. 1. Components of a communication system.

For biological systems, communication may be defined in terms of adaptiveness, i.e., maximizing an individual's reproductive success, for both emitter and receiver as implied by Marler (1967) and Otte (1974). That is, both signaler and recipient (responder) must benefit from the interaction. Such a definition allows restriction of cases to co-evolved systems in which both participants enhance their respective fitnesses. Many acoustic interactions in which only one individual benefits in this Darwinian sense, e.g., recognition of an approaching predator, location of prey and echo-location, may be disregarded. The overt sign of the operation of a bio-communicative system is the repeated observation of the production of a distinctive type of signal by one individual and the corresponding adaptive behavioral response by another individual. It must be noted, however, that a response in the receiving individual may be limited to neural processing (assessment, comparison, decision not to respond), secular changes in endocrine profiles, e.g., reproductive maturation, or displaced in time, so that a direct cause-effect relationship may not always be obvious.

Acoustic communication in anuran amphibians has been reviewed by Bogert (1960), Blair (1963, 1968), Paillette (1971) and Straughan (1973). A considerable amount of new information has been assembled since these reviews were completed, however, and there has been a very important shift in approach with increased emphasis on the evolutionary significance of interactions at the level of the individual, rather than of populations and species (Williams 1966; Otte 1974; Alexander 1975). Because most anurans use air (the atmosphere) as the channel for sound communication, only such cases will be considered in this review. Furthermore, signals that operate at very close range, i.e., less than one body length, or with physical contact between emitter and receiver (courtship and identification of sex and reproductive state), will be largely disregarded.

The characteristics of acoustic energy and the properties of air as a communication channel provide a set of constraints on the potential range of signals that could be generated. These possible signals, in turn, will be further constrained by the biological limitations and adaptive compromises of the emitter and receiver subsystems, and the extent of utilization of the channel by other co-existent bio-communication systems. Hence, a consideration of all four components of the communication system is necessary in order to allow development of an integrated evolutionary approach to acoustic communication in anurans.

THE REPERTOIRE

In biological systems, the common repertoire of the emitter and receiver will be derived from such sources as a shared genetic background, convergent evolution, or cultural transmission. For frogs, it seems that the basic structure of the signal and contextual stimulus field (releaser) are determined largely by inheritance. Such a conclusion is reached through a consideration of the results of studies of acoustic signals of hybrids (Blair 1956; Gerhardt 1974a; Littlejohn et al. 1971; Schneider and Eichelberg 1974; Zweifel 1968), which are generally mosaic or intermediate in structure. In most species of anurans, parent-offspring contact usually ceases before hatching, so there is little opportunity for cultural transmission of specific information about the way in which a signal should be structured, and the context in which it should be produced or assessed. In this case, a maturing individual would require either a genetically-specified template against which all sounds subsequently heard may be compared (both emitter and receiver) and the appropriate motor pattern developed through practice (emitter), or else a fully specified genetic program that allows the signal or appropriate response to be generated without the need for an external model. Deprivation (Kaspar Hauser) experiments would clarify this issue, but they have yet to be reported for anurans.

The following is a provisional classification of possible types of signals used in long-range communication that may be recognized in the acoustic repertoires of anurans. The list is based on both observational and experimental studies; supporting references will be provided where appropriate in other sections of the paper.

1. Advertisement Call. This name was suggested by Wells (in press) and can be applied to the signal previously referred to as the mating call (*sensu lato*). This signal, which is generally produced by a breeding male, may have two important functions in reproduction (Bogert 1960):

a) Announcement of occupied territory to other males of the same or different species; and

b) Attraction of a conspecific female to the calling male (= mating call in the strict sense).

These functions may be combined into one monophasic signal so that the sex of the receiver and the context determine the type of response; or else they may be separated between the components of a biphasic call, as in *Eleutherodactylus coqui* (Nairns and Capranica 1976). It is possible that the biphasic advertisement calls of some other species, e.g., *Geocrinia victoriana* (Littlejohn and Martin 1964), also have these two functions partitioned between the two components, but previous studies do not seem to have been organized to test for this possibility. Thus far, most experimental investigations have been directed at assessing the specificity of the female-attracting function of the advertisement call (Forester 1973; Gerhardt 1974b; Littlejohn and Watson 1974; Straughan 1975), with territorial aspects only now beginning to be studied in detail (Allan 1973; Whitney and Krebs 1975).

2. Territorial Call. This signal is produced by a resident male in response to an advertisement call received above a critical threshold of intensity (Allan 1973; Capranica 1965; Hawe 1970; Rosen and Lemon 1974; Snyder and Jameson 1965). Below this threshold the advertisement call of a nearby male may be heard and responded to with an advertisement call, so that there is an alternation of advertisement calls (Loftus-Hills 1971a, 1974; Rosen and Lemon 1974). The territorial call may be produced as a result of acoustic stimulation alone, and thus could occur when the intruder was some distance from the resident. This term is used in the original sense of Bogert (1960).

3. Encounter Call. This term was suggested by McDiarmid and Adler (1974) and Wells (in press) for all signals evoked by territorial interactions. There may be at least two levels of acoustic

interaction, however, long-range and short-range, with distinctive signals being produced in each category. It is suggested that the term "encounter call" might be restricted to short-range interactions. Furthermore, an encounter call also could be elicited by visual and vibrational stimuli associated with the closeness of the intruder. The monophasic call of Hyla regilla would be included in this category while the trill would be a territorial call (Allan 1973).

4. Reciprocation Call. This name is suggested for the signal given by breeding females of some species in response to the advertisement call of a conspecific male, and which allows the pair to converge through alternation of signals and acoustic orientation. Dixon (1957) described such a sequence for Tomodactylus, and Wells (in press) referred to some other possible examples. In this situation, males presumably do not hold territories.

A classification and summary of the characteristics of these four types of signal are presented in Table 1.

THE EMITTER SUBSYSTEM

The structure and function of the physical components of the anuran emitter subsystem have been described by McAlister (1959, 1961), Martin (1971, 1972), Martin and Gans (1972) and Gans (1973), with emphasis on North American species of Bufo. Watkins et al. (1970) have commented on the significance of the vocal sac in call production. The neural control of vocalization has been reviewed by Schmidt (1973). The following general account of the production of acoustic signals is based on these previously mentioned works.

During long-range communication, the pneumatic system of the anuran is sealed; the mouth, and usually the nares, are closed, so that air is driven from the lungs through the larynx, and into the mouth cavity and vocal sac or sacs. The energy for sound production is provided by the body wall musculature, which is under the influence of outputs from a central neural program and determines the gross temporal structure of the signal, i.e., duration, repetition rate, envelope-shape. Air may be driven forward steadily during a call, or a smaller parcel may be moved forward and backward through the larynx. This latter process could be important in very long call sequences. There is normally no sound production during the return air flow.

As it passes forward through the larynx, the air stream is first modified by the vocal cords. These structures vibrate passively and act as an oscillator that generates the fundamental frequency of the call. This tone (or a harmonic) becomes the

Table 1. Long-range acoustic signaling repertoire of anurans: A provisional classification

	1. ADVERTISEMENT CALL (MALE)	
BIPHASIC CALL	(PHASE I a) Announcement of occupied territory to other males	FUNCTIONS MAY BE COMBINED IN A MONOPHASIC CALL
	(PHASE II b) Attraction of conspecific breeding female (= "mating call")	
	2. TERRITORIAL CALL (MALE)	
	Higher acoustic threshold, long-range territorial response to acoustic signal of approaching male	FUNCTIONS MAY BE COMBINED IN SAME SIGNAL
	3. ENCOUNTER CALL (MALE)	
	Lower acoustic threshold, close-range response to acoustic signal of approaching male, or to vibrational and visual stimuli generated by approaching male	
	4. RECIPROCATION CALL (FEMALE)	
	Produced in response to advertisement call of male, and facilitating mutual convergence for breeding. In this case males would not maintain a territory.	

carrier (dominant) frequency and may be further modified by a second pair of membranes, the arytenoid cartilages, which modulate the amplitude of the carrier to produce a series of pulses. The arytenoid cartilages may vibrate passively at their natural frequency, or be driven by the laryngeal muscles under central neural control.

The thin-walled, elastic vocal sac acts as a balloon radiator (pulsating sphere) and through impedance matching, couples the closed emitter system to the atmosphere, thus facilitating the transfer of acoustic energy into the channel.

The fundamental frequency of the advertisement call is determined mainly by the mass, length and tension of the vocal cords (Martin 1972). There is a direct relationship between fundamental frequency and size of larynx, or of the emitter (as body length), so that smaller species have higher carrier frequencies (Martin 1972). Although considered unlikely by Watkins _et al_. (1970), the later work of Martin (1972) indicates that the vocal sac has resonant properties. Controlled deflation experiments on the vocal sac of _Bufo cognatus_ resulted in a detuning of the dominant frequency, with acoustical energy being spread across more sidebands; the dominant frequency did not shift, however (Martin 1972).

THE CHANNEL

The following factors contribute to transmission losses (reduction in signal strength) through the atmosphere:

1. Spherical divergence (inverse square law) results in attenuation at the rate of 6 dB (sound pressure level (SPL) reduced to one-half) for each doubling of the length of the transmission path, and affects all frequencies equally. Beaming, ducting, diffraction and refraction may lead to lower losses than expected on theoretical grounds (Wiener and Keast 1959), but more empirical observations are necessary.

2. Molecular absorption of sound energy causes further losses; these increase at higher frequencies, higher temperatures and lower relative humidities (Embleton _et al_. 1976; Harris 1966).

3. Ground impedance becomes very important as an attenuating factor at higher frequencies, particularly when emitter and receiver are close to the ground surface (Embleton _et al_. 1976).

4. The nature of the substrate and vegetation also influences transmission loss, with different frequencies being affected by different substrates (Aylor 1972; Morton 1975).

Losses attributable to factors 2, 3 and 4 are combined as excess attenuation, i.e., in addition to that due to spherical divergence. Because higher frequencies are most affected, transmission at lower frequencies would appear more efficient, especially below 4 kHz. Thus, the atmosphere acts as a low pass filter, and an upper frequency limit is set by the properties of air as a channel. The determination of a lower limit, and hence of the potential transmission bandwidth, will be considered in a later section.

THE RECEIVER SUBSYSTEM

The structure and function of the anuran receptor system, including central elements and processing, have been reviewed by Capranica (1976). Comparative studies of reception have been made by Loftus-Hills and Johnstone (1970), Loftus-Hills (1973a), and Lombard and Straughan (1974). The following summary is derived mainly from these papers.

In most frogs, acoustic energy is received at the tympanic membrane and conducted to the inner ear by the columellar complex. In frogs that lack a tympanic membrane and middle ear, but employ acoustic communication (e.g., Pseudophryne semimarmorata), the areas of reception and the conductive pathway are uncertain, but probably involve the suprascapular-opercular link (Loftus-Hills 1973b).

The tympanic membrane and columella together act as an impedance-matching device between the air and the perilymphatic fluid (Capranica 1976), but the function of the tympanic membrane as a sound collector is uncertain. Frishkopf et al. (1968) found no relationship between size of tympanum and auditory sensitivity in Rana catesbeiana. Loftus-Hills (1973a), however, obtained a significant correlation between tympanum width (and hence area) and sensitivity to frequencies used in long-range communication for 15 species of about 8 dB/mm ($r = 0.76$ for a partial correlation analysis minus the effect of frequency). Loftus-Hills (1973a) also observed that there was no significant correlation between best auditory frequency and tympanum width; and Lombard and Straughan (1974) found that removal of the tympanum, while resulting in 20 dB loss of sensitivity, did not affect the tuning curves in Hyla regilla and H. versicolor. Hence, it is unlikely that the tympanum is a tuned (resonant) structure.

The principal sites of acoustic transduction in the inner ear are the amphibian papilla and the basillar papilla. Recent neurophysiological studies on the auditory nerve (see Capranica 1976) have indicated that there are three peaks of frequency sensitivity; the low (less than 500-600 Hz) and medium (500-1000 Hz) peaks are associated with the amphibian papilla, and the high frequency (the frequency depending on the species) with the basillar papilla. One exception so far found is Scaphiopus couchi, which lacks a middle peak of sensitivity (Capranica and Moffat 1975).

The low frequency sensitivity is probably concerned with the detection of signals transmitted through the substrate, e.g., associated with an approaching predator, and with contact communicative stimuli, e.g., release calling, with conduction to the inner ear by the suprascapular-opercular link (Capranica 1976). The middle and high frequency peaks clearly are associated with the reception of air-borne acoustic stimuli, particularly intraspecific signals, with which there is a close correlation to their carrier frequencies (Capranica 1976; Gerhardt 1976; Loftus-Hills 1973a).

There is an inverse relationship between hearing threshold of the upper frequency peak and the frequency of this peak, i.e., the higher the frequency, the lower the sensitivity, with the attenuation being about 20 dB/octave for the 15 species of anurans studied by Loftus-Hills (1973a). The threshold value is also associated with the size of the animal, with smaller species having higher thresholds, i.e., being less sensitive (Loftus-Hills 1973a); size of the tympanum may be influencing the relationship, however, rather than the efficiency of neural transduction being frequency-dependent.

The degree of tuning (sharpness or selectivity) of the high frequency component is comparable to that of other terrestrial vertebrates, with Q values (determined by dividing the best frequency by the bandwidth at 10 dB below the level of the best frequency) between 1.0 and 4.0 (Capranica 1976). Hearing thresholds in the high frequency band are generally higher than in the low frequency band, often by more than 30 dB (e.g., Ranidella parinsignifera, Loftus-Hills 1973a).

In summary, smaller species may suffer significant disadvantages in long-range acoustical communication. Smaller resonating structures and impedance-matching devices mean that higher carrier frequencies must be used in communication, and greater transmission losses suffered because of higher levels of excess attenuation, particularly through ground impedance. The efficiency of communication will nevertheless be relative to the size and mobility of the organisms concerned. Small, ground-dwelling frogs may effectively use the acoustic channel for communication, but on a greatly reduced scale.

The use of lower frequencies for detection of predators and for contact or vibrational communication will set a lower frequency limit at about 400 Hz for long-range signaling. The effective bandwidth for long-range acoustic communication in anurans will thus be about 400-4000 Hz. This conclusion is supported by the consideration of carrier frequencies in mating calls of 106 species of anurans, made by Loftus-Hills and Johnstone (1970), in which about 87 percent fell between 1 kHz and 4 kHz.

SIGNAL STRENGTH AND PATTERN OF RADIATION

Because of the very short rise times and pulse durations, transients, and different rates and degrees of amplitude modulation that may occur in anuran acoustic signals, standard sound level meters are unsuitable for measuring maximum (peak) sound pressure levels (SPL). The low inertia of an electron beam in an oscilloscope, or an impulse sound level meter, must be used to obtain accurate measurements of peak signal levels.

Few measurements of peak sound pressure levels in anuran signals are presently available. Loftus-Hills and Littlejohn (1971a) and Loftus-Hills (1973b) provided values obtained in the field and with a calibrated oscilloscope for advertisement calls of seven species of Australian frogs. These values were expressed as equivalent to root mean square levels, and a correction factor of 3 dB must be added to them to obtain true peak levels (Gerhardt 1975). Gerhardt (1975) determined true peak sound pressure levels for advertisement calls of 21 species of North American anurans in the field by using an impulse sound level meter. Both sets of measurements were made at a distance of 50 cm and relative to 0 dB = 2 x 10^{-5} N/m^2. Sound pressure level values for 28 species range from 90.5 dB for Ranidella signifera to 113 dB for Rana virgatipes, with a mode at 105-109 dB (12 species).

Loftus-Hills and Littlejohn (1971a) used measuring microphone heights between 15 and 25 cm (grazing angles of 16°42' to 26°36'). Gerhardt (1975) placed the measuring microphone horizontally and in line with the snout of the frog (presumably about 3-5 cm for frogs calling on the ground). For both situations, losses due to ground impedance are not significant for frequencies below 4 kHz at 50 cm (Embleton et al. 1976). There should not be any additional losses due to excess attenuation for measurements obtained from frogs calling at elevated sites in open vegetation.

Loftus-Hills and Littlejohn (1971a) found that smaller species have less intense signals, presumably because of lower surface areas of vocal sacs and less powerful generators (body-wall muscles). Gerhardt (1975), however, did not find such a relationship; Bufo quercicus (body length 19-33 mm, Conant 1975) and Bufo americanus (body length 51-89 mm, Conant 1975) have peak SPL values of 105.5 dB. Loftus-Hills and Littlejohn (1971a) also noted that, for species of comparable size, impulsive (explosive) calls, and calls with full amplitude modulation, i.e., 100 percent, at relatively low pulse rates were more intense by about 3 dB than calls with gradual changes in amplitude, lower levels of amplitude modulation, or higher pulse rates. Advertisement calls of smaller frogs were found to belong in the former category.

The only published measurements of sound pressure fields of calling anurans have been provided by Gerhardt (1975) for three North American species. Hyla crucifer was found to have a uniform sound field for the advertisement call. Propagation was directional in Hyla chrysoscelis, which has a front-to-back difference of 5 dB at 50 cm, and less so in Bufo americanus (1-3 dB).

INFLUENCE OF TEMPERATURE

The effect of temperature of emitter on signal structure has been studied in detail by Zweifel (1959, 1968), Schneider (1967, 1968, 1974), Lörcher (1969), Schneider and Nevo (1972), and Nevo and Schneider (1976). Other studies of acoustic signals usually have included correction factors for effects of temperature, derived from least squares linear regression analyses, as part of the study of geographic variation, but have not treated this variable to any extent (Blair 1955; Forester 1973; Fouquette 1975; Littlejohn 1965a; Ralin 1968). Consideration of these data allows generalization about the effects of temperature on signal structure. Signal duration, when it does not consist of a single short pulse or damped oscillation, is strongly and negatively correlated with temperature, while pulse repetition rate and call repetition rate are strongly and positively correlated with temperature. The number of pulses in a signal and the carrier (or fundamental) frequency show a variable picture with either a weak relationship (positive or negative) or no significant correlation. These observations further suggest that signal components that are determined by active physiological (neural and muscular) processes, e.g., duration, pulse repetition rate, and call repetition rate, have high temperature coefficients, whereas those resulting from the passive properties of the emitter system, e.g., carrier frequency, are uncorrelated or only slightly correlated with temperature.

The only published study of the effects of temperature on the receiver subsystem is Loftus-Hills' (1973a). He determined the hearing threshold for one male of Litoria ewingi on two successive days at 20°C and 9.0°C, and obtained essentially identical hearing curves and thresholds. This finding is consistent with the temperature stability of the dominant frequency in the advertisement call of L. ewingi (Littlejohn 1965a, 1976).

Because of the high temperature coefficients for some temporal components in signals, and their significance in coding, it must be postulated that a central discriminator in the receiver will have a similar temperature coefficient. It is also possible that homologous systems exist in emitter and receiver: one to determine the value of the temporal component, and the other to provide a template against which the received message can be compared.

TRANSMISSION RANGE

The maximum distance over which emitter and receiver can communicate will depend on the following factors:

1. Carrier frequency;
2. Intensity of signal;
3. Sensitivity of receiver;
4. Grazing angle of transmitter and receiver;
5. Noise level;
6. Nature of substrate and vegetation; and
7. Meteorological conditions.

Loftus-Hills (1971b) and Loftus-Hills and Littlejohn (1971a) have estimated maximum effective ranges of propagation for eight species of Australian anurans based on carrier frequencies, advertisement call intensities, receptor thresholds at the appropriate carrier frequency of the advertisement call, spherical divergence, and excess attenuation due to molecular absorption. Values range from 2.1 m for Ranidella parinsignifera to 135 m for Litoria dentata. These estimates may require further reduction if allowance is made for losses due to ground impedance (Embleton et al. 1976). Such losses become significant for carrier frequencies above 1 kHz and with horizontal propagation by a source that is close to the ground. For example, if the source and the receiver are 3 cm above the ground, then the loss for a 2 kHz signal will be about 6 dB at 1.5 m, and about 20 dB at 3 m (estimates based on Fig. 2, Embleton et al. 1976).

THE STRUCTURE AND DIFFERENTIATION OF LONG-RANGE ACOUSTIC SIGNALS WITHIN THE REPERTOIRE OF A SPECIES

Very little information is available on the structure and pattern of differentiation of signals within the acoustic repertoire of a species. Thus far, most of the detailed studies have been directed towards the species-specific, mate-attracting function of the male advertisement call, particularly as a premating reproductive isolating mechanism. Investigations of the role of acoustic signals in interactions between males seem to have concentrated on behavioral aspects, and seldom have sufficient physical data been provided to allow adequate comparisons of the different signals involved, e.g., territorial calls and encounter calls.

An arbitrary dichotomy can be made between spectral and temporal components of acoustic signals, based on the periodicity of the carrier frequency relative to the duration of the signal. Thus, two potential categories of coding system are available for communication.

The use of spectral characteristics requires the presence of a frequency analyzer in the acoustic pathway to resolve differences and decode the appropriate messages. Neurophysiological studies have shown that the anuran acoustic receptor system can resolve up to three spectral peaks: low and middle frequencies in the amphibian papilla and high frequency in the basillar papilla (Capranica 1976). Thus, there is the basis for an elementary frequency analyzer.

Capranica (1965) found that two frequency components (less than 500 Hz, and 1000-1800 Hz) were necessary in a synthetic advertisement (mating) call to evoke an acoustic response in males of *Rana catesbeiana*. He also found that sufficient energy between 400 and 700 Hz would inhibit these responses. These findings were correlated with the results of physiological studies of receptor characteristics carried out by Frishkopf and Goldstein (1963) on the ear of *R. catesbeiana*. Gerhardt (1974c) showed that the spectrum of the advertisement call of *Hyla cinerea* is bimodal, with frequency components of 900-1100 Hz and 2700-3300 Hz. While phonotaxis by breeding conspecific females could be obtained to synthetic calls containing either of these spectral components, the response was enhanced when both frequency bands were present in the stimulus (Gerhardt 1974c). Furthermore, the addition of spectral components at 1800 and 2100 Hz rendered a synthetic call less effective in eliciting phonotaxis (Gerhardt 1974c). The functional significance of this middle-band inhibition has yet to be determined. Later studies (Gerhardt 1976) indicated that the low-frequency component suffers less attenuation during transmission through the atmosphere, thus facilitating distance communication; as the distance shortens, then the high-frequency component is perceived, and presumably confers greater directionality.

Definite evidence for spectral differentiation between signals is confined to the male-repelling (1.1-1.3 kHz) and mate-attracting components (1.7-2.2 kHz) in the biphasic advertisement call of *Eleutherodactylus coqui* (Nairns and Capranica 1976). In this case, specificity of response is determined by sexual dimorphism in the receptor tuning curves rather than by frequency discrimination.

The bulk of direct and circumstantial evidence indicates that temporal differentiation, e.g., signal duration, repetition rate, and degree and rate of amplitude modulation, is the most common form of coding, as would be expected from the limited powers of frequency resolution displayed by the anuran ear. This temporal information can be extracted from the carrier by demodulation in the otic capsule (peripheral processing), and transmitted to the auditory areas of the brain for further evaluation (central processing).

The structure and function of two acoustic signals associated with interactions between breeding males of Ranidella signifera were determined by Hawe (1970) through physical analysis and playback experiments. At 12°C, the advertisement call has a mean duration of 104 msec, and a mean pulse repetition rate of 33 pulses/sec, while the corresponding values for the territorial call are 230 msec and 55 pulses/sec. The dominant frequencies are similar (Littlejohn, unpublished obs.).

Allan (1973) has described the structure and function of three calls in the repertoire of Hyla regilla in the context of male-male interactions. Durations and pulse repetition rates of the three signals used in playback experiments are as follows: diphasic (advertisement call)--250 msec, 101 pulses/sec; trill (territorial call)--580 msec, 54 pulses/sec; monophasic (encounter call)--200 msec, 146 pulses/sec. Dominant frequencies are similar and fall between 2000 and 2250 Hz. Temperatures at the recording sites were similar (air: 17.C-18.5°C; water: 20.5-21.5°C).

There is clearly some redundancy in these sets of signals. Until more studies are carried out, however, including controlled playback experiments using synthetic signals, there is little more that can be said on the subject of coding.

LOCALIZATION OF THE EMITTER

Marler (1967) and Feng et al. (1976) have discussed possible mechanisms of sound localization. Such a process involves two components: distance and direction. Assessment of intensity gradients should allow distance determination within the scope of most exteroceptors, while the binaural detection of differences in intensity, phase and time of arrival at the receptors can provide directional cues. Thus, an acoustic signal should be structured to maximize or minimize locatability of the source, depending on its function.

Localization by differences in intensity is best when the wavelengths are shorter than the interaural distance, and when head shadowing is maximal, whereas detection of phase differences is best when the wavelengths are longer than the interaural distance. Localization based on differences in time of arrival requires fast rise and decay times (abrupt beginnings and endings), short durations and moderate repetition rates, and transient frequencies to allow accurate timing. An ideal signal with minimal locatability would be transient free (slow rise and decay times), of pure tonal composition, and of a frequency with a wavelength too short for detecting phase differences and too long for assessment of intensity differences (Marler 1967).

Constraints for anurans will rest with their limited frequency analysis; temporal structure, however, should be amenable to optimization.

If one assumes that the advertisement call should provide maximal, and the territorial call minimal, locational cues, then this should be reflected at least in temporal structure. As indicated earlier, very little information is available on the structure of signals within the acoustic repertoire of anurans for which function has been verified. Consideration of the structure of these calls in *Hyla regilla* (Allan 1973, Littlejohn 1971, Straughan 1975) indicates that each call includes potential directional cues in having fast rise times and distinct pulsing. Until further comparative studies are made, however, no generalization is possible.

Patterns of orientation indicate that an effective system of acoustic localization exists in anurans (Axtell 1959; Feng *et al*. 1976; Littlejohn and Watson 1974). Behavioral and neurophysiological studies of sound localization in *Hyla cinerea* and *H. gratiosa* suggest that directionality is based on interaural cues, possibly head shadowing of the contralateral ear (resulting in attenuation of about 25 dB), and central detection of small differences in phase, as little as 100 microsec (Feng *et al*. 1976). Hill and Boyan (1976) described an effective system of acoustic localization in crickets that is based on phase-lag pressure augmentation on the inner surface of the ipsilateral tympanic membrane. Sound also reaches the ipsilateral membrane by way of the contralateral tympanic membrane and a tracheal link; this longer pathway is about half a wave length when the source is at right angles to the longitudinal alignment of the receiver. This arrangement results in an increase of sensitivity of about 15 dB in the ipsilateral ear, relative to the contralateral ear when the angle of incidence is 90°, and of about 10 dB at 0°, i.e., head-on. Strother (1959) suggested a similar system for anurans, with conduction from the contralateral ear by way of the Eustachian tubes and pharyngo-oral cavity, but he did not provide a mechanism. Clearly this area warrants further investigation.

DESIGN FEATURES OF ACOUSTIC SIGNALS

Moles (1963), Marler (1967), Konishi (1970), Loftus-Hills (1971b) and Martin (1971) have discussed the structural and functional aspects of acoustic signals as design features.

The ideal properties of an acoustic signal, i.e., those that optimize specificity, range of propagation, and localization of the emitter, will depend on the function of the sound. Thus, for a mate-attracting signal, high specificity and locatability, and maximum effective range would be expected. In contrast, signals associated with territorial behavior might have low specificity (to repel all potential competitors for space and channel bandwidth, regardless of species), short range (to conserve energy), and minimum locatability (to reduce the possibility of accurate positioning of the resident or intruder).

Such optimal design features may not be fully realized in natural systems because (1) two functions are often combined within one signal, e.g., advertisement call; (2) of the selective constraints and limitations in anatomical structures; and (3) of their prior evolution and utilization by other sympatric species. Selection should favor diversification in regard to specificity, but convergence in processes maximizing range and locatability (Loftus-Hills 1971b).

ACOUSTIC INTERACTIONS

The following components may be recognized as resources used in sound communication:

1. The optimal atmospheric channel bandwidth between 400 Hz and 4000 Hz;

2. Transmission time;

3. Suitable transmission stations for effective propagation of signals;

4. Adequate space for the establishment of an individual distance sufficient to allow emitter distinctness; and

5. Temporal coding patterns below the limit set by the shortest damping period of emitter structures and the lowest frequency of neural fusion in the auditory pathway.

It is usual for individuals belonging to several taxa to utilize a common, restricted area for acoustic signaling associated with reproduction. Hence, for a particular set of signaling species, there is the potential for interference as the emitters (usually males) strive to optimize their utilization of the resources of communication to secure matings.

Two levels of interference between signalers may be recognized: intraspecific and interspecific. Intraspecific interactions will involve direct competition for mates, or for those resources that will increase the probability of securing mates, e.g., calling sites, and the favoring through selection of those individuals that have the most effective signals (distinctness, range, ease of location, aspects indicative of suitability as a mate, etc.). This sexual selection should result in decreased variation between males (assuming that frogs are promiscuous) as an optimum is reached. Evolutionary trends based on the "runaway effect" of female sexual selection (Fisher 1958) may also occur, but it is unlikely that these paths would be contrary to the maintenance of maximum efficiency of transmission. Alexander (1975) has discussed evolutionary aspects of acoustic communication at the level of individual fitness.

Interspecific interactions will involve only the resources of communication, i.e., competition for bandwidth, temporal codes, transmission time and calling sites. In this context, the signals of non-conspecifics constitute noise, and may represent a selective pressure to depart from the optimal acoustic design features. Interspecific interactions should thus lead to divergence in acoustic behavior and signal structure because of presence of these acoustic, but not reproductive, competitors.

Following Littlejohn (1965b) and Littlejohn and Martin (1969), the following means of reducing acoustic interference are suggested to result from selection for communicative efficiency:

1. Spectral stratification through the partitioning of the premium atmospheric bandwidth into a number of non-overlapping frequency bands;

2. Spatial separation through (a) aggregation of individuals into single-species assemblages, or (b) the utilization of distinctive, species-characteristic calling sites by individuals;

3. Temporal partitioning of the potential transmission period at one of three levels: (a) into species-specific breeding seasons, (b) into definite non-overlapping time-periods of the daily cycle, or (c) through alternation, antiphony, etc., within the same daily time-period; and

4. Differentiation of species-specific coding patterns of advertisement calls would be expected when common transmission frequency bands and times were employed so that reproductive efficiency would be maintained. Signals associated with territorial behavior, however, might converge (or remain the same if already similar), or responsiveness become generalized, under these conditions.

Loftus-Hills (1971b) gave three possible explanations for diversification of carrier frequencies in acoustic signals:

"1. Frequency carries information regarding species identification, and differences have been selected for in this context;

2. Frequency differences have evolved to minimize acoustic interference between species calling at common breeding sites; and

3. Frequency differences have evolved incidentally, or indirectly, and are due to factors such as body size for which there may be direct selection for ecological reasons."

Consideration of advertisement calls of closely-related, sympatric, synchronically-breeding taxa indicates that carrier frequencies are usually similar, with ranges of variation overlapping extensively (Fouquette 1975; Gerhardt 1973; Littlejohn 1964, 1965a; Ralin 1968; Straughan 1975). In other cases there may be considerable differences in carrier frequency, but temporal patterns are also distinctive (Blair 1955; Littlejohn and Martin 1965; A. A. Martin 1972). Cases of signal differentiation based solely on carrier frequency are presently not known, but this situation could be due to the small number of detailed analyses of signal structure. It appears that the constraints imposed by selection for optimal body size (Nevo and Schneider 1976) or optimal carrier frequency may be so great or that other mechanisms are employed to minimize acoustic interference, that transmission frequencies remain similar.

Straughan (1975) discussed the limitations of coding systems based solely on spectral differences ("channel isolation") in which receptors lack a frequency discriminator. In these cases, heterospecific signals of different frequencies and sufficient intensity could be detected and non-adaptive responses elicited with consequent loss of fitness to the receiver.

It is now clear that adequate space for effective signaling ("individual distance") is actively sought and maintained by males of most species that have prolonged breeding seasons, and that such males will behave agonistically in this context. In these situations, the advertisement call of the resident serves to announce occupied space, and the territorial call provides a "warning" to the intruder. Such situations are common within the reproductive biology of prolonged breeders, as is evidenced by the distinctive signals and complex behavior associated with territoriality. The amount of energy invested in this activity gives a measure of the importance of space as a resource for breeding males; and the return may be seen in enhanced signaling efficiency (distinctness and effective range of an individual's call) and the consequent higher probability of securing matings.

Because of the restriction of most species to nocturnal breeding, territories presumably are re-established each night. Hence, there must be a period of spatial consolidation at the start of each calling session. In some species there is a period of vigorous territorial calling at dusk (Allan 1973; Wells 1976) before advertisement calling begins, suggesting the establishment of temporary spatial distributions.

Likewise, interspecific interactions could occur between males of different species that breed at the same site, if they initially required identical transmission stations. Interspecific interactions could lead to the evolutionary development of one of three possible states:

1. A shift in transmission time;

2. Specialization into different microhabitats, e.g., calling in water, on banks, in vegetation; and

3. Apportioning the same microhabitat through continued spatial interactions.

The first two states involve the development of non-interactive solutions. The last involves continuing interactions, in which case territorial signals might be expected to converge (or remain the same in cognate species), or males might show territorial responses to a wider range of acoustic stimuli, providing that they could be heard, i.e., were loud enough, or close to the optimal carrier frequency used in intraspecific communication.

Even with the establishment of an optimal spatial distribution, there are still problems of maximizing intelligibility if males are relatively close in terms of signal intensities. To transmit randomly, or with independent regularity, means that a proportion of the signals will be overlapped by those of neighbors, and so reduced in intelligibility or directionality.

Accordingly, it would seem to increase their fitness if adjacent individuals timed transmissions so that overlap was minimized. Thus, calling out of phase, as in alternation, antiphony, duets, trios, etc., with one individual as a pacemaker, leader, and the others as successive followers in an extended hierarchy, will maximize the signaling efficiencies of all members of the set, and so tend to even out their competitiveness. It is not clear whether the leader or the last follower is fitter, in the sense of securing more mates, for this will depend on the duration and repetition rate of the signal of the more dominant individual, and on the mechanism of orientation by females. In sustained alternating, however, both signallers should be equivalent with respect to this type of interaction. This topic has been discussed in some detail by Alexander (1975) and Wells (in press) from an evolutionary point of view.

The autumn and spring breeding assemblages at Yan Yean, 34 km NNE of Melbourne, have been studied in some detail and will be discussed briefly as examples of the ways in which interspecific acoustic interference might be minimized.

Two terrestrially-ovipositing species breed in the autumn (April-May): Geocrinia victoriana and Pseudophryne semimarmorata. While the temporal structures of advertisement calls of these species are quite distinctive, the carrier frequencies overlap extensively at around 2600 Hz (Littlejohn and Martin 1969). Furthermore, they often form a mosaic spatial distribution with calling males of each species in close proximity. Hence there is a potentially serious problem of acoustic jamming because of the common transmission frequencies. Littlejohn and Martin (1969), through field playback experiments, demonstrated the P. semimarmorata, which has a short, pulsed call, is inhibited from calling by the longer calls of nearby G. victoriana if received at an intensity of 80 dB or more. P. semimarmorata calls soon after the G. victoriana has concluded a call sequence. Thus, temporal partitioning seems to be the principal mechanism operating to reduce interference between these two species.

Seven species constitute the spring chorus (September-November): Limnodynastes dumerili, L. peroni, L. tasmaniensis, Litoria ewingi, L. raniformis, L. verreauxi and Ranidella signifera. The general spatial distribution of calling males is indicated in figure 2. Diagrammatic representations of dominant frequencies and principal temporal codes in advertisement calls are shown in figure 3. Where calling positions and carrier frequencies are similar, then phasing of calls may occur, but this aspect has not been adequately investigated.

DIVERGENCE AND CONVERGENCE IN ACOUSTIC SIGNALING SYSTEMS

The evolution of a communication element, i.e., a mutually-adaptive message-response unit, may occur incidentally or through the direct action of selection in the context of information transfer. Furthermore, the process may be viewed at two levels: intraspecific--involving members of the same reproductive set, and interspecific--involving members of different reproductive sets. Intraspecific interactions should lead to reduced between-individual variation and to optimization of effectiveness of communication

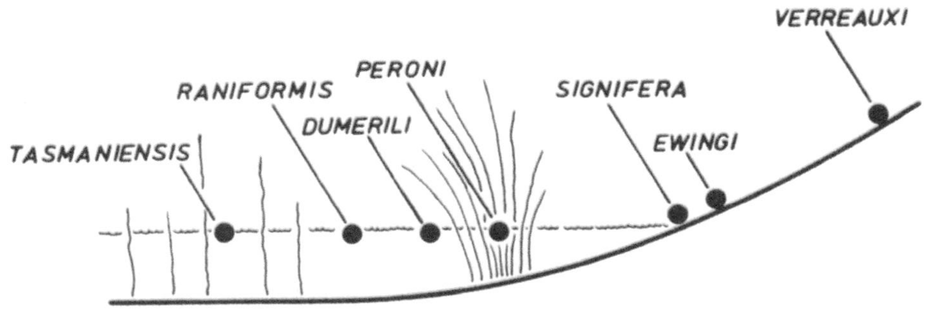

Fig. 2. General spatial distribution of calling males of seven species of anurans (Limnodynastes dumerili, L. peroni, L. tasmaniensis, Litoria ewingi, L. raniformis, L. verreauxi and Ranidella signifera) in a spring chorus at Yan Yean, Victoria (based on unpublished observations).

through reproductive competition. Interspecific interactions may be of two types: (1) communicative, as in territorial conflicts, and could lead to convergence of communicative elements, or to reduced specificity of signals and responses; or (2) non-communicative, as in selective mating behavior, when heterospecific signals would constitute noise, and divergent evolution would be expected to occur as a result of selection for specificity and distinctness.

The speciation process is of interest in that the development of the sympatric state requires disruption of the original communication system, with selection favoring non-communication between members of differentiated daughter, i.e., cognate, populations. As speciation approaches completion, most, or all, of the signals of one cognate population will begin to constitute an actual or potential source of noise to individuals of the other cognate population. A signal will be transformed into noise as the mutual benefits to the interactants are reduced through the loss of reproductive fitness that arises from genetic incompatibility and the production of adaptively inferior hybrids.

Conversely, the communication systems of previously isolated cognate populations, although effectively differentiated, may no longer constitute noise if inter-breeding occurs and the resulting hybrids are of equal or greater fitness than the parental individuals. Such could be the case where previously allopatric cognate populations meet on an ecotone, and the hybrids, because of their adaptation to an intermediate environment, are of greater fitness than the parental individuals in transitional habitats. The initiation and maintenance of a narrow hybrid zone between Geocrinia laevis and G. victoriana in south-eastern Australia has been

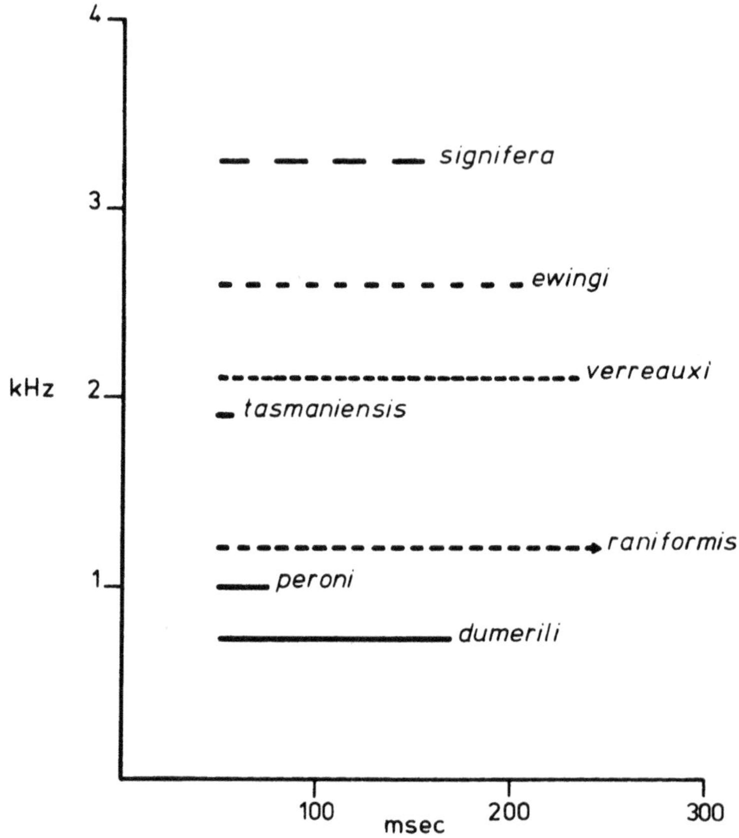

Fig. 3. Diagrammatic representation of dominant frequencies and principal temporal coding patterns in advertisement calls of seven species of anurans (*Limnodynastes dumerili*, *L. peroni*, *L. tasmaniensis*, *Litoria ewingi*, *L. raniformis*, *L. verreauxi* and *Ranidella signifera*) in a spring chorus at Yan Yean, Victoria (based on Littlejohn 1964, 1965a, b, unpublished observations; and Littlejohn and Martin 1965).

explained in this way by Littlejohn and Watson (1973, 1976).

Comparisons of the signaling systems of sympatric cognate species may provide useful information about the degree of distinctness that is necessary to allow efficient intraspecific communication in complex acoustic communities, and on the evolution of species-specific temporal codes. This area has been extensively investigated from a different point of view in the study of the origin and operation of differences in mating calls (= advertisement calls, in part) of sympatric species in the context of reproductive isolation. Littlejohn (1969) reviewed differences in the temporal structures of advertisement calls of extensively sympatric cognate species. He found that pulse rates were often distinctive, with mean values differing by a factor of two or more, and with a gap between the extremes of ranges of variation. Subsequent investigations of species pairs in which there is differentiation in pulse rates of advertisement calls has confirmed this factor (Fouquette 1975; Littlejohn 1971).

Although many studies of the structure of advertisement calls in congeneric sympatric species pairs have been made, experimental verification of the role of pulse rates in providing the species specificity of the female phonotactic response has been carried out on only two species pairs: Litoria ewingi and L. verreauxi (Loftus-Hills and Littlejohn 1971b) and Hyla cadaverina and H. regilla (Straughan 1975).

The dynamics of the evolution of genetically-determined coding systems probably are difficult to trace because of the very long time spans involved. The essential stages of the process may be reconstructed, however, by studying signal structure (emitter) and behavioral responses (receiver) in cognate pairs of species that are presumed to have expanded their geographic ranges from disjunct allopatry to extensive sympatry. Interactive (sympatric) and non-interactive (allopatric) components are available for study in such cases. This spatial arrangement may be equated to a temporal sequence by assuming that the allopatric populations represent pre-interactive states, and the sympatric populations represent interactive or post-interactive states. Grant (1972) commented on the dangers of making such assumptions, particularly as the allopatric populations might just as easily be interpreted as representing post-interactive populations that have undergone evolutionary release, and converged towards an optimal state.

Three such situations have been adequately investigated within the Anura, in the context of reproductive character displacement (sensu Brown and Wilson 1956). They may also be seen, however, as representing communicative interactions in which species-specific coding patterns have arisen, or been refined. In Gastrophryne carolinensis and G. olivacea, dominant frequencies, call durations, and pulse rates are more distinctive in sympatry than in allopatry (Awbrey 1965; Blair 1955). In Litoria ewingi and L. verreauxi (Littlejohn 1965a; Littlejohn and Loftus-Hills 1968; Loftus-Hills and Littlejohn 1971b) and in Pseudacris nigrita and P. triseriata feriarum (Fouquette 1975), there is difference of pulse rates in sympatry to the extent that ranges of variation do not overlap, but are separated by a considerable gap.

Displacement of means, and absence of overlap of ranges of variation in temporal components, provide prima facie evidence for the development of species-specific codes. Shifts in dominant frequencies, however, may arise as incidental effects of other selective processes, such as ecological interactions, that influence body size, and hence carrier frequency. The assumed functional significance of these differences in signal structure requires verification through the experimental analysis of the associated receiver specificity, but few data are available. Specificity of female phonotactic response has been assessed in some sympatric populations of Litoria ewingi and L. verreauxi, and found to be consistent with the trends in pulse rates of advertisement calls (Littlejohn and Loftus-Hills 1968). Experimental documentation of the assumption that specificity rests with the differences in pulse rate in these species was carried out by Loftus-Hills and Littlejohn (1971b), who used synthetic signals in controlled two-choice discrimination tests of female phonotaxis. Awbrey (1965), using synthetic signals, attempted to determine which mating-call components provided cues for the specificity of phonotactic response by females of Gastrophryne olivacea. Because of a lack of consistency between results obtained with synthetic mating calls of G. carolinensis and G. olivacea and those obtained with recorded natural mating calls, however, he was unable to draw any conclusions.

These studies show how quantitative shifts in temporal codes might occur, and provide a measure of the minimum amount of difference required to ensure that efficient intraspecific communication is achieved. Only one of the cases so far analyzed illustrates how a qualitative change in temporal coding might occur. In Litoria verreauxi the degree of amplitude modulation is progressively reduced with increasing depth of sympatry, so that calls of some individuals from extreme sympatry are almost completely lacking in pulse modulation (Littlejohn 1965a). Presumably there are parallel shifts in the response systems of the receivers, but the efficacy of unmodulated advertisement calls in attracting females of L. verreauxi has not been tested.

Cases of reproductive character displacement involving the development of species-specific codes are rare, and in most of the species pairs that have been studied, acoustic signals are just as distinctive in allopatry as in sympatry (Blair 1974). An incidental origin of coding systems might therefore be a favored hypothesis. Other mechanisms of speciation than the classical geographic model may have operated, however, such as sympatric speciation, or speciation through small peripheral populations that subsequently became fully sympatric, interacted with the parental species, and developed distinctive codes during this phase. The new species could then undergo an expansion of range to include areas allopatric to the parental species, and could thus carry the post-interactive, modified coding pattern throughout its geographic distribution. Otte (1974) provided a discussion of this type of evolutionary process.

Likewise, efficient, specific coding systems eventually operating between cognate species may have an indirect origin. If, during an allopatric phase, each daughter population were exposed to a different set of soniferous species (i.e., different acoustic environments), and acoustic signals diverged sufficiently during this period, then there might be no communicative interaction in a subsequent sympatric phase (Littlejohn 1965a, 1969).

REFERENCES

Alexander, R. D. 1975. Natural selection and specialized chorusing behavior in acoustical insects. IN: D. Pimentel (ed.), Insects, Science and Society, pp. 35-77. Academic Press, New York.

Allan, D. M. 1973. Some relationships of vocalization to behavior in the Pacific treefrog, Hyla regilla. Herpetologica 29:366-371.

Awbrey, F. T. 1965. An experimental investigation of the effectiveness of anuran mating calls as isolating mechanisms. Ph. D. Thesis, Univ. Texas, Austin.

Axtell, R. W. 1959. Female reaction to the male call in two anurans (Amphibia). Southwest. Nat. 3:70-76.

Aylor, D. 1972. Noise reduction by vegetation and ground. J. Acoust. Soc. Am. 51:197-205.

Blair, W. F. 1955. Mating call and stage of speciation in the Microhyla olivacea-M. carolinensis complex. Evolution 9: 469-480.

Blair, W. F. 1956. The mating calls of hybrid toads. Texas J. Sci. 8:350-355.

Blair, W. F. 1963. Acoustic behaviour of Amphibia. IN: R. G. Busnel (ed.), Acoustic Behaviour of Animals, pp. 694-708, 803-804. Elsevier, Amsterdam.

Blair, W. F. 1968. Amphibians and reptiles. IN: T. A. Sebeok (ed.), Animal Communication, pp. 289-310. Indiana Univ. Press, Bloomington.

Blair, W. F. 1974. Character displacement in frogs. Am. Zool. 14:1119-1125.

Bogert, C. M. 1960. The influence of sound on the behavior of amphibians and reptiles. IN: W. E. Lanyon and W. N. Tavolga (eds.), Animal Sounds and Communication, pp. 137-320. Am. Inst. Biol. Sci. Publ. 7, Washington, D. C.

Brown, W. L., and E. O. Wilson. 1956. Character displacement. Syst. Zool. 5:49-64.

Capranica, R. R. 1965. The Evoked Vocal Response of the Bullfrog. Research Monograph, No. 33. M. I. T. Press, Cambridge, Mass.

Capranica, R. R. 1976. The auditory system. IN: B. Lofts (ed.), Physiology of the Amphibia, Vol. III, pp. 443-466. Academic Press, New York.

Capranica, R. R., and A. J. M. Moffat. 1975. Selectivity of the peripheral auditory system of spadefoot toads (Scaphiopus couchi) for sounds of biological significance. J. Comp. Physiol. 100:231-249.

Cherry, C. 1957. On Human Communication. M. I. T. Press, Cambridge, Mass.

Conant, R. 1975. A Field Guide to Reptiles and Amphibians of Eastern and Central North America (Second Edition). Houghton Mifflin Co., Boston.

Dixon, J. R. 1957. Geographic variation and distribution of the genus Tomodactylus in Mexico. Texas J. Sci. 9:379-409.

Embleton, T. F. W., J. E. Piercy and N. Olson. 1976. Outdoor sound propagation over ground of finite impedance. J. Acoust. Soc. Am. 59:267-277.

Feng, A. S., H. C. Gerhardt and R. R. Capranica. 1976. Sound localization behavior of the green treefrog (Hyla cinerea) and the barking treefrog (H. gratiosa). J. Comp. Physiol. 107: 241-252.

Fisher, R. A. 1958. The Genetical Theory of Natural Selection (Second Edition). Dover Publications, New York.

Forester, D. C. 1973. Mating call as a reproductive isolating mechanism between Scaphiopus bombifrons and S. hammondi. Copeia 1973:60-67.

Fouquette, M. J. 1975. Speciation in chorus frogs. I. Reproductive character displacement in the Pseudacris nigrita complex. Syst. Zool. 24:16-23.

Frishkopf, L. S., R. R. Capranica and M. H. Goldstein. 1968. Neural coding in the bullfrog's auditory system. A teleological approach. Proc. I. E. E. E. 56:969-980.

Frishkopf, L. S., and M. H. Goldstein. 1963. Responses to acoustic stimuli from single units in the eighth nerve of the bullfrog. J. Acoust. Soc. Am. 35:1219-1228.

Gans, C. 1973. Sound production in the Salientia: Mechanism and evolution of the emitter. Am. Zool. 13:1179-1194.

Gerhardt, H. C. 1973. Reproductive interactions between Hyla crucifer and Pseudacris ornata (Anura: Hylidae). Am. Midl. Nat. 89:81-88.

Gerhardt, H. C. 1974a. The vocalizations of some hybrid treefrogs: Acoustic and behavioural analyses. Behaviour 49: 130-151.

Gerhardt, H. C. 1974b. Behavioral isolation of the treefrogs, Hyla cinerea and Hyla andersonii. Am. Midl. Nat. 91:424-433.

Gerhardt, H. C. 1974c. The significance of some spectral features in mating call recognition in the green treefrog (Hyla cinerea). J. Exp. Biol. 61:229-241.

Gerhardt, H. C. 1975. Sound pressure levels and radiation patterns of the vocalizations of some North American frogs and toads. J. Comp. Physiol. 102:1-12.

Gerhardt, H. C. 1976. Significance of two frequency bands in long distance vocal communication in the green treefrog. Nature 261:692-694.

Grant, P. R. 1972. Convergent and divergent character displacement. Biol. J. Linn. Soc. 4:39-68.

Harris, C. M. 1966. Absorption of sound in air versus humidity and temperature. J. Acoust. Soc. Amer. 40:148-159.

Hawe, S. M. 1970. Calling behaviour and territoriality in males of two species of Crinia (Anura: Leptodactylidae). Unpublished B. Sc. Honours Research Report, Department of Zoology, University of Melbourne.

Hill, K. G., and G. S. Boyan. 1976. Directional hearing in crickets. Nature 262:390-391.

Johnston, T. D. 1976. Theoretical considerations in the adaptation of animal communication systems. J. Theoret. Biol. 57: 43-72.

Konishi, M. 1970. Evolution of design features in the coding of species-specificity. Am. Zool. 10:67-72.

Littlejohn, M. J. 1964. Geographic isolation and mating call differentiation in Crinia signifera. Evolution 18:262-266.

Littlejohn, M. J. 1965a. Premating isolation in the Hyla ewingi complex (Anura: Hylidae). Evolution 19:234-243.

Littlejohn, M. J. 1965b. Vocal communication in frogs. Aust. Nat. Hist. 15:52-55.

Littlejohn, M. J. 1969. The systematic significance of isolating mechanisms. IN: Systematic Biology. Proceedings of an International Conference, pp. 459-482. National Academy of Sciences, Washington, D. C.

Littlejohn, M. J. 1971. A re-appraisal of mating call differentiation in Hyla cadaverina (= Hyla californiae) and Hyla regilla. Evolution 25:98-102.

Littlejohn, M. J. 1976. The Litoria ewingi complex (Anura: Hylidae) in south-eastern Australia. IV. Variation in mating-call structure across a narrow hybrid zone between L. ewingi L. paraewingi. Aust. J. Zool. 24:283-293.

Littlejohn, M. J., and J. J. Loftus-Hills. 1968. An experimental evaluation of premating isolation in the Hyla ewingi complex (Anura: Hylidae). Evolution 22:659-663.

Littlejohn, M. J., and A. A. Martin. 1964. The Crinia laevis complex (Anura: Leptodactylidae) in south-eastern Australia. Aust. J. Zool. 12:70-83.

Littlejohn, M. J., and A. A. Martin. 1965. Mating call structure in three sympatric species of Limnodynastes (Anura: Leptodactylidae). Copeia 1965:509-511.

Littlejohn, M. J., and A. A. Martin. 1969. Acoustic interaction between two species of leptodactylid frogs. Anim. Behav. 17: 785-791.

Littlejohn, M. J., and G. F. Watson. 1973. Mating-call variation across a narrow hybrid zone between Crinia laevis and C. victoriana (Anura: Leptodactylidae). Aust. J. Zool. 21:277-284.

Littlejohn, M. J., and G. F. Watson. 1974. Mating call discrimination and phonotaxis by females of the Crinia laevis complex (Anura: Leptodactylidae). Copeia 1974:171-175.

Littlejohn, M. J., and G. F. Watson. 1976. Effectiveness of a hybrid mating call in eliciting phonotaxis by females of the Geocrinia laevis complex (Anura: Leptodactylidae). Copeia 1976:76-79.

Littlejohn, M. J., G. F. Watson and J. J. Loftus-Hills. 1971. Contact hybridization in the Crinia laevis complex (Anura: Leptodactylidae). Aust. J. Zool. 19:85-100.

Loftus-Hills, J. J. 1971a. Neural correlates of acoustic behaviour in the Australian bullfrog Limnodynastes dorsalis (Anura: Leptodactylidae). Z. Vergl. Physiol. 74:140-152.

Loftus-Hills, J. J. 1971b. Auditory Function and Acoustic Communication in Anuran Amphibians. Ph. D. Thesis, University of Melbourne, Melbourne.

Loftus-Hills, J. J. 1973a. Comparative aspects of auditory function in Australian anurans. Aust. J. Zool. 21:353-367.

Loftus-Hills, J. J. 1973b. Neural mechanisms underlying acoustic behaviour of the frog Pseudophryne semimarmorata (Anura: Leptodactylidae). Anim. Behav. 21:781-787.

Loftus-Hills, J. J. 1974. Analysis of an acoustic pacemaker in Strecker's chorus frog Pseudacris streckeri (Anura: Hylidae). J. Comp. Physiol. 90:75-87.

Loftus-Hills, J. J., and B. M. Johnstone. 1970. Auditory function, communication and brain-evoked response in anuran amphibians. J. Acoust. Soc. Am. 47:1131-1138.

Loftus-Hills, J. J., and M. J. Littlejohn. 1971a. Mating-call sound intensities of anuran amphibians. J. Acoust. Soc. Am. 49:1327-1329.

Loftus-Hills, J. J., and M. J. Littlejohn. 1971b. Pulse repetition rate as the basis for mating call discrimination by two sympatric species of Hyla. Copeia 1971:154-156.

Lombard, R. E., and I. R. Straughan. 1974. Functional aspects of anuran middle ear structures. J. Exp. Biol. 61:71-93.

Lörcher, K. 1969. Vergleichende bio-akustische untersuchungen an der rot- und gelbbauchunke, Bombina bombina (L.) und Bombina *v. variegata* (L.). *Oecologia* *3*:84-124.

Marler, P. 1967. Animal communication signals. *Science* *157*:769-774.

Martin, A. A. 1972. Studies in Australian Amphibia. III. The *Limnodynastes dorsalis* complex (Anura: Leptodactylidae). *Aust. J. Zool.* *20*:165-211.

Martin, W. F. 1971. Mechanics of sound production in toads of the genus *Bufo*: Passive elements. *J. Exp. Zool.* *176*:273-294.

Martin, W. F. 1972. Evolution of vocalization in the genus *Bufo*. IN: W. F. Blair (ed.), *Evolution in the Genus Bufo*, pp. 279-309. Univ. Texas Press, Austin.

Martin, W. F., and C. Gans. 1972. Muscular control of the vocal tract during release signaling in the toad *Bufo valliceps*. *J. Morphol.* *137*:1-27.

McAlister, W. H. 1959. The vocal structures and method of call production in the genus *Scaphiopus* Holbrook. *Texas J. Sci.* *11*:60-77.

McAlister, W. H. 1961. The mechanics of sound production in North American *Bufo*. *Copeia* *1961*:86-95.

McDiarmid, R. W., and K. Adler. 1974. Notes on territorial and vocal behavior of neotropical frogs of the genus *Centrolenella*. *Herpetologica* *30*:75-78.

Moles, A. 1963. Animal language and information theory. IN: R. G. Busnel (ed.), *Acoustic Behaviour of Animals*, pp. 112-131. Elsevier, Amsterdam.

Morton, E. S. 1975. Ecological sources of selection on avian sounds. *Amer. Nat.* *109*:17-34.

Nairns, P. M., and R. R. Capranica. 1976. Sexual differences in the auditory system of the tree frog *Eleutherodactylus coqui*. *Science* *192*:378-380.

Nevo, E., and H. Schneider. 1976. Mating call pattern of green toads in Israel and its ecological correlate. *J. Zool.* (Lond.) *178*:133-145.

Otte, D. 1974. Effects and functions in the evolution of signaling systems. *Ann. Rev. Ecol. Syst.* *5*:385-417.

Paillette, M. 1971. Communication acoustique chez les amphibians anoures. J. de Psychol. Norm. et. Pathol. 1971:327-351.

Ralin, D. B. 1968. Ecological and reproductive differentiation in the cryptic species of the Hyla versicolor complex (Hylidae). Southwest. Nat. 13:283-299.

Rosen, M., and R. E. Lemon. 1974. The vocal behavior of spring peepers, Hyla crucifer. Copeia 1974:940-950.

Schmidt, R. S. 1973. Central mechanisms of frog calling. Am. Zool. 13:1169-1177.

Schneider, H. 1967. Rufe und rufverhalten des laubfrosches, Hyla arborea arborea (L.). Z. Vergl. Physiol. 57:174-189.

Schneider, H. 1968. Bio-akustische untersuchungen am Mittelmeerlaubfrosch. Z. Vergl. Physiol. 61:369-385.

Schneider, H. 1974. Structure of the mating calls and relationships of the European tree frogs (Hylidae, Anura). Oecologia 14:99-110.

Schneider, H., and H. Eichelberg. 1974. The mating call of hybrids of the fire-bellied toad and yellow-bellied toad (Bombina bombina (L.), Bombina v. variegata (L.), Discoglossidae, Anura). Oecologia 16:61-71.

Schneider, H., and E. Nevo. 1972. Bio-acoustic studv of the yellow-lemon tree frog, Hyla arborea savignyi Audouin. Zool. Jb. Physiol. 76:497-506.

Shannon, C. E., and W. Weaver. 1949. The Mathematical Theory of Communication. Univ. Illinois Press, Urbana.

Snyder, W. F., and D. L. Jameson. 1965. Multivariate geographic variation of mating call in populations of the Pacific tree frog (Hyla regilla). Copeia 1965:129-142.

Straughan, I. R. 1973. Evolution of anuran mating calls. Bioacoustical aspects. IN: J. A. Vial (ed.), Evolutionary Biology of the Anurans, pp. 321-327. Univ. Missouri Press, Columbia.

Straughan, I. R. 1975. An analysis of the mechanisms of mating call discrimination in the frogs Hyla regilla and H. cadaverina. Copeia 1975:415-424.

Strother, W. F. 1959. The electrical response of the auditory mechanism in the bullfrog (Rana catesbeiana). J. Comp.

Physiol. Psychol. 52:157-162.

Watkins, W. A., E. R. Baylor and A. T. Bowen. 1970. The call of Eleutherodactylus johnstonei, the whistling frog of Bermuda. Copeia 1970:558-561.

Wells, K. D. 1976. Territorial Behaviour of the Green Frog (Rana clamitans). Ph. D. Thesis, Cornell Univ., Ithaca.

Wells, K. D. In press. The social behaviour of anuran amphibians: A review. Anim. Behav.

Whitney, C. L., and J. R. Krebs. 1975. Spacing and calling in Pacific tree frogs, Hyla regilla. Can. J. Zool. 53:1519-1527.

Wiener, F. M., and D. N. Keast. 1959. Experimental study of the propagation of sound over ground. J. Acoust. Soc. Am. 31: 724-733.

Williams, G. C. 1966. Adaptation and Natural Selection. Princeton Univ. Press, Princeton.

Zweifel, R. G. 1959. Effect of temperature on call of the frog, Bombina variegata. Copeia 1959:322-327.

Zweifel, R. G. 1968. Effects of temperature, body size, and hybridization on mating calls of toads, Bufo a. americanus and Bufo woodhousei fowleri. Copeia 1968:269-285.

ACOUSTIC BEHAVIOR AND PHYSIOLOGY OF VOCALIZATION IN THE EUROPEAN TREE FROG, *Hyla arborea* (L.)

Hans Schneider

Zoological Institute, University of Bonn
Poppelsdorfer Schloss, D-53 Bonn, Germany

It was about ten years ago that bio-acoustic methods were first systematically applied to the study of the European anurans (Schneider 1966a, b). The results that have since been obtained are numerous, for so short a span of time (Heinzmann 1970; Heusser 1970; Lörcher 1966, 1969; Obert 1971, 1974, 1975a, b, c; Schmid 1976; Schneider 1973; Schneider and Eichelberg 1974; Wahl 1969; Weber 1974, 1975, 1976; Weber and Schneider 1971).

The anuran fauna of Europe, as compared with other regions, comprises remarkably few species--only thirteen in Central Europe, though these are distributed among five families. With so few species in a given area, it becomes possible to investigate certain questions in all the native species. The structure of the sound-producing apparatus, for example, is known for all thirteen Central European species, and the effect of temperature upon the mating calls has been analyzed in most of the species for which such experiments would be informative. Aside from this broad program, certain species have been used to clarify special questions of bio-acoustics. Most notable of these are the fire-bellied toad, *Bombina bombina* (L.), and the tree frog, *Hyla arborea* (L.), both admirably suited and rewarding objects for such studies. In the case of the fire-bellied toad interest has been directed chiefly toward the hormonal control of calling activity (Obert 1973), whereas the studies of the tree frog have been concerned with the external influences on calls and calling behavior and, more particularly, with the nervous control of the sound-producing apparatus. This report surveys the present state of research on the tree frog.

The work has been focused chiefly upon the Central European tree frog, Hyla arborea arborea (L.) and has been supplemented by comparative studies of Hyla a. sarda (de Betta) on islands in the western Mediterranean, Hyla a. kretensis Ahl on Crete, Hyla a. molleri Bedriaga in Portugal, Hyla meridionalis Boettger in southern France and on the Canary island Tenerife, and Hyla a. savignyi Audouin in Israel (Fig. 1).

I. CALLING BEHAVIOR

The Central European tree frog is distributed over a wide range. It extends from northern Spain in the south to southern Sweden in the north and to the Urals in the east. The frog is found mainly in permanent bodies of water, varying in size but characterized by abundant plant growth. Until a few years ago tree frogs were numerous in Central Europe, but because of destruction or alteration of the natural biotopes there has recently been

Fig. 1. The locations in which frogs have been studied: 1, Bebenhausen, southwestern Germany (H. a. arborea); 2, Domaine d'Anghione, Corsica (H. a. sarda); 3, Chania, Crete (H. a. kretensis); 4, Coimbra, Portugal (H. a. molleri); 5, Salin de Giraud, the Camargue, France (H. meridionalis); 6, Punta del Hidalgo, Tenerife (H. meridionalis); 7, Saar, Israel (H. a. savignyi).

a rapid population decline which still continues today. In southern Europe the frog populations are still large; they occupy not only the natural waters but also irrigation systems, rice paddies and cisterns.

The onset of the annual period of calling and reproduction is critically affected by the seasonal temperature changes. The activity periods begin earliest in southern Europe; their onset is later the further north a population lives. There are exceptions among populations of regions with an especially favorable climate. On the island of Crete the tree frogs are already calling actively in February, and on Corsica, in Portugal and in southern France intense activity is underway in March. By contrast, the calling periods of the tree frogs native to southwestern Germany do not begin until the last days of April or early May. In Israel and on Tenerife calling begins earlier than in Europe--in January or even in December.

Like most of the European species of Anura, with the exception of the Discoglossidae, the tree frogs have only one calling period per year, and it can last for several months.

The daily calling phase of a male in readiness for mating begins with the onset of twilight. Each day when the air temperature exceeds the lower threshold for calling, the tree frogs call. In the spring the weather sometimes changes suddenly, with a fall in atmospheric pressure and a drop in temperature that may be quite severe. During these days the frogs cease to call and begin again only when the weather has improved.

Since the tree frogs never begin to call before evening, the daily cycle of calling activity is dependent not only on temperature but also on light intensity. This was in fact confirmed by observations, extending over two years, of the onset times of calling and chorus behavior (Schneider 1971a). The data obtained for a population of ca. 40 animals in southwestern Germany in 1970 are shown in Fig. 2. The tree frogs were first observed to call on May 4; the calling period ended on July 13. There were a few days during this time when the frogs did not call, since the weather was cool and the air temperature had already fallen below the lower threshold for calling by evening. The tree frogs never began to call until the light intensity had decreased to 250 lux, and usually waited until there was considerably less light. Choruses (simultaneous calling by at least ten frogs) were heard only when it was much darker. The light intensities permitting calling were reached about 7:45 in the evening at the beginning of the calling period and toward 9:00 in June and July.

Analysis of the data showed a clear dependence of the daily onset of calling upon light intensity, but no effect of tempera-

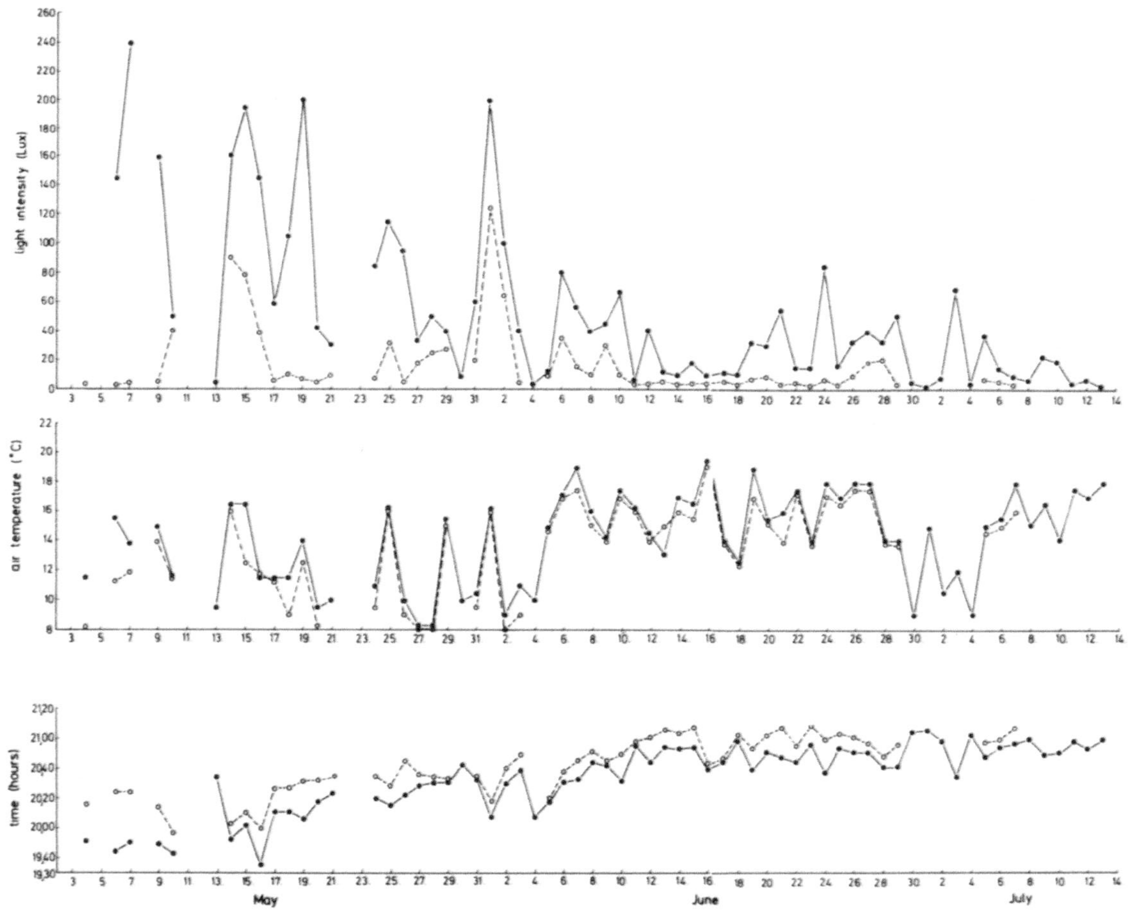

Fig. 2. Hyla a. arborea: the onset of calling and chorus behavior by a population in southwestern Germany in 1970. ●——● light intensity, air temperature and time of day when calling began; o---o the same data for chorus onset. From H. Schneider, Oecologia (Berl.) 8, 310-320 (1971). By permission of Springer-Verlag, Heidelberg.

ture. Temperature affects calling only in that it must be above the lower threshold if the frogs are to show any calling activity at all.

A second relationship derivable from these data was between the onset of calling or chorus behavior and the date. This correlation can best be expressed in terms of the end of civil twilight (defined as the time when the sun is 6° below the horizon) in the same place over the same period; the two variables are plotted vs. the date in Fig. 3. The curve showing the end of civil twilight is almost exactly centered in the dot pattern representing the onset of calling. Only in July, when the curve turns downward, does it depart from the dot pattern; soon after this desynchronization the frogs end their calling period. This correlation of calling onset with time of year is a sure indication that there exists a changing state of readiness to call, which is under endogenous control and is an effective factor in calling behavior.

Other factors, such as atmospheric pressure and amount of precipitation, were not monitored during these observations, but they quite probably do play a role; for example, after it has rained the chorus behavior of the tree frogs is more intense. The additional evidence provided by Blankenhorn (1972) was therefore particularly welcome; he also examined the meteorological variables affecting the onset and duration of calling by *Hyla arborea*, with somewhat different results. He concluded that the critical factors determining the beginning of calling were the water temperature at 7:00 p.m., the atmospheric pressure and the amount of precipitation on the preceding day, while that affecting duration of the chorus was the water temperature at midnight.

At the onset of the daily calling phases calling is sporadic at first; one frog begins, and others gradually join in. From time to time general pauses occur between the series of calls. Later in the evening the number of calling frogs increases more rapidly, as does the intensity of calling activity. When the activity is at its peak, the tree-frog chorus is no longer interrupted by pauses. It often happens that for a few minutes at a time two members of the chorus, usually neighboring frogs, call in alternation, so that the calls of one frog occupy the intervals in the partner's calling. Once the peak of the daily calling phase has been passed, the end is signalled by the reoccurence of general pauses, at first lasting only a minute or so. As soon as one frog resumes calling others join in, but with increasing delays. The pauses become longer and the bursts of calling shorter, until the pauses predominate. By this time the frogs are already leaving the pond, so that the volume of the chorus steadily declines.

The duration of the choruses in the evening depends primarily on the temperature, in particular with respect to whether and when

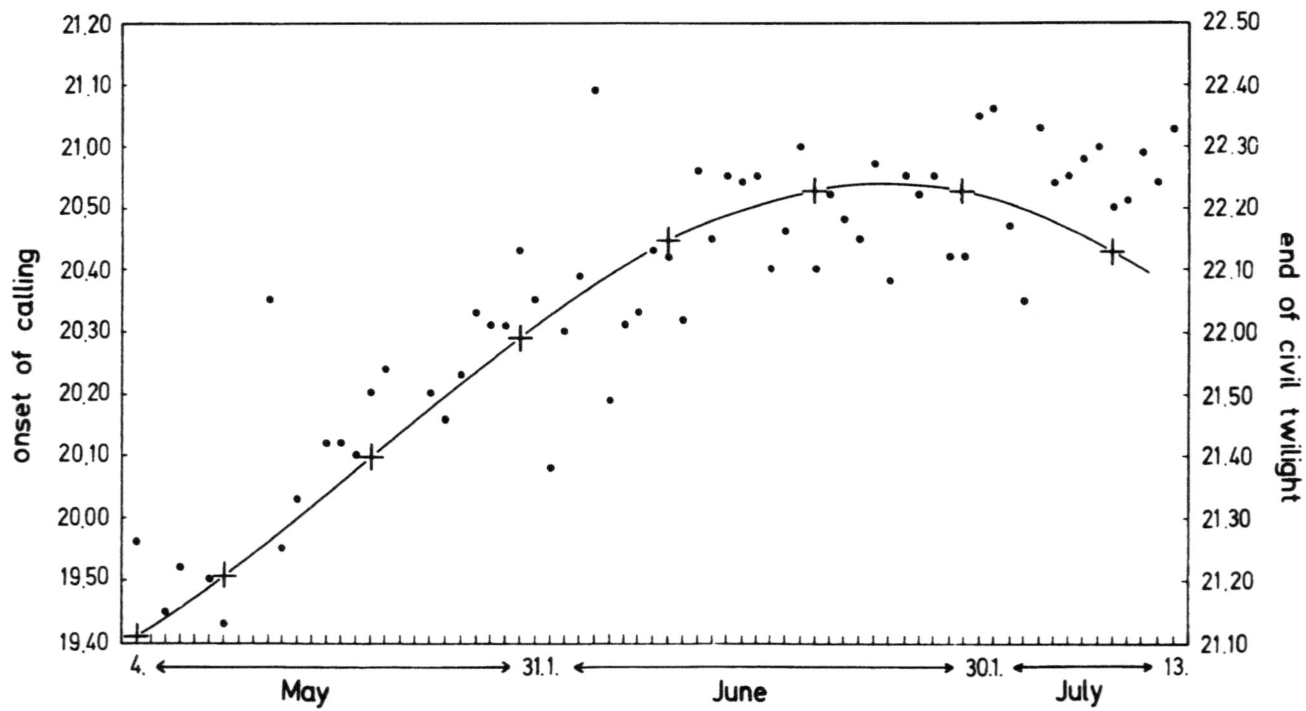

Fig. 3. Hyla a. arborea: the onset of calling by a population in southwestern Germany in 1970, compared with the end of civil twilight at 45° north latitude during the same time (+ —— +); the ordinates represent the time of day. From H. Schneider, Oecologia (Berl.) 8, 310-320 (1971). By permission of Springer-Verlag, Heidelberg.

the lower threshold for calling is reached (Schneider 1967). During the first part of the calling period, in May, the temperature is not very high (usually about 10-12° C) when calling begins in the evening, and it continues to fall rapidly. On clear days there is greater heating during the day, but more rapid cooling in the evening, than when the sky is overcast. The lower threshold for calling is at an air temperature of 8° C, with some slight variation among individual frogs. Calling activity is reduced even before this temperature is reached.

Because of the fall in air temperature, in May and June calling is terminated between 9:30 and 11:00 in the evening. Later in the calling period the temperature rises higher during the day, remains higher in the evening and falls more slowly. The greatest air temperature measured at the observation pond at the time of onset of calling was 20.5° C. During the hours that followed, the temperature fell only to between 16° and 14° C; the lower threshold for calling was never reached. However, the frogs called only until about midnight. Even so there was intensive calling activity, the choruses then came to a relatively rapid end. Only on a few days did individual frogs, usually two or three at most, continue to call sporadically until morning. The Mediterranean tree frogs in the Camargue regularly end their choruses considerably later when the weather is favorable, often not ceasing to call until dawn.

The calling sites chosen by the frogs are at the very edge of the water; the frogs usually sit in such a way that the hindlegs and some of the underside of the body are barely covered by water. Most of the tree frogs, then, are to be found at the margins of a pond. If there are islands of dry land in the pond or fields of algae or other objects floating on the surface, the frogs also make use of these as calling sites. It is rare for a frog to call while completely on land, but after it has rained, when soil and vegetation are still wet, they may sit as much as a meter away from the pond to call. Almost without exception the frogs display their green coloration during the night.

The frogs adopt a characteristic posture while calling. The calls are produced by expiration of air from the lungs into the vocal sac, which expands still further in the process. The vocal sac of the Mediterranean tree frog contains little air during the intervals; it gradually swells during the calls, reaching the maximal expansion at the end of the calls.

During the day the tree frogs stay out of the pond, although if tall plants are growing in the water some of the frogs remain on these. Shortly before and during the twilight hours, the frogs return to the water to call. It is only on evenings when the air temperature barely exceeds the lower threshold for calling that relatively few frogs come to the calling site. Occasionally individual frogs begin to call while still at some distance from the water and sometimes while they are still sitting on trees and shrubs.

Whereas the tree frogs can easily be located in the evening by listening to the calls, they are well hidden during the day. Only when sunbathing do they sit in exposed places. But a 20-day preliminary effort to determine their daytime quarters proved successful. By monitoring the signals from transmitters attached to the tree frogs it was possible to locate the frogs and trace their movements. Some of the animals chose the same hiding and calling places several days in succession, following approximately the same routes between the two places each day. Frogs altered their daytime locations more often than their calling sites.

The temperature ranges that the European tree frogs exhibit calling activity are different. The lower threshold is higher in Hyla a. arborea (8° C) than for Hyla meridionalis (5-6° C). On the other hand, the upper threshold is difficult to determine at the calling grounds because the air cools so rapidly in the evening. According to observations in the laboratory the upper threshold of Hyla a. arborea and of Hyla meridionalis seem to be about 22° C, since at this temperature the frogs often sit at their calling sites in typical calling posture, but rarely produce calls. Other tree frogs are still very active in calling at higher air temperatures--for example, Hyla a. sarda at 24-25° C, and Hyla a. savignyi at 26-28° C.

II. THE CALLS

1. The Mating Call

Tree frogs produce their mating calls in series, which in the Central European species usually comprises 15-30 calls. At the peak of the calling period these series may be very long, each containing several hundred calls. The calls within a series are separated by intervals; that between the first and second calls is always the longest. The intervals following the next few calls are progressively shorter, and from the fourth or fifth call on they remain roughly equal in duration (Schneider 1966a, 1967).

Each call is formed of sound pulses, the number and amplitude-modulation of which distinguish the calls of the different species and subspecies (Fig. 4). The call of Hyla a. arborea quite consistently comprises nine pulses ($\bar{x} = 9.1 \pm 0.4$, $n = 77$). The few calls that depart from this average have no less than seven and no more than eleven pulses; the variation is often due to the

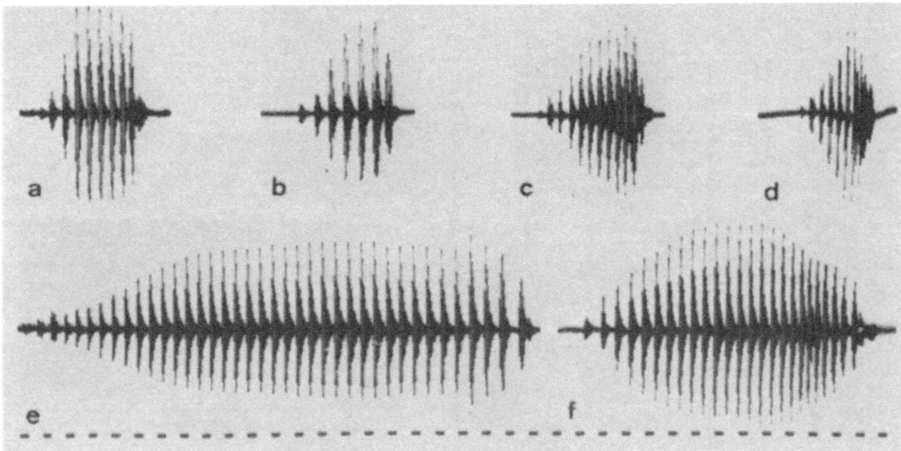

Fig. 4 a-f. Oscillographs of mating calls of European tree frogs at 12° C. a, *Hyla a. arborea*; b, *Hyla a. kretensis*; c, *Hyla a. molleri*; d, *Hyla a. sarda*; e, *Hyla meridionalis*; f, *Hyla a. savignyi*. Time marks, 50 Hz.

interpolation of small pulses at the beginning or end of the call, or the merging of pulses within the call.

The amplitude of the first pulse is always very small. The following pulses grow steadily in amplitude until a maximum is reached and diminish again toward the end of the call. The amplitude is not always modulated smoothly. The next-to-last pulse may be the largest of all, so that there is a sharp reduction of amplitude at the last pulse. Each pulse has the form of a strongly damped oscillation. The calls of the tree frog are very loud. Although no sound-pressure measurements have been made, their intensity is well illustrated by the fact that from a distance of 50-75 cm one finds them painful to the ear after only a few minutes of listening. With respect to frequency composition, the calls may be described as noise with most of the sound energy concentrated around 1-1.5 and 2-2.5 kHz (Fig. 5).

Because of these characteristics, the mating call is excellently suited to the role it plays in auditory communication; it is extraordinarily loud, sharply differentiated in pattern and highly stereotyped.

The number of pulses in the calls of the different species of European tree frogs varies (Fig. 4; Schneider 1974). *Hyla a. molleri* has slightly more (10.5 ± 0.5; n = 75) pulses per call than *Hyla a. arborea*, while there are fewer, on the average, in

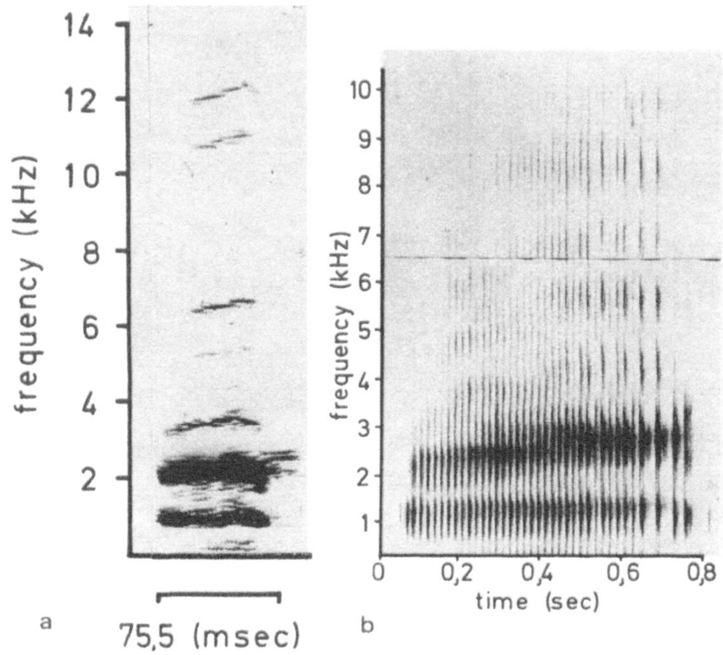

Fig. 5 a-b. Sonogram of a mating call of Hyla a. molleri (a) and Hyla meridionalis (b). Filter band with 45 Hz. From H. Schneider, Z. vergl. Physiol. 61, 369-385 (1968) and Oecologia (Berl.) 14, 99-110 (1974). By permission of Springer-Verlag, Heidelberg.

the calls of Hyla a. kretensis (8.2 ± 0.6; n = 67) and Hyla a. sarda (8.7 ± 1.2; n = 57). Only in the case of Hyla a. sarda does the reduction of pulse number result in a shortening of the call. Hyla a. savignyi, by contrast, has a call containing considerably more pulses; their number varies between 15 and 26. This relatively large range of variation results from interindividual differences. The number of pulses in the calls of a given tree frog is very uniform. The calls of Hyla meridionalis in the Camargue (Schneider 1968) and on Tenerife comprise the most pulses, 36 to 48. There are thus about five times as many pulses in its call as in that of Hyla a. arborea.

2. The Territorial Call

The territorial call is given comparatively often, particularly while the frogs are approaching the water in the evening twilight. When two frogs choose calling sites only a short distance (up to about 15 cm) apart, they give territorial calls. If one of the frogs has been established on its site for some time, it produces territorial calls as soon as the newcomer begins to call. If neither has become established, they both call. When a rival is calling, Mediterranean tree frogs, which give very long mating calls, often actually stop in the middle of such a call and proceed to give a territorial call instead (Fig. 6). Under the influence of the territorial calls the tree frogs move apart until the distance between them is about 30 cm. In southern Europe, where it often happens that many tree frogs move to the same body of water to call, I have seen smaller distances between calling frogs. Occasionally these frogs were separated by as little as 10 cm.

Since tree frogs tend so strongly to give territorial calls in response to the mating calls of their rivals, it seems likely that neighboring frogs tell how far apart they are by the loudness of the calls they hear. Experiments to test this are now in progress.

The territorial calls, like the mating calls, are composed of pulses varying in amplitude. Both calling rate and the number of pulses per call can change, so that the duration of the territorial calls is more variable than that of the mating calls. The territorial calls of the different European tree frogs closely resemble one another.

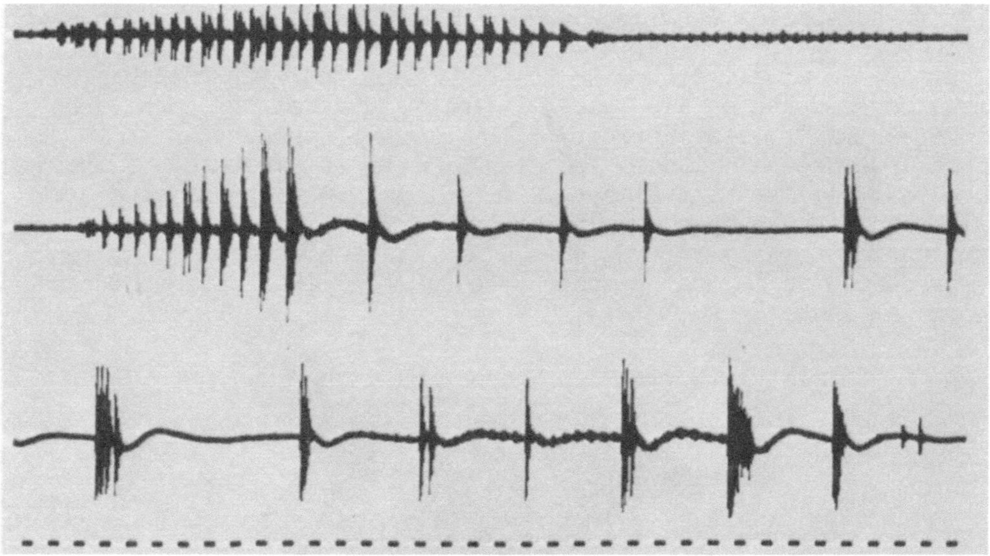

Fig. 6. Hyla meridionalis: the mating call of a neighboring rival (top) is answered by an established frog, which breaks off the mating call it has begun (middle) to give a territorial call (middle and bottom). Time marks, 50 Hz.

3. The Release Call

I suspected that Central European tree frogs are capable of giving this call but have not yet been able to prove that they do, since the territorial calls have the effect of keeping the males so far apart that one ordinarily never seizes and tries to clasp another. In Southern Europe, however, the density of males at the mating ground is so great that they do come into direct contact. In such an encounter the clasped male emits a series of release calls. These calls are very similar to territorial calls.

4. The Distress Call

This striking call is relatively uncommon. It is a piercing cry that varies in duration, intensity and pitch (Fig. 7). One can hear it occasionally when calling frogs are caught.

III. THE EFFECT OF TEMPERATURE ON THE MATING CALLS

The influence of temperature on the mating calls of the European tree frogs has been a focal point of my research in recent years. Collaboration with E. Nevo has made it possible to include Hyla a. savignyi in the comparison of mating calls (Schneider and Nevo 1972). Within the range of distribution of each species or subspecies I selected one suitable population, and in each I recorded the mating calls of up to 85 frogs. The recordings were made at widely differing temperatures; wherever possible, these covered the entire range of temperatures at which the tree frogs exhibit calling activity. The value of such experiments depends upon the method used to determine the temperature. Temperature measurements were always made next to or on the calling animals, since the temperature does not change uniformly in space or time during a daily calling phase, and it is not uncommon to find a difference of several degrees at a given time at different places around a pond.

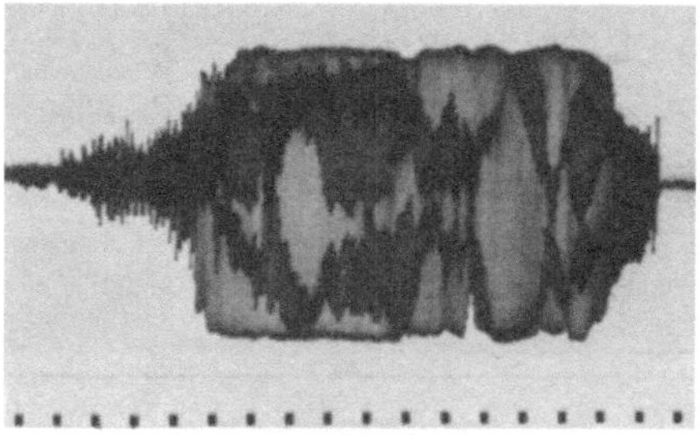

Fig. 7. Hyla a. savignyi: oscillograph of a distress call. Time marks, 50 Hz. From A. Knorr, 1976.

It is the temperature of the air that is most critical, since the frogs are almost completely out of the water while calling. Only when the air temperature falls very quickly, producing a sudden large difference between air and water, is the frog's body at a somewhat higher temperature than the air.

Under the influence of rising temperature there is a shortening of both calls and intervals of all the tree frogs investigated (Fig. 8), so that the number of calls per unit time increases.

All the regressions derived for the data are described by linear equations. As temperature rises so does the fundamental frequency of the calls, but since the frequency spectrum is so broad this effect is not very apparent to the ear.

As determined by the computed regression lines, the calls of the Central European tree frog average 94.95 msec in duration at 10° C and 64.95 msec at 20° C (Fig. 9 and Table 1). Thus a temperature increase of 10° C shortens the calls by about one-third and halves the intervals (Fig. 10; Schneider 1967).

Although the tree frog living on Crete is a separate subspecies and the straight-line distance between the two populations studied is about 1800 km, the mating calls of these two tree frogs are essentially identical in both call duration (Fig. 11) and interval length (Fig. 12; Schneider 1974). For example, the calculated difference in call duration at 20° C was 0.98 msec and at 10° C only 0.32 msec (cf. Table 1).

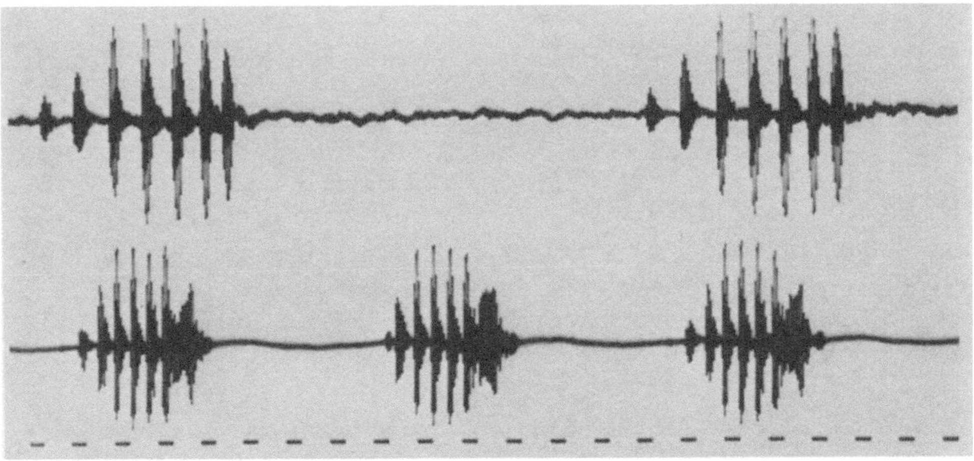

Fig. 8. Oscillographs of mating calls from two series, one (above) produced at 9° C and the other at 22.5° C. Time marks, 50 Hz. From H. Schneider, Praktikum der Verhaltensforschung, Stuttgart, 1971. By permission of Gustav Fischer Verlag, Stuttgart.

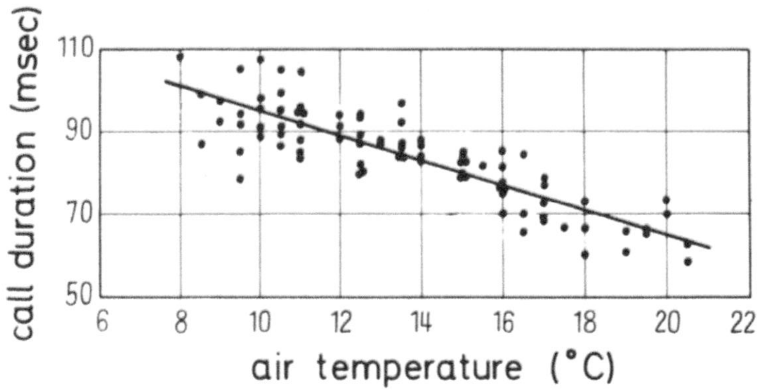

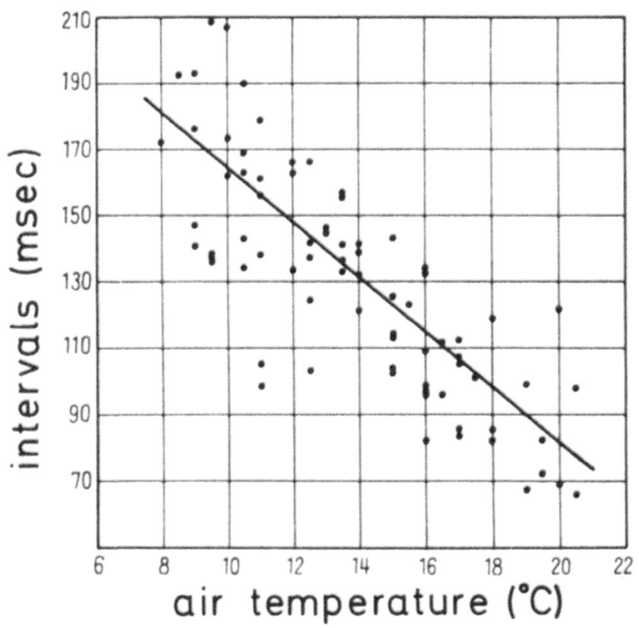

Figs. 9 and 10. Hyla a. arborea: the effect of temperature upon call duration (Fig. 9) and the intervals between calls (Fig. 10). From H. Schneider, Z. vergl. Physiol. 57, 174-189 (1967). By permission of Springer-Verlag, Heidelberg.

In the case of Hyla a. molleri the differences are rather greater. At 10° C the calls are longer (98.69 msec) on the average than those of Hyla a. arborea, but at 20° C they are shorter (53.79 msec). The regression line computed for Hyla a. molleri (Fig. 11 and Table 1), accordingly, has a greater slope. The t-test for difference between the regression coefficients b_1 and b_2

Table 1. Analysis of the effect of temperature on calling in European Hylids: Regression equations and results of the F-tests (y = dependent variables, x = temperature). After Schneider, 1974.

	N	Regression equation	F-test
Hyla a. arborea			
call duration	84	y = 124.95 - 3.00x	234.76***
intervals	84	y = 239.90 - 8.02x	117.55***
Hyla a. sarda			
call duration	71	y = 68.91 - 0.79x	78.33***
intervals	71	y = 151.48 - 2.89x	44.58***
Hyla a. kretensis			
call duration	67	y = 123.33 - 2.87x	255.15***
intervals	66	y = 228.52 - 6.87x	105.61***
Hyla a. molleri			
call duration	75	y = 143.59 - 4.49x	297.55***
intervals	73	y = 270.17 - 9.55x	49.41***
Hyla a. savignyi			
pulses/call	81	y = 21.42 - 0.10x	5.36*
call duration	83	y = 186.80 - 2.59x	61.89***
intervals	83	y = 256.89 - 4.18x	14.86***
Hyla meridionalis (Camargue)			
pulses/call	75	y = 47.05 - 0.51x	11.74***
call duration[a]	35	y = 768.10 - 25.95x	626.77***
intervals[a]	24	y = 3406.63 - 135.97x	17.62***
Hyla meridionalis (Tenerife)			
call duration	83	y = 613.32 - 12.10x	9.47**
intervals	83	y = 6601.60 - 302.60x	56.49***

[a]Calculations based on data of a single male.
***Significant at $p \leq 0.001$; **significant at $p \leq 0.01$; *significant at $p \leq 0.05$.

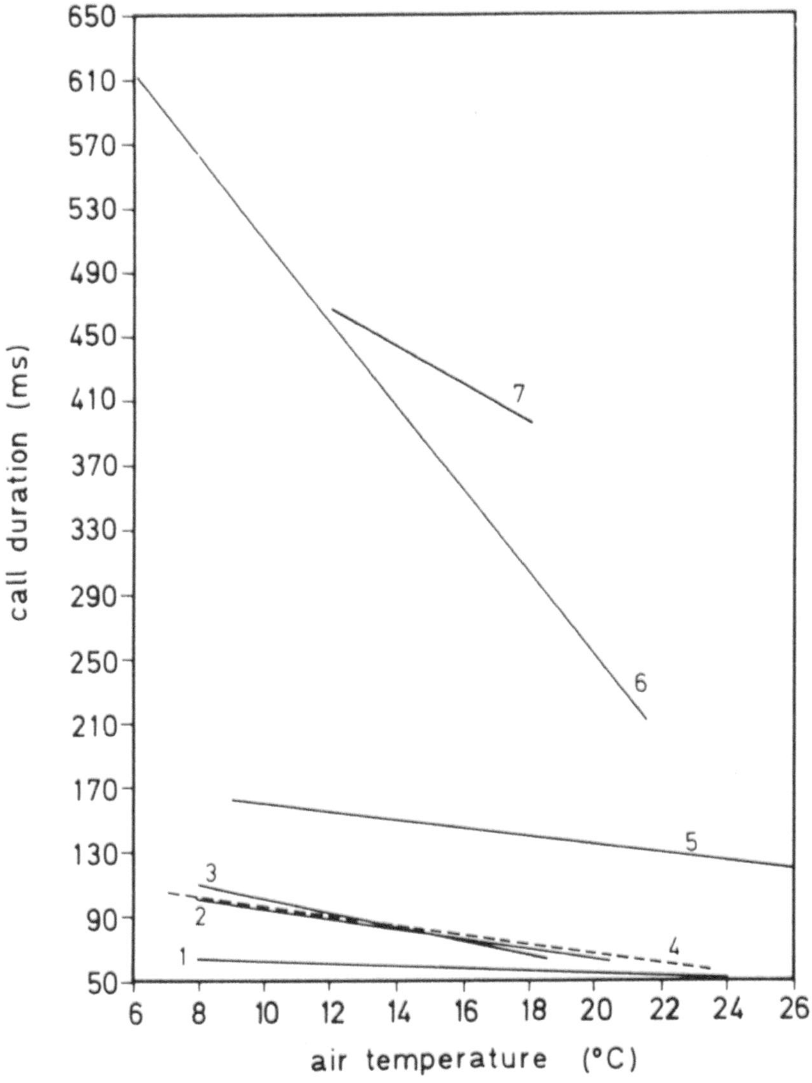

Fig. 11. The effect of temperature upon call duration in the various tree frogs studied. 1, *Hyla a. sarda*; 2, *Hyla a. arborea*; 3, *Hyla a. molleri*; 4, *Hyla a. kretensis*; 5, *Hyla a. savignyi*; 6 and 7, *Hyla meridionalis*. From H. Schneider, Oecologia (Berl.) 14, 99-110 (1974). By permission of Springer-Verlag, Heidelberg.

($b_1 \hat{=}$ *Hyla a. molleria*; $b_2 \hat{=}$ *Hyla a. arborea*) is highly significant; $t = 4.534$. Computations of confidence interval using the F-statistic for probability level 0.05 show a statistically significant,

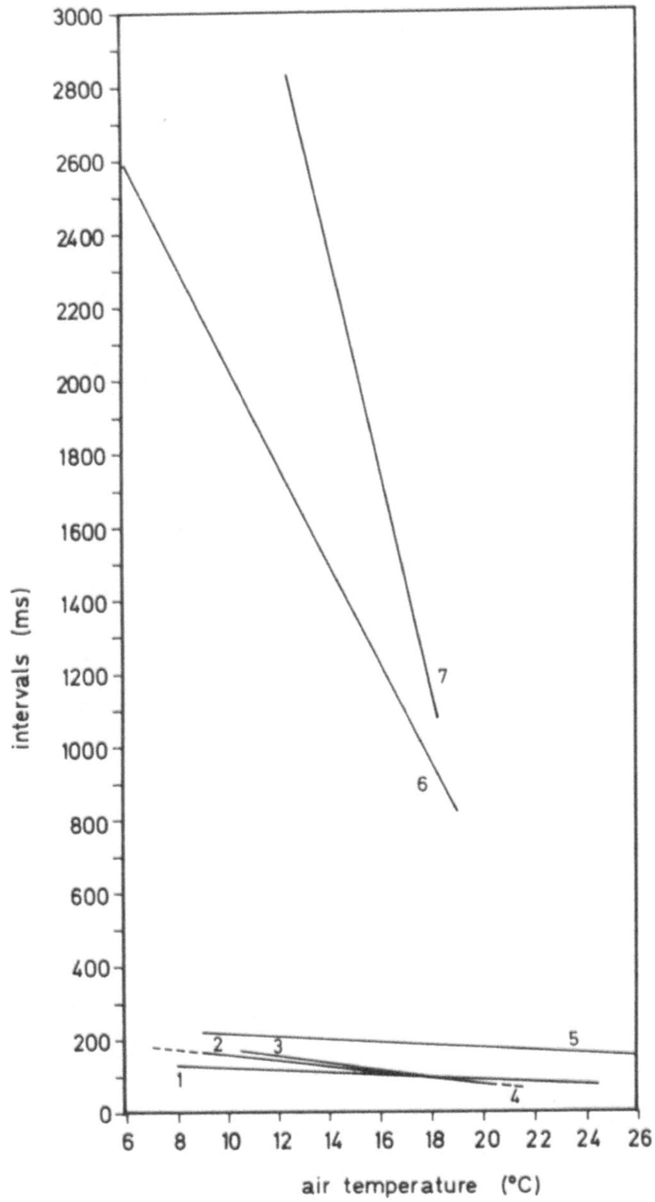

Fig. 12. The effect of temperature upon interval length in the various tree frogs studied. 1, *Hyla a. sarda*; 3, *Hyla a. molleri*; 4, *Hyla a. kretensis*; 5, *Hyla a. savignyi*; 6 and 7, *Hyla meridionalis* From H. Schneider, Oecologia (Berl.) 14, 99-110 (1974). By permission of Springer-Verlag, Heidelberg.

though very small, difference in the duration of the calls of the two tree frogs only at air temperatures of less than 10.5° C and more than 15° C. The intervals (Fig. 12 and Table 1) are slightly longer than those of Hyla a. arborea at temperatures below 15° C, but not at higher temperatures. Despite the statistical significance of these differences, one may say that the mating calls of Hyla a. molleri and Hyla a. arborea resemble one another very closely.

The shortest calls and intervals are those of Hyla a. sarda. As an extension of my 1974 observations I have since recorded calls at temperatures around 24° C. The influence of temperature is expressed in just the same way in this range. Since the effect of temperature on the calls and intervals of Hyla a. sarda is less than on those of the other frogs, it is only in the lower temperature range that a clear difference exists between the calls of Hyla a. sarda and Hyla a. arborea and hence those of Hyla a. kretensis and Hyla a. molleri. The confidence intervals ($F_{0.05}$) for interval length overlap at temperatures of 14.5° C or more, and those for the duration of the call begin to overlap at 18° C.

In accordance with the greater number of pulses in the call of Hyla a. savignyi, its duration is longer than that of the call of Hyla a. arborea (Schneider and Nevo 1972), and the intervals are also longer (Figs. 11 and 12 and Table 1). Nevertheless, there is a close correlation between the mating calls of these two frogs in that the computed regression lines are nearly parallel; the effect of temperature is the same with respect to both the calls and the intervals (Schneider 1974).

The mating call of the Mediterranean tree frog (Hyla meridionalis) is not only longer than those discussed so far, but is also more subject to the influence of temperature. Both of these properties suggest that this tree frog is a species distinct from the others. This is of interest because Boettger, who examined some preserved specimens from Tenerife in 1874, regarded the Mediterranean tree frog as a subspecies (Hyla arborea meridionalis) of the Central European species. As early as 1884 Héron-Royer noted certain differences between the two, as a result of which he renamed the Mediterranean form Hyla barytonus to call attention to the voice as its most prominent characteristic. Mertens and Müller (1940, 1966) and Mertens and Wermuth (1960), however, gave the Mediterranean frog the status of a subspecies. In 1967 Paillette pointed out that at 16° C the mating call of the Mediterranean tree frog is about five times as long as that of Hyla a. arborea. Detailed investigation (Schneider 1968) has shown that the calls of the Mediterranean tree frog in the Camargue average 612.4 msec in duration at 6° C and only 197.2 msec at 22° C (Fig. 11). The repetition rate is very low, since at 6° C the intervals last

2590.8 msec on the average, as compared with 823.2 msec at 19° C (Fig. 12). These interval values are computed from an equation derived after the original work was published in 1968; this new equation was published by Schneider (1974) and is given in Table 1, but the corresponding curve was not included in Figure 11 of the 1974 publication.

An increase in temperature also decreases the number of pulses in each call of Hyla meridionalis, from an average 43.99 pulses/call at 6° C to 36.85 at 20° C (Fig. 13). The number of pulses in the call of Hyla a. savignyi is reduced by increased temperature, although the change is slight. A 10° C rise in temperature lowers the average number of pulses in the call by one. Temperature does not affect the number of pulses in the mating calls of the other tree frogs (Fig. 13).

The European tree frogs closely resemble one another in both size and coloration. For this reason, the mating call is an important distinguishing characteristic. The call of the Mediterranean tree frog differs considerably from that of the Central European form. According to Flindt et al. (1968), the serum albumins of these two frogs also differ. Near Arcachon both Mediterranean and Central European frogs are found, but no hybrids are produced (Paillette 1967). All these findings justify the conclusion that the Mediterranean form is a separate species.

This research is currently being extended by a study of the mating calls of the tree frog on Tenerife. Because the temperature on Tenerife varies little throughout the day, its effect upon the mating calls is known for only a small range of temperatures. The results show quite clearly that both the calls (Fig. 11 and Table 1) and the intervals (Fig. 12 and Table 1) are similar in duration to those of the tree frog native to southern France, although evaluation of the confidence intervals shows that there are differences. I am cautious about accepting these differences until data have been obtained for other temperature ranges.

From these results I conclude that Hyla meridionalis is not a close relative of Hyla arborea. Since the former species lives in North Africa as well as in southern France and on Tenerife, I infer that its nearest relatives are frogs native to Africa.

IV. THE SOUND-PRODUCING APPARATUS

1. The Larynx

The specialization of the tree frog for calling is reflected in the fact that it has the largest larynx, in comparison with its

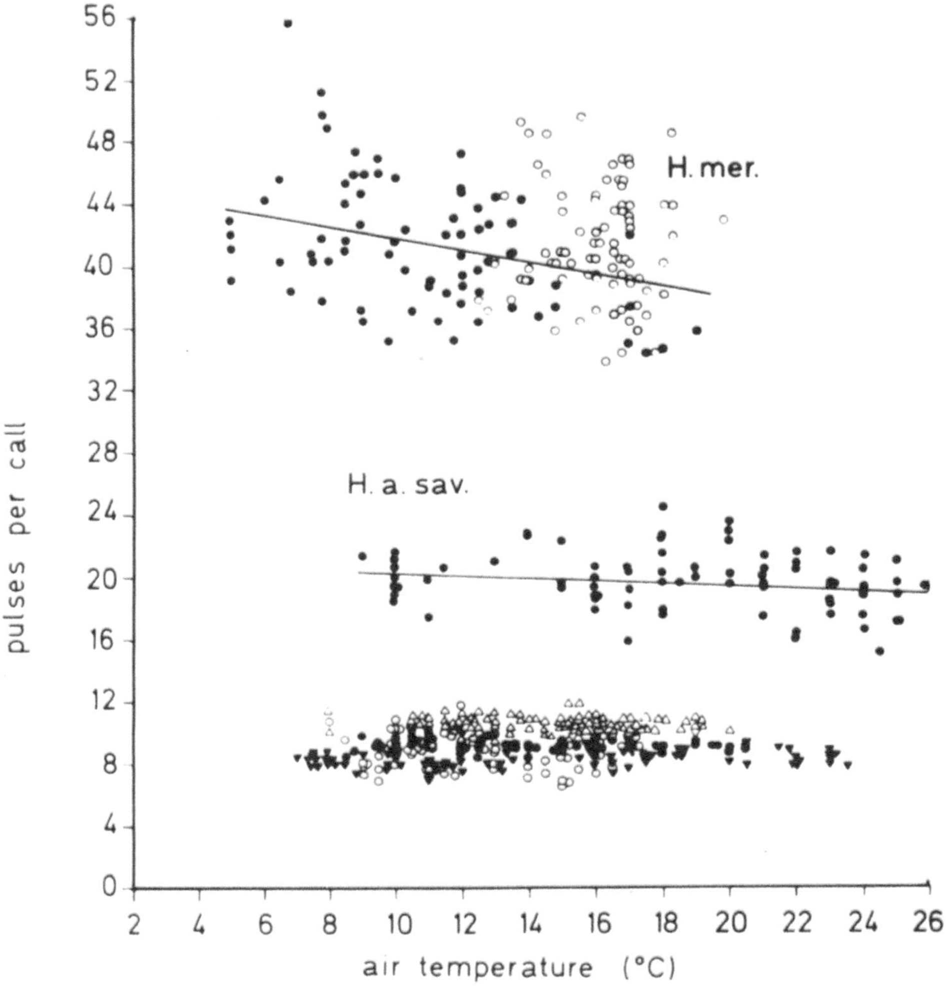

Fig. 13. The effect of temperature upon the number of pulses per call in the various tree frogs studied. H. mer., *Hyla meridionalis*: ● Camargue, o Tenerife; H. a. sav., *Hyla a. savignyi*; ● *Hyla a. arborea*; o *Hyla a. sarda*; ▼ *Hyla a. kretensis*; Δ *Hyla a. molleri*. From H. Schneider, Oecologia (Berl.) 14, 99-110 (1974). By permission of Springer-Verlag, Heidelberg.

body size, of all the Central European anurans; moreover, it is only in the male that the larynx is so well developed. The larynx of a 48-mm-long male *Hyla arborea* measures 9.5 mm in its dorsoventral dimension. That of the female is only half or even one-third as large.

The skeleton of the larynx consists of a ring-shaped pars crico-trachealis and two large arytenoid cartilages (Schneider

1970). The pars crico-trachealis is attached to the two postero-medial processes of the hyoid cartilage; these ossified projections extend diagonally up and back to form a fork around the larynx (Fig. 14). The two arytenoid cartilages abut rostrally against the pars crico-trachealis. Along the posterior edges of the arytenoids lie the dorso-ventrally oriented ligamentous vocal cords (Fig. 15), attached to the arytenoids at either end and by way of a frenulum in the middle. Vocal cords are also developed in female frogs.

The musculature of the larynx is relatively massive and consists of the M. dilatator, the sphincter muscles posterior to it (M. sphincter anterior, M. sphincter posterior) and the M. hyolaryngeus (Fig. 16). These muscles open the aditus during respiration via the lungs and during calling, and tense or relax the vocal cords.

As comparative examination of Hyla meridionalis has shown (Schneider 1970), this form does not differ with respect to the

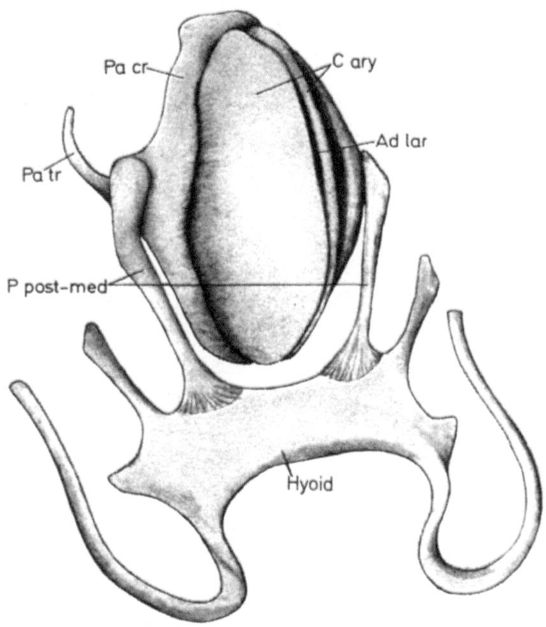

Fig. 14. Hyla a. arborea: the laryngeal skeleton of a male frog. Ad lar, aditus laryngis; C ary, arytenoid cartilage; P post-med, postero-medial process; Pa cr, Pa tr, pars cricoidea and pars trachealis of the crico-tracheal cartilage. From H. Schneider, Z. Morph. Tiere 66, 299-309 (1970). By permission of Springer-Verlag, Heidelberg.

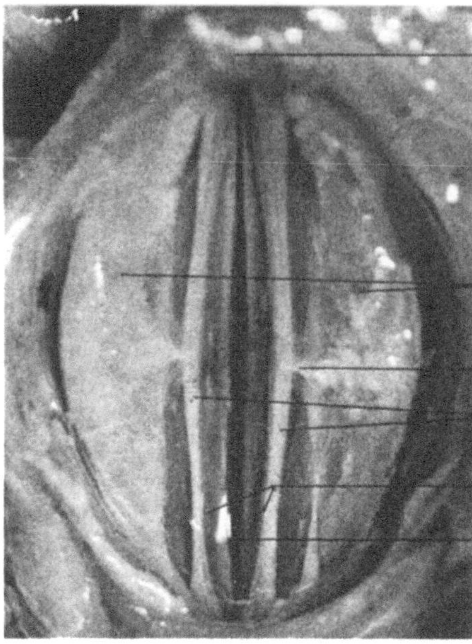

Fig. 15. Hyla a. arborea: caudal aspect of the larynx of a male, showing the vocal cords. Fre, frenulum; Li voc, vocal cords; Ri gl, rima glottidis; Ad lar, aditus laryngis; C ary, arytenoid cartilage; P post-med, postero-medial process; Pa cr, pars cricoidea. From H. Schneider, Z. Morph. Tiere 66, 299-309 (1970). By permission of Springer-Verlag, Heidelberg.

size of the larynx in relation to the body size, and differs little or not at all in structure of the laryngeal skeleton and musculature. The morphological differences between the larynx of the fire-bellied toad and that of the yellow-bellied toad (two species that are closely related), for example, are considerably greater (Schmid 1976).

2. The Fine Structure of the Laryngeal Muscles

Since Hyla a. arborea produces calls at a very high rate--6.82 calls per second at 20° C--the laryngeal muscles must operate rapidly. Recent studies (Eichelberg and Schneider 1973; Manz 1975) have shown that the laryngeal muscles of the male are superbly adapted to their task in both morphology and physiology. This becomes evident when these muscles are compared with a muscle of the hind limb and with the laryngeal muscles of the female.

The four laryngeal muscles of the male tree frog are similar in structure. The fibers are thin, as are the myofibrils; the mean diameter of the latter is 220 μm (Fig. 17). The sarcoplasmic

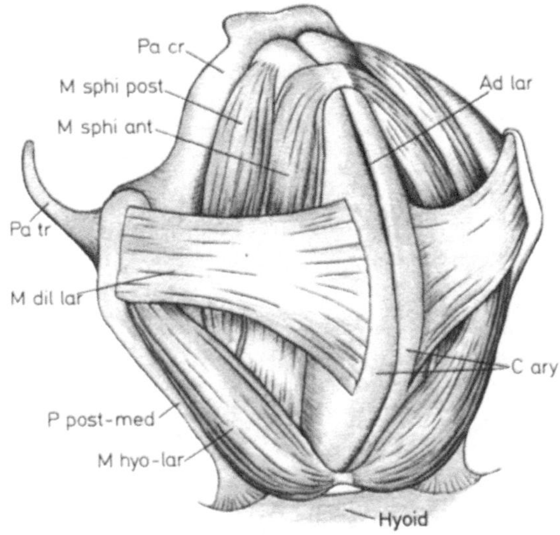

Fig. 16. Hyla a. arborea: the laryngeal musculature of the male. M dil lar, M. dilatator laryngis; m hyo-lar, M. hyo-laryngeus; M sphi ant, M. sphincter anterior; M sphi post, M. sphincter posterior. Other abbreviations as in Fig. 14. From H. Schneider, Z. Morph. Tiere 66, 299-309 (1970). By permission of Springer-Verlag, Heidelberg.

reticulum is massively developed. Triads regularly appear at the level of the Z lines and are distinguished by small central vesicles and large lateral vesicles. The very numerous mitochondria are usually aggregated in groups and may even form bands. The mitochondria are of conspicuous size and contain many cristae. All the laryngeal muscles are rich in fat. The lipoid spheres are usually immediately adjacent to the mitochondria. In addition to the lipoids, the laryngeal muscles of the male also contain considerable quantities of glycogen. The number of motor end plates is remarkably high; their substructure offers an example of the typical motor end plate arrangement (Fig. 18). The fold apparatus in the region of the muscle membrane is highly developed. In the axon the synaptic vesicles are very abundant, as are the mitochondria; the latter are noticeably smaller than those in the muscle fibers.

In order to see which of these properties represent specializations of the laryngeal muscles of the male tree frog, Eichelberg and Schneider (1973) compared them with those of the hind-foot extensor (M. plantaris longus). In the latter muscle the fibers are about ten times as thick as those of the laryngeal muscles, the

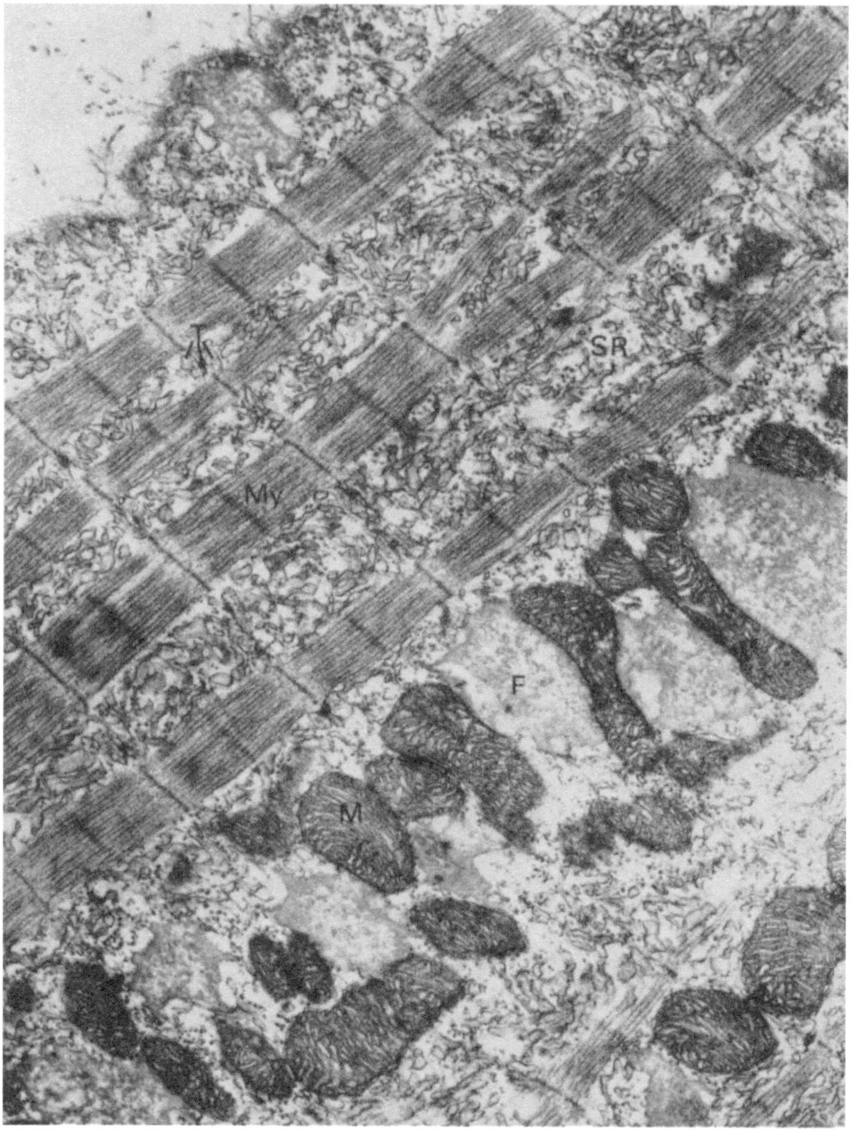

Fig. 17. Hyla a. arborea: longitudinal section of the M. dilatator of a male. F, fat; M, mitochondria; My, myofibrils; SR, sarcoplasmic reticulum; T, triad. Mag. 28,000X. From H. Eichelberg and H. Schneider, Z. Zellforsch. 141, 223-233 (1973). By permission of Springer-Verlag, Heidelberg.

myofibrils are three times as thick, and the number of motor end plates is much smaller, scarcely more than 1/25 of that of the laryngeal muscle (Fig. 19). The sarcoplasmic reticulum is sparse,

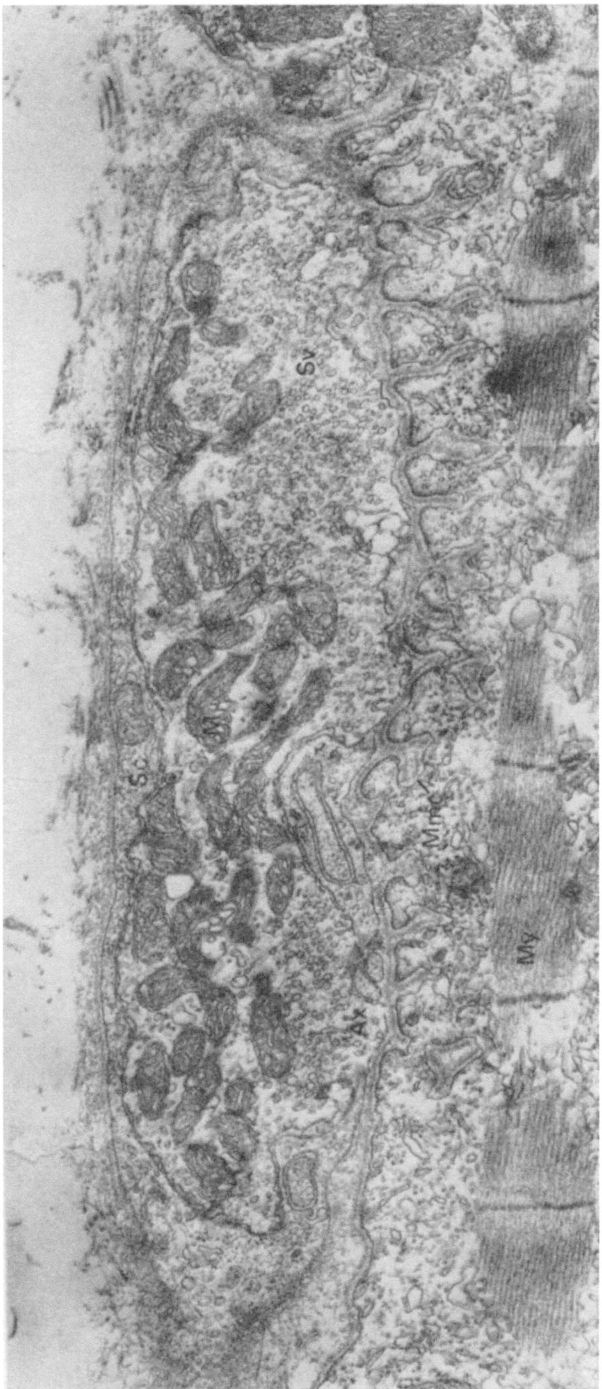

Fig. 18. *Hyla a. arborea*: motor end plate in a laryngeal muscle of a male. Ax, axon; M, mitochondrion; Mmc, muscle-membrane complex; My, myofibrils; Sc, Schwann cell; Sv, synaptic vesicle. Mag. 19,000X. Generously provided by H. Eichelberg.

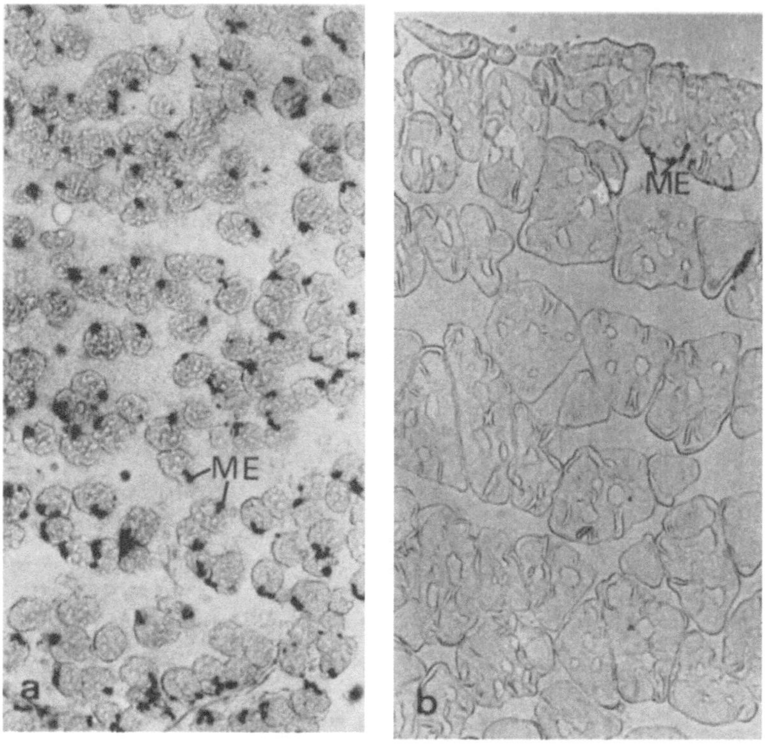

Fig. 19. Hyla a. arborea: frozen section through a laryngeal muscle (a) and the M. plantaris longus (b) of a male, stained to show the motor end plates (ME). Mag. 80X. From H. Eichelberg and H. Schneider, Z. Zellforsch. 141, 223-233 (1973). By permission of Springer-Verlag, Heidelberg.

and so are the mitochondria. Lipoids are almost entirely lacking (Fig. 20). Glycogen is present as an energy store, but there is considerably less of it than in the laryngeal muscles.

The special structure of the laryngeal muscles of the male tree frog is also revealed by comparison with the equivalent muscles of the female, which does not give mating calls (Eichelberg and Schneider 1974). The larynx of the female also has four pairs of muscles, but these are poorly developed. The two sphincter muscles in particular are usually difficult to locate. All the muscles have a glassy, transparent appearance. Three fiber types can be distinguished on the basis of differences in diameter: very thin fibers like those in the laryngeal muscles of the males, fibers of intermediate diameter, and large-diameter fibers like those in the limb muscle. These thick fibers are the most numerous and the very

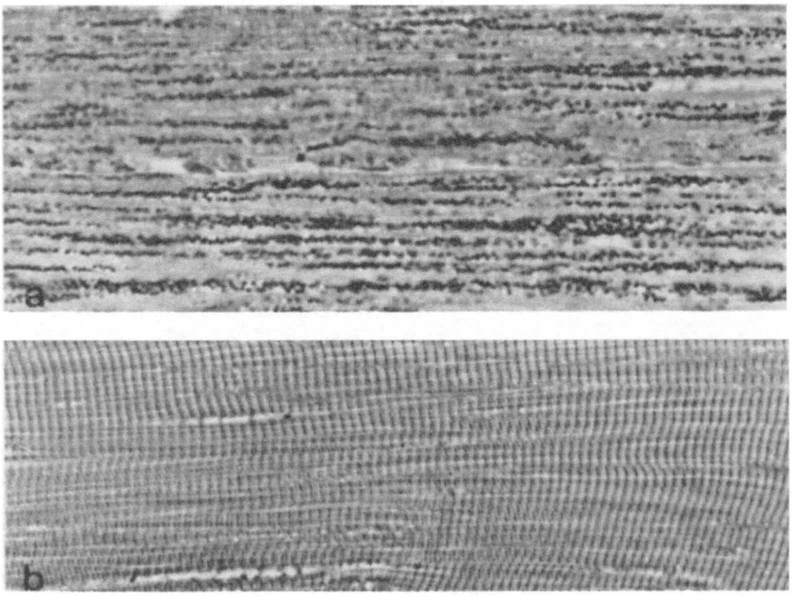

Fig. 20. Hyla a. arborea: semithin longitudinal section through a fat-containing M. sphincter anterior (a) and the fat-free M. plantaris longus (b). Mag. 1000X. From H. Eichelberg and H. Schneider, Z. Zellforsch. 141, 223-233 (1973). By permission of Springer-Verlag, Heidelberg.

thin fibers are least common; the fibers are present in roughly the proportions 9:3:1. Characteristics typical of the three types can also be found in the fine structure of the fibers. Only the very thin fibers contain large quantities of sarcoplasmic reticulum, mitochondria, motor end plates and lipoids. In the fibers of intermediate diameter there are fewer of these components, and their numbers are least in the largest fibers. For example, in a standard area of the thin fibers there were 724 motor end plates, whereas in the same area intermediate fibers contained 277 end plates and large fibers only 22 end plates.

The presence of lipoids throughout the laryngeal muscles of the male suggests that they do not represent fat stores, but rather a direct source of energy for the muscle. My colleagues Eichelberg and Obert (1976) confirmed that this is indeed true of the fire-bellied toad. Results obtained with the toad are likely to apply to the tree frog as well, since the fine structure of the laryngeal musculature is the same in males of the two anuran genera. Fire-bellied toads can be brought into continuous calling activity by administration of pregnant mare serum, so that it is possible to test whether the energy stores in the laryngeal muscles are depleted by prolonged calling. After thirteen days of calling activity a large fraction of the fibers in the laryngeal muscles is almost free of fat (Fig. 21). The breakdown of the fatty substances does not result from administration of the hormone to elicit calling, since other toads that received equal amounts of hormone but were prevented from calling by castration have laryngeal muscles containing the normal amount of fat. Breakdown of the fatty substances occurs only after calling has been continued for some time. After electrical stimulation of the M. dilatator for four hours at 80 stimuli per second (ambient temperature, 20° C) there is no detectable reduction in lipoid content, though the glycogen content is reduced.

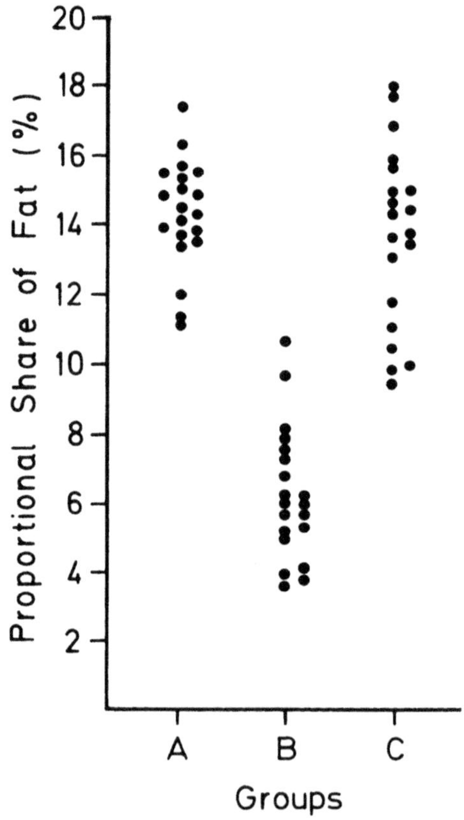

Fig. 21. Bombina bombina: the fat content in the M. dilatator laryngis in males of three experimental groups, as follows: A, no calling activity; B, after 13 days of hormone-elicited calling; C, treated with hormone but no calling (castrates). Each dot represents one random sample. From H. Eichelberg and H.-J. Obert, Cell Tiss. Res. 167, 1-10 (1976). By permission of Springer-Verlag, Heidelberg.

From these results Eichelberg and Obert concluded that the glycogen in the laryngeal muscles is sufficient only for relatively short periods of muscle activity and that it is supplemented during calling by mobilization of the fatty substances in the muscles.

3. The Fusion Frequencies of the Laryngeal Muscles

Analysis of their fine structure identifies the laryngeal muscles of *Hyla a. arborea* as fast muscles in every respect. Manz (1975) determined the stimulus frequencies at which the muscles go into tetanus. Since the activity of the muscles is strongly temperature-dependent, the experiments covered a range of temperatures.

The freshly excised larynx of a tree frog was placed in a dish of Ringer solution. The stimulus electrode was in contact with a muscle and the indifferent electrode dipped into the Ringer solution. A transducer converted the contractions of the muscles into voltage fluctuations that were recorded on an oscilloscope. The stimuli were square pulses lasting 1 msec, given in sequences lasting from 70 to 320 msec, depending on the frequency of the pulses. Fusion frequencies were determined for the temperature range 6-24° C in steps of 2° C. Manz tested the four laryngeal muscles of both males and females and compared these with the M. plantaris longus of the hind limb.

At 6° C the fusion frequencies of the laryngeal muscles of male tree frogs were between 30 and 50 Hz; with increasing temperature they rose rapidly, reaching 240-260 Hz at 24° C (Fig. 22). The regression curves computed for this range of temperatures are quadratic. Within the smaller range 8-22° C they are linear; this is the temperature range in which the frogs exhibit calling activity, with calls and intervals that are also linearly related to temperature. The curves for the four laryngeal muscles lie so close together that a common regression line can be derived (Fig. 23).

The fusion frequencies of the limb muscle are much lower than those of the laryngeal muscles. At 6° C 20 stimuli per second, and at 24° C 120 stimuli per second, suffice to put this muscle into a smooth tetanus (Fig. 23).

The four laryngeal muscles of the female tree frog are also physiologically uniform, but they are not very fast. Their fusion frequencies are even lower than those of the limb muscle. At 10° C fusion is achieved with 30 stimuli per second, while at 20° C only 80 stimuli per second are required (Fig. 23).

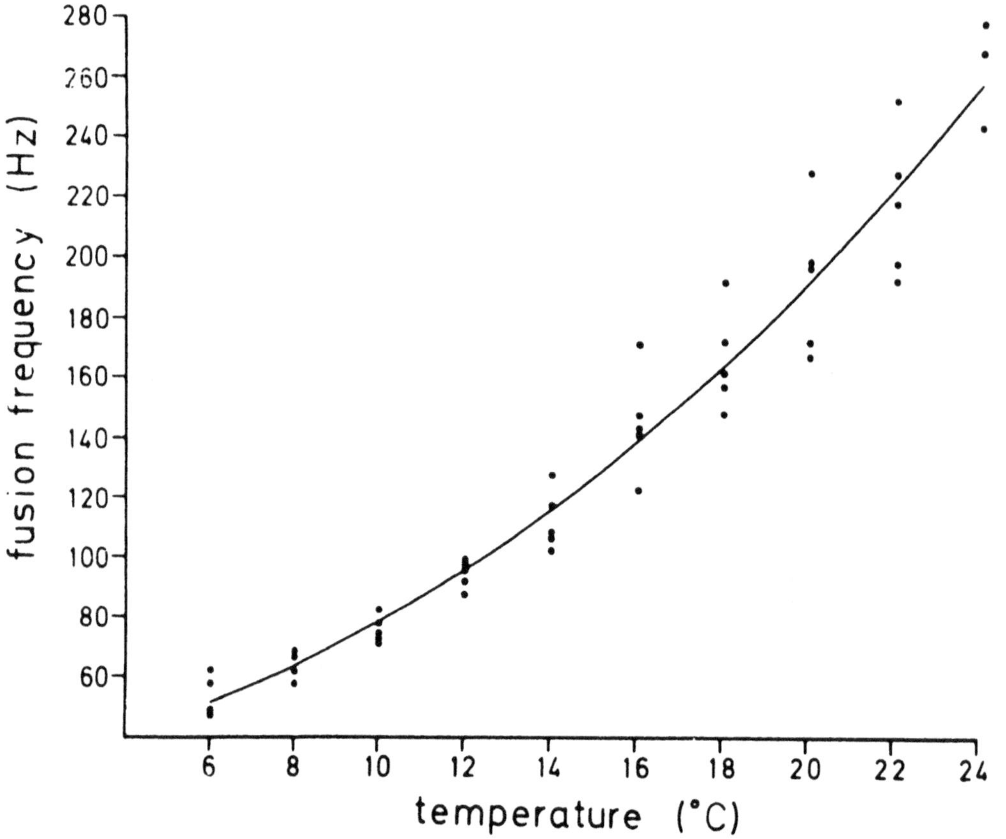

Fig. 22. Hyla a. arborea: the effect of temperature on the fusion frequency of the M. dilatator laryngis of a male. From R. Manz, Zool. Jb. Physiol. 79, 221-245 (1975). By permission of VEB Gustav Fischer Verlag, Jena.

In his experiments, Manz was chiefly concerned with the bioacoustics of the Anura, but the results are of considerable relevance in general. For the first time precise data have been obtained about the physiological properties of a number of morphologically different muscles of several anuran species, in particular as regards their temperature dependence.

V. THE CENTRAL NERVOUS MECHANISMS INVOLVED IN CALLING

The structure of the larynx enables male Hyla a. arborea to produce very brief mating calls in rapid succession. In view of the fact that the laryngeal structure is nearly the same in the Mediterranean tree frog as in the Central European form, whereas the mating calls differ greatly, the specificity of the mating

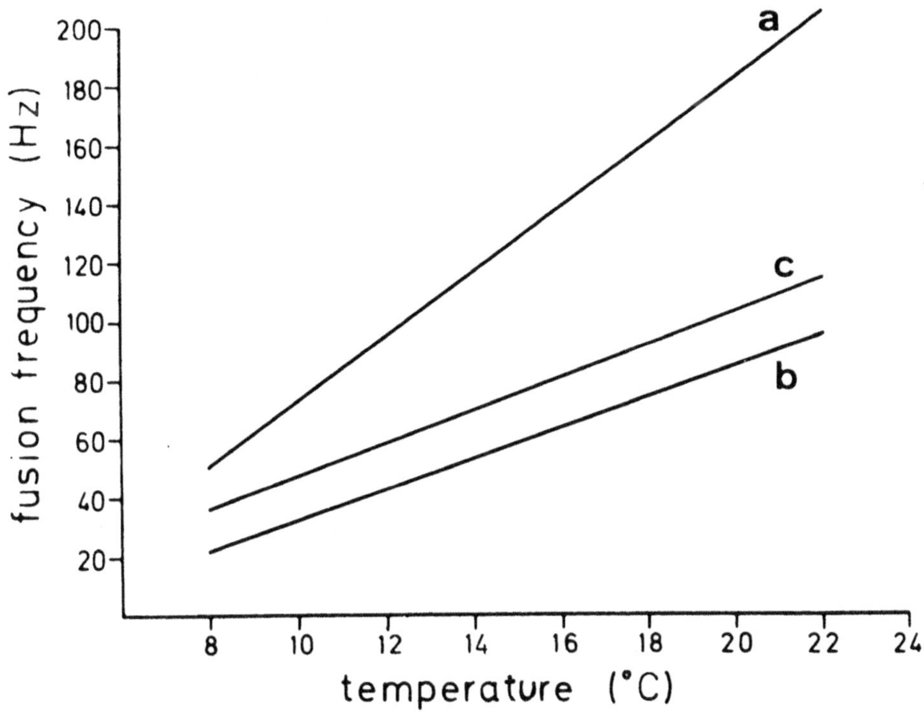

Fig. 23. Hyla a. arborea: the fusion frequencies of the laryngeal muscles of male (a) and female (b) frogs and of the M. plantaris longus of male and female frogs (c). From R. Manz, Zool. Jb. Physiol. 79, 221-245 (1975). By permission of VEB Gustav Fischer Verlag, Jena.

calls in this case must depend upon the controlling mechanisms in the central nervous system. The situation in tree frogs, then, is different from that in Bufo viridis viridis. Central European Bufo v. viridis (Lörcher and Schneider 1973) give mating calls of higher fundamental frequency than the toads of the same species living in Israel (Nevo and Schneider 1976). The toads of these two ranges differ in body length (the Central European Bufo v. viridis is smaller), and the difference in fundamental frequency is thus due to the difference in size of the larynx, which depends upon body size.

The central nervous mechanisms underlying calling by the tree frog are currently being investigated. Electrical stimulation of the brain can provide information as to the function and anatomical location of the centers involved. In preparation for such experiments Knorr and Schneider (1975) constructed an atlas of the brain of Hyla a. arborea, using a coordinate system by which all points in the brain can be defined and located for stimulation.

For the experiments themselves Knorr (1976) chose Hyla a. savignyi. The stimulating electrodes were steel pins sharpened electrolytically to a tip diameter of 30-50 μm and insulated to the tip with Insl-X. Electrodes were chronically implanted in the brains of 68 free-moving tree frogs. Fourteen other animals were stimulated by electrodes that could be moved vertically during the experiment; these frogs were fixed in a head-holder and immobilized by transection of the limb nerves (for details of this difficult method see Knorr 1976). The electrodes were advanced by a micromanipulator in steps of 1 μm. The metal plate on which the frogs rested served as the indifferent electrode.

The stimuli were monophasic negative square pulses of 0.5-3.5 V, 5 msec in duration, providing a current of 0.7-2.0 μA; they were given at a frequency of 100 Hz. The sounds elicited by these stimuli were recorded on magnetic tape and analyzed via oscillograms and sonograms. The sites of stimulation in the brain were identified histologically. The spots were marked at the end of stimulation by applying a positive voltage to the electrode so as to pass direct current of about 0.5 μA for 15 sec. Fe^{3+} ions were thus deposited in the tissue surrounding the electrode tip, so that the area could be stained by the Berlin-blue reaction with 2 per cent potassium ferrocyanide.

In his investigation Knorr concentrated on stimulation of the preoptic region. This area seemed most promising, particularly since Schmidt successfully stimulated there in the brains of Rana pipiens and Bufo americanus . On the basis of these results and of recordings from the midbrain and medulla of Rana pipiens, Hyla cinerea and Hyla versicolor, Schmidt developed a model of the mechanisms underlying calling (Schmidt 1965, 1966, 1968, 1971, 1974).

The preoptic region is also the critical area for control of the mating and release calls in Hyla a. savignyi. It is the most anterior and most ventral region of the diencephalon (Fig. 24). It is bounded by the telencephalon beneath the anterior commissure; the third ventricle lies dorsal to it, and its caudal boundary is the optic chiasm. It is subdivided into three large nuclei: the anterior preoptic nucleus, with numerous large somata, is followed by the smaller magnocellular nucleus and finally by the posterior preoptic nucleus, which also contains many somata.

Electrical stimulation anywhere in the nuclear area of the preoptic region, particularly the anterior preoptic and magnocellular nuclei, as well as in certain parts of the hypothalamus, elicits a behavior pattern that Knorr termed "readiness to call" and that precedes the actual calling activity. It appears with a latency of at least 40 sec. The first sign of readiness to call

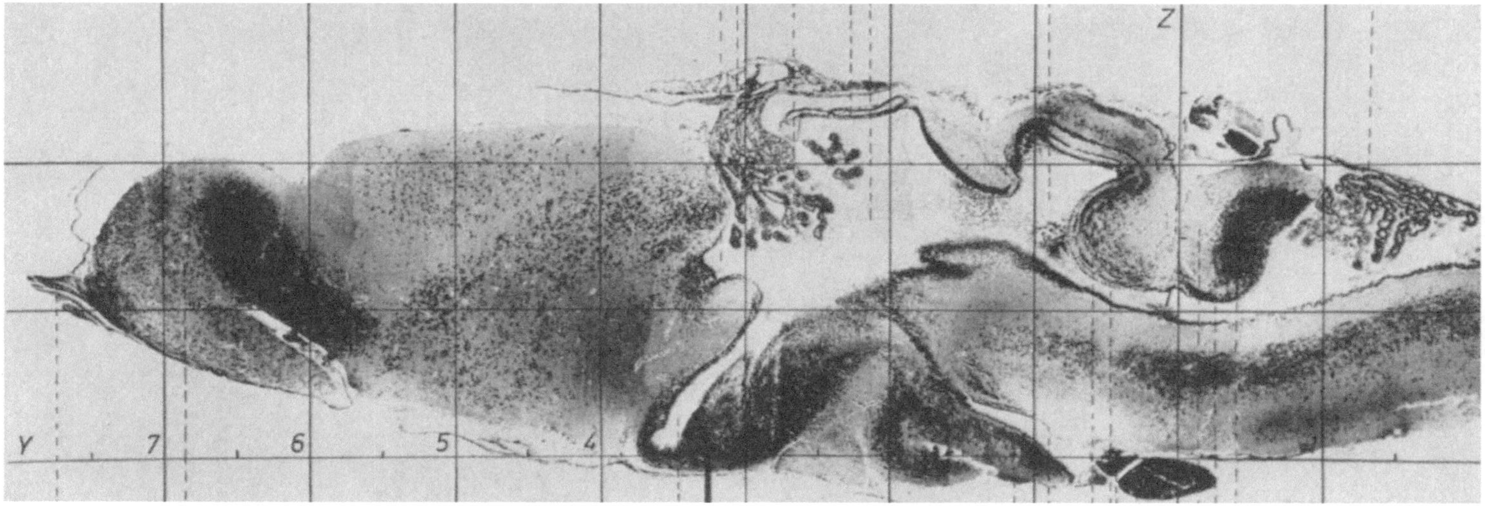

Nu. praeopt.

Fig. 24. Hyla a. arborea: longitudinal section through the brain of a male at the midline. Nu praeopt., preoptic nucleus. From A. Knorr and H. Schneider, Bonner zool. Beitr. 26, 370-382 (1975). By permission of the publisher.

is an acceleration of oral respiration; the tree frog becomes restless, adopts the calling posture and fills the vocal sac and lungs with air. If the mating calls of their own species are played back to animals in this state, they respond by giving mating calls themselves. Some of the frogs turn toward the sound source and try to approach it. At all the loci shown in Fig. 25, stimulation produced readiness to call as its sole result; even when stimulation was prolonged it did not elicit mating calls. Such a state of readiness also appears under natural conditions during the mating season.

Stimulation in the preoptic nuclei can result also in actual calling activity. The mating call was elicited by stimulation at twelve points in the brain. Seven of these are in the anterior preoptic nucleus, three in the posterior dorsal part of the posterior nucleus, and two between these nuclei, in the vicinity of the fibers emerging from the anterior preoptic nucleus (Fig. 26). No calls were produced by stimulation in the magnocellular nucleus.

The latency between stimulus onset and calling amounted to some minutes. During this time the frog first entered the state of readiness, which became progressively more intense until the mating calls were given (Fig. 27). Electrically elicited mating calls rarely occurred in isolation; they were ordinarily produced in series, as they are under natural conditions, especially when the anterior preoptic nucleus was stimulated. Depending on the site stimulated, the series may comprise a small (2-3), intermediate (6-9), or large (17-22) number of calls. All the calls elicited at a particular site have the same structure, whereas those associated with different stimulus sites vary in structure. Most of the calls are composed of distinct pulses with the same duration as in the natural mating call, but the gradual rise and fall in amplitude characteristic of the natural call is often lacking (Fig. 28). The frequency spectrum is noisy, extending to 8 kHz with energy maxima between 1.0 and 2.5 kHz. This, too, is in good agreement with the spectrum of the natural mating call. The intervals between the calls at room temperature are usually between 190 and 200 msec long.

When stimulation is stopped, calling activity first comes to an end, and then the state of readiness gradually disappears. If the end of the stimulus falls during a series of calls, the frog continues to call for a while. When stimulation is resumed, readiness to call builds up again and eventually gives rise to calling. Depending on the site of stimulation readiness may remain high, so that additional series of calls are produced spontaneously. While calling, the stimulated tree frogs maintain the characteristic calling posture.

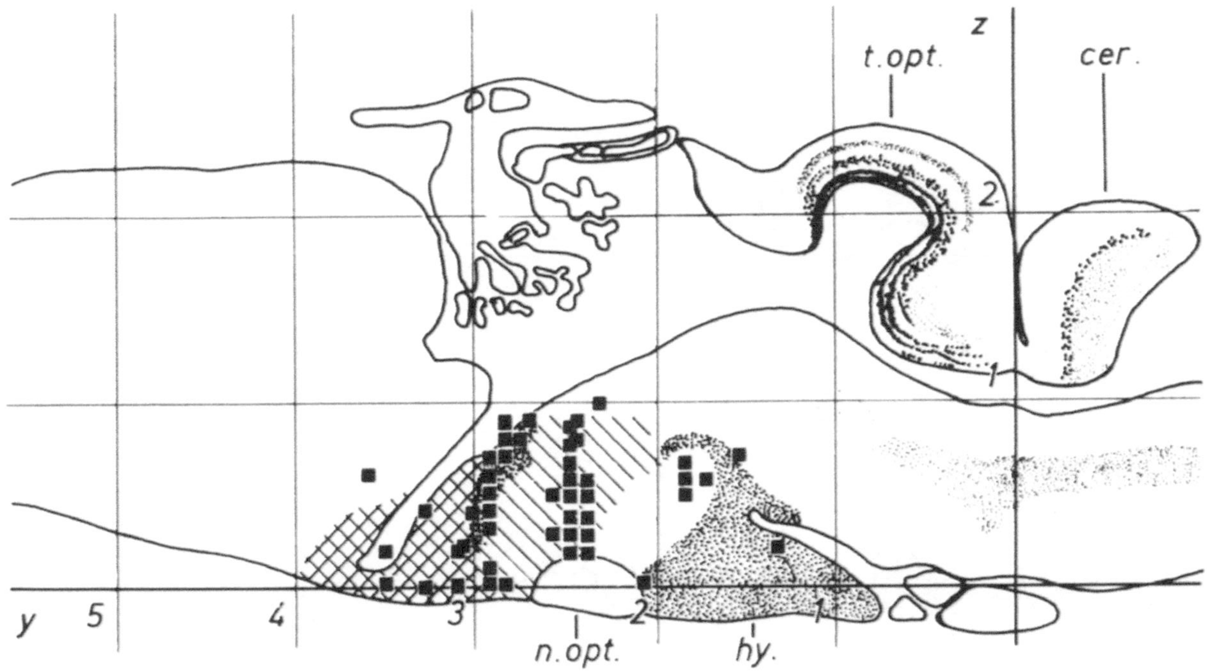

Fig. 25. Hyla a. **savignyi**: diagrammatic longitudinal section through the brain ca. 100 µm from the midline. Electrical stimulation at the indicated sites produced a state of readiness for calling. XX anterior preoptic nucleus; ⁘ magnocellular nucleus; \\\ posterior preoptic nucleus. From A. Knorr, dissertation, 1976.

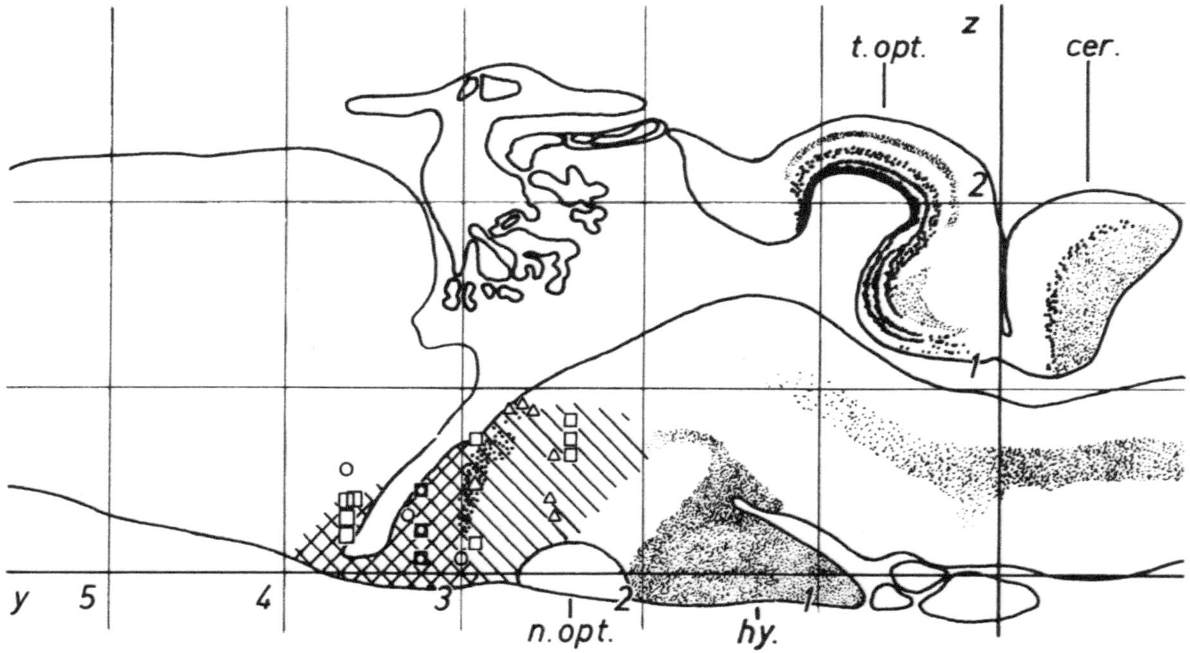

Fig. 26. Hyla a. savignyi: diagrammatic longitudinal section through the brain of a male, ca. 100 µm from the midline. Electrical stimulation at the indicated sites elicited the following behavior: mating calls (□), egg-laying posture (○), mating calls plus egg-laying posture (⊡), mucus production (Δ). From A. Knorr, dissertation, 1976.

Fig. 27. Hyla a. savignyi: calling posture of a male, elicited by electrical stimulation. From A. Knorr, dissertation, 1976.

By stimulating the brains of male tree frogs electrically, Knorr was also able to elicit a posture that Eibl-Eibesfeldt (1954) observed among female tree frogs during copulation and termed the "signal position." Just before depositing eggs, the female arches her back deeply downward--a sign to the male that he should discharge the sperm. This signal position resulted from stimulation at four points, three of them in the anterior preoptic nucleus and the fourth immediately in front of it. All these sites also elicited prolonged calling activity, and an increase of 0.3 V in the stimulus was sufficient to produce the signal position. Stimulation

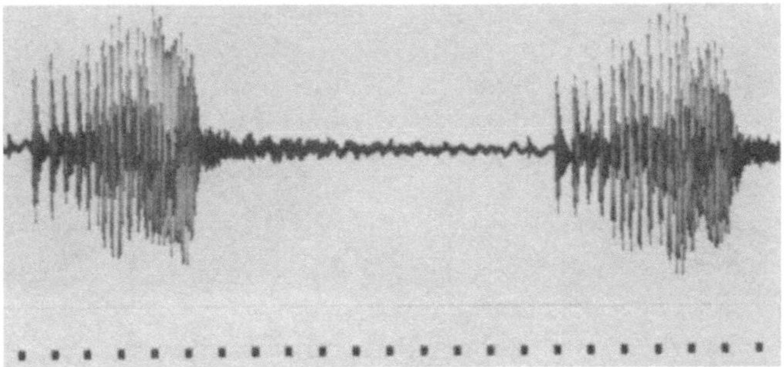

Fig. 28. Hyla a. savignyi: oscillograph of two mating calls elicited by electrical stimulation at a single site in the preoptic region. From Knorr, dissertation, 1976.

of the same loci in the brains of females also caused them to adopt the position for egg-laying. In response to voltages only slightly higher the female produced quiet vocalizations in series that resembled the mating calls of the males. The low intensity of these sounds is probably associated with the fact that the female larynx is less well equipped for sound production.

Calls and calling behavior are very conspicuous and distinctive characteristics of _Hyla arborea_ that are based on both the morphological and the physiological organization of the animal. The form they take reflects the great significance of acoustic signals in tree-frog communication. Though it may lie far in the future, a thorough understanding of communication by these animals is a goal worth striving for.

REFERENCES

Blankenhorn, H. J. 1972. Meteorological variables affecting onset and duration of calling in _Hyla arborea_ L. and _Bufo calamita_ Laur. _Oecologica_ (Berl.) _9_:223-234.

Boettger, O. 1874. Reptilien von Marocco und von den canarischen Inseln. II. Amphibia. _Abh. Senck. nat. Ges._ _9_:186-190.

Eibl-Eibesfeldt, I. 1953. Vergleichende Verhaltensstudien an Anuren. _Z. Tierpsychol._ _9_:338.

Eichelberg, H., and H. Schneider. 1973. Die Feinstruktur der Kehlkopfmuskeln des Laubfrosches, _Hyla arborea arborea_ (L.) im Vergleich zu einem Skelettmuskel. _Z. Zellfrosch._ _141_:223-233.

Eichelberg, H., and H. Schneider. 1974. The fine structure of the larynx muscles in female tree frogs, _Hyla a. arborea_ L. (Anura, Amphibia). _Cell Tiss. Res._ _152_:185-191.

Eichelberg, H., and H. J. Obert. 1976. Fat and glycogen utilization in the larynx muscles of fire-bellied toads (_Bombina bombina_ L.) during calling activity. _Cell Tiss. Res._ _167_:1-10.

Flindt, R., H. Hemmer and R. Jaeger. 1968. Das Serumeiweißbild mitteleuropäischer Anuren. _Zool. Jb. Physiol._ _74_:155-163.

Heinzmann, U. 1970. Untersuchungen zur Bio-Akustik und Ökologie der Geburtshelferkröte, _Alytes o. obstetricans_ (Laur.). _Oecologia_ (Berl.) _5_:19-55.

Héron-Royer. 1884. Note sur une forme de rainette nouvelle pour la faune française (_Hyla barytonus_). _Bull. Soc. zool. France_ _9_:221-238.

Heusser, H. 1970. Paarungs- und Befreiungsruf der Erdkröte, Bufo bufo bufo (L.). Z. Tierpsychol. 27:894-898.

Knorr, A. 1976. Reizphysiologische Untersuchungen über die zentralnervöse Kontrolle des Paarungsrufes und einiger weiterer Verhaltensweisen beim Laubfrosh Hyla arborea savignyi Audouin. Dissertation, Univ. Bonn.

Knorr, A., and H. Schneider. 1975. Beitrag zur Morphologie des Gehirns von Hyla arborea arborea (L.) (Hylidae, Anura). Bonner zool. Beitr. 26:370-382.

Lörcher, K. 1966. Einfluss der Wassertemperatur auf die Paarungsrufe der Unken. Naturwissenschaften 53:559-560

Lörcher, K. 1969. Vergleichende bio-akustische Untersuchungen an der Rot- und Gelbbauchunke, Bombina bombina (L.) und Bombina v. variegata (L.). Oecologia (Berl.) 3:84-124.

Lörcher, K., and H. Schneider. 1973. Vergleichende bio-akustische Untersuchungen an der Kreuzkröte, Bufo calamita (Laur.), und der Wechselkröte, Bufo v. viridis (Laur.). Z. Tierpsychol. 32: 506-521.

Manz, R. 1975. Die Fusionsfrequenzen der Kehlkopfmuskeln und eines Beinmuskels in Abhängigkeit von der Temperatur bei europäischen Froschlurchen (Anura). Zool. Jb. Physiol. 79:221-245.

Mertens, R., and L. Müller. 1940. Die Amphibien und Reptilien Europas. Abh. Senck. nat. Ges. 451:1-56.

Mertens, R., and L. Müller. 1966. Liste der Amphibien und Reptilien Europas. Abh. Senck. nat. Ges. 41:1-62.

Mertens, R., and H. Wermuth. 1960. Die Amphibien und Reptilien Europas. Frankfurt/M: W. Kramer.

Nevo, E., and H. Schneider. 1976. Mating call pattern of green toads in Israel and its ecological correlate. J. Zool., Lond. 178:133-145.

Obert, H.-J. 1971. Morphologie von Hoden und Hypophyse der Erdkröte (Bufo b. bufo) während der Fortpflanzungsperiode. Zool. Jb. Physiol. 76:185-197.

Obert, H.-J. 1973. Untersuchungen zur hormonalen Steuerung der Ruf- und Paarungsaktivitat der Rot- und Gelbbauchunke Bombina bombina (L.) und Bombina v. variegate (L.) Zool. Jb. Physiol. 77:166-198.

Obert, H.-J. 1974. Untersuchungen zur hormonalen Steuerung der Rufaktivität von Fröschen und Kröten der Familien Ranidae, Discoglossidae, Hylidae und Bufonidae (Rana temporaria, Bombina variegata, Hyla arborea, Bufo bufo). Zool. Jb. Physiol. 78:219-241.

Obert, H.-J. 1975a. Investigations into the significance of testicles and interrenal gland in the hormonal control of mating call activity in the common frog Rana temporaria (L.). Zool. Jb. Physiol. 79:246-261.

Obert, H.-J. 1975b. Zur hormonalen Steuerung der Differenzierungshöhe von Cutisbildungen bei Anuren. Zool. Jb. Anat. 94:319-338.

Obert, H.-J. 1975c. The dependence of calling activity in Rana esculenta Linné 1758 and Rana ridibunda Pallas 1771 upon exogenous factors (Ranidae, Anura). Oecologia (Berl.) 18:317-328.

Paillette, M. 1967. Valeur taxinomique des émissions sonores chez Hyla (Amphibiens, Anoures) de la faune française. C. R. Acad. Sci. (Paris), Sér. D, 264:1626-1628.

Schmid, E. 1976. Vergleichende Untersuchungen am Kehlkopf heimischer Froschlurche. Dissertation, Univ. Bonn.

Schmidt, R. S. 1965. Larynx control and call production in frogs. Copeia 1965:143-147.

Schmidt, R. S. 1966. Central mechanisms of frog calling. Behaviour 26:252-285.

Schmidt, R. S. 1968. Preoptic activation of frog mating calling. Behaviour 30:239-257.

Schmidt, R. S. 1971. A model of the central mechanisms of male anuran acoustic behavior. Behaviour 34:288-317.

Schmidt, R. S. 1974. Neural correlates of frog calling. J. Comp. Physiol. 88:321-333.

Schneider, H. 1966a. Die Paarungsrufe einheimischer Froschlurche (Discoglossidae, Pelobatidae, Bufonidae, Hylidae). Z. Morph. Ökol. Tiere 57:119-136.

Schneider, H. 1966b. Bio-Akustik der Froschlurche. Ein Bericht über den gegenwärtigen Stand der Forschung. Stuttgarter Beitr. Naturkde. 152:1-16.

Schneider, H. 1967. Rufe und Rufverhalten des Laubfrosches, Hyla arborea arborea (L.). Z. vergl. Physiol. 57:174-189.

Schneider, H. 1968. Bio-akustische Untersuchungen am Mittelmeerlaubfrosch. Z. vergl. Physiol. 61:369-385.

Schneider, H. 1970. Morphologie des Larynx von Hyla a. arborea (L.) und Hyla meridionalis Boettger (Amphibia, Anura). Z. Morph. Tiere 66:299-309.

Schneider, H. 1971a. Die Steuerung des täglichen Rufbeginns beim Laubfrosch, Hyla arborea arborea (L.). Oecologia (Berl.) 8: 310-320.

Schneider, H. 1971b. Beobachtungen und Versuche zum Verhalten der Froschlurche. IN: A. W. Stokes (ed.), Praktikum der Verhaltensforschung. Stuttgart: Gustav Fischer Verlag.

Schneider, H. 1973. Die Paarungsrufe einheimischer Ranidae (Anura, Amphibia). Bonner zool. Beitr. 24:51-61.

Schneider, H. 1974. Structure of the mating calls and relationships of the European tree frogs (Hylidae, Anura). Oecologia (Berl.) 14:99-110.

Schneider, H., and E. Nevo. 1972. Bio-acoustic study of the yellow-lemon treefrog, Hyla arborea savignyi Audouin. Zool. Jb. Physiol. 76:497-506.

Schneider, H., and H. Eichelberg. 1974. The mating call of hybrids of the fire-bellied toad and yellow-bellied toad (Bombina bomina (L.), Bombina v. variegata (L.), Discoglossidae, Anura). Oecologia (Berl.) 16:61-71.

Wahl, M. 1969. Untersuchungen zur Bio-Akustik des Wasserfrosches Rana esculenta (L.). Oecologia (Berl.) 3:14-55.

Weber, E. 1974. Vergleichende Untersuchungen zur Bio-akustik von Discoglossus pictus, Otth 1837 und Discoglossus sardus, Tschudi 1837 (Discoglossidae, Anura). Zool. Jb. Physiol. 78:40-84.

Weber, E. 1975. Die Veränderung der Befreiungsrufe bei sechs europäischen Anuren nach Ausschaltung von Kehlkopfmuskeln (Amphibia). Zool. Jb. Physiol. 79:311-320.

Weber, E. 1976. Die Veränderung der Paarungs- und Revierrufe von Hyla arborea savignyi Audouin (Anura) nach Ausschaltung von Kehlkopfmuskeln. Zool. Jb. Physiol. (in press).

Weber, E., and H. Schneider. 1971. Artspezifische Paarungsrufe bei Discoglossus pictus Otth, 1837 und Discoglossus sardus Tschudi, 1837 (Discoglossidae, Anura). Z. Naturforsch. 26b: 1055-1057.

AUDITORY PROCESSING OF VOCAL SIGNALS IN ANURANS

Robert R. Capranica

Section of Neurobiology and Behavior and
School of Electrical Engineering
Cornell University, Ithaca, New York 14853

Acoustic signaling between animals involves a rather remarkable cooperative communication channel. It enables rapid transfer of information over a considerable distance in which the sender and receiver need not be in visible contact, such as through dense foliage or in total darkness. The "information" within an acoustic signal is conveyed by the temporal and spectral pattern of sound pressure that the sender must control in order to transmit a "meaningful" message. The receiver, in turn, must possess a receptor apparatus that is sensitive to these incoming pressure variations as well as a nervous system that can decode them. But communication rarely occurs in isolation. In general there is an ambient background of other sounds from other sources that we collectively refer to as "noise" and that interferes with detection of those signals of interest. This is a universal problem that all animals face in communicating with other members of their own species.

Let us begin by asking a question of strategy. Suppose an animal is interested in detecting a particular set of sensory signals. How should it go about it? We can consider two extreme strategies. One possibility would be to evolve an initial selectivity within the sense organs themselves. That is, "design" the receptor organs and peripheral nervous system to be specialized for only a small set of signals, so that the peripheral sensory system acts as a "sensory filter." The advantage of this strategy is that the central nervous system, namely the brain, would have to sort out and interpret only a small number of communication signals. So a simpler central nervous system could be effective. But clearly there is a disadvantage. The animal would be restricted only to certain events in the environment, and many other events would have to go undetected.

The other extreme possibility would be to design the sense organs and peripheral nervous system so that they could detect every type of communication signal imaginable. Then nothing would be missed, but the brain must process an enormous array of signals and somehow sort them all out. This strategy requires an exceedingly complicated brain and poses a serious obstacle. Animals are of limited size and thus must possess a finite number of neurons in their central auditory nervous systems.

So what have animals done in the course of signal evolution? They have compromised. Some, particularly lower vertebrates, have opted to emphasize selective processing immediately within their receptor and peripheral nervous systems. Others, such as ourselves, have chosen to place greater emphasis on sensory processing in the central nervous system. This trade-off between peripheral vs. central processing is an interesting one; it gives us some insight into how different animals have evolved their own strategies in communicating in a noisy environment.

For the past several years I have been interested in acoustic communication in animals and the way that meaningful sounds are processed in the auditory system. Most of this work has involved anurans for a variety of reasons. First of all, sound communication plays a very significant role in their lives. They are highly vocal animals, yet they possess a small set of signals in their repertoire, such as mating calls, release calls, territorial calls, distress calls, warning calls, etc. (Bogert 1960). Furthermore, it is possible to conduct behavioral studies of their ability to discriminate between various features in some of the sounds that they employ, particularly mating calls. And finally, their nervous system is comparatively simpler than higher vertebrates so that it might be possible, at least in my lifetime, eventually to understand the neural basis for "recognition" of sounds of biological significance.

It has been estimated that there are about 2600 species and sub-species of frogs and toads in the world today (Cochran 1961). Males of most species are vocal and possess an inflatable vocal sac that serves to broadcast their advertisement calls. Most species possess a simple subgular sac beneath their throats whereas others, particularly ranids, may possess paired lateral vocal sacs (Liu 1935). The advantage of a particular vocal sac configuration is not obvious; it could possibly play a role in providing females with appropriate directional cues. It seems as though the signal properties of the mating calls depend primarily on active laryngeal mechanisms and not on the vocal sac itself

(Martin 1971). Thus the vocal sac serves chiefly as an efficient mechanism to couple energy from the animal to the environment. The intensity of a male's call can be quite high, of the order of 100-110 decibels sound pressure level in many species (relative to .0002 dyne/cm^2) (Loftus-Hills and Littlejohn 1971; Gerhardt 1975). Males must therefore expend a considerable amount of energy in broadcasting such intense signals.

The mating advertisement call of each species is distinct and serves as a species identification signal. In a number of studies it has been shown that females are selectively attracted to the calls of males of their own species (e.g., Littlejohn and Michaud 1959; Blair and Littlejohn 1960; Gerhardt 1973). Thus the male's call provides a premating isolating mechanism to minimize cross-breeding between sympatric species (Blair 1958). The basis for this selectivity resides in the distinctive spectral and temporal features in a male's call. Obviously the auditory system of an anuran must be sensitive to these species-specific features to enable such selective discrimination. In the following discussion, suppose we consider three different species: the bullfrog *Rana catesbeiana*, the cricket frog *Acris crepitans*, and the Puerto Rican treefrog *Eleutherodactylus coqui*. For each species, we first will summarize behavioral studies that have identified the important features in the male's call and then consider the sensitivity of their auditory systems to these features. As we shall see, not only can the anuran's peripheral auditory sensitivity be species-specific, but it can also be geographically specific and even sexually specific.

Rana catesbeiana

Male bullfrogs, like males of many anuran species, will respond vocally to the mating call of another male of their species. This evoked calling behavior provides a basis for identifying the distinctive species-specific features in a male bullfrog's call. Capranica (1965, 1968) maintained several colonies of bullfrogs in the laboratory and systematically broadcasted a variety of natural and synthetic calls to them while monitoring their vocal responses. He found that males would respond only to certain stimuli and thereby identified the distinctive features in the male's mating call that give rise to species identification.

Figure 1 shows the mating call of an adult male which would evoke a high degree of calling from other male bullfrogs. We find that it possesses certain features which experiments showed were necessary to evoke a good response. It has a bimodal spectrum, with an energy peak in a low-frequency region around 200-300 Hz, a dip in the mid-frequency region around 500-700 Hz, and a broad concentration of energy in the higher-frequency region centered

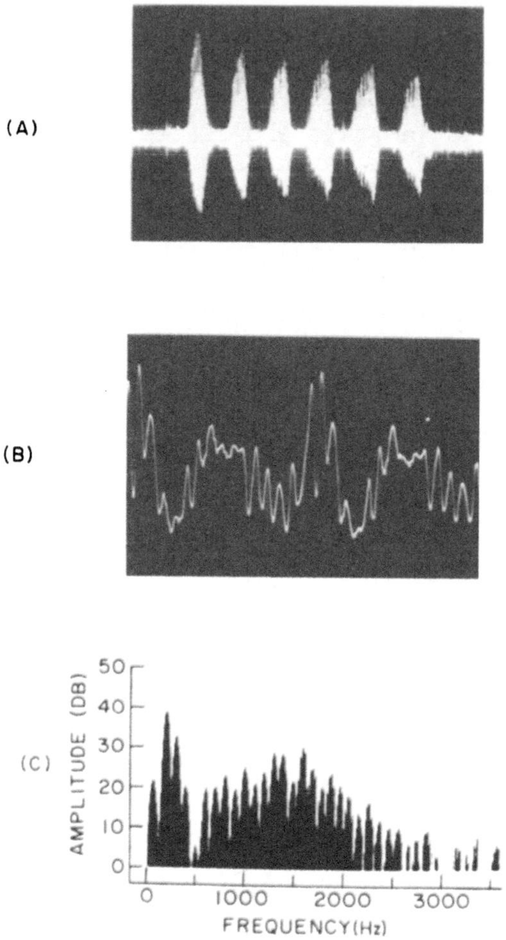

Fig. 1. Analysis of a field-recorded mating call of an adult male bullfrog. (A) Mating call, comprised of 6 similar croaks, in its entirety. (B) Representative expansion of the croak waveform showing the waveshape in a single fundamental period. (C) Spectrographic section taken through the middle of a croak. The harmonic structure of the spectrum reflects the quasiperiodic nature of the croak waveform (from Capranica 1966).

around 1400-1500 Hz. Furthermore note that the croak waveform is pulsatile with a periodicity of about 100/sec. These features are distinctive for the bullfrog. A mating call lacking one or more of these features will be less effective and the male will not be as successful in competition with other males. A young juvenile male's call would not elicit significant calling from other males even though he might participate in a chorus. His call does not possess a low-frequency peak around 200 Hz. Rather the lower peak of energy in his call occurs around 700-800 Hz (Capranica 1965). As he matures, presumably because the vocal cords and supporting structures in his larynx become more massive, his ability to produce such low-frequency energy improves and he gradually achieves an adult call and the status to initiate chorusing. This ability is a result of maturity which occurs normally with an increase in size.

At this point a consideration of the auditory apparatus of the bullfrog would be useful. A schematic diagram of the middle and inner ear is shown in figure 2. The large tympanic membrane

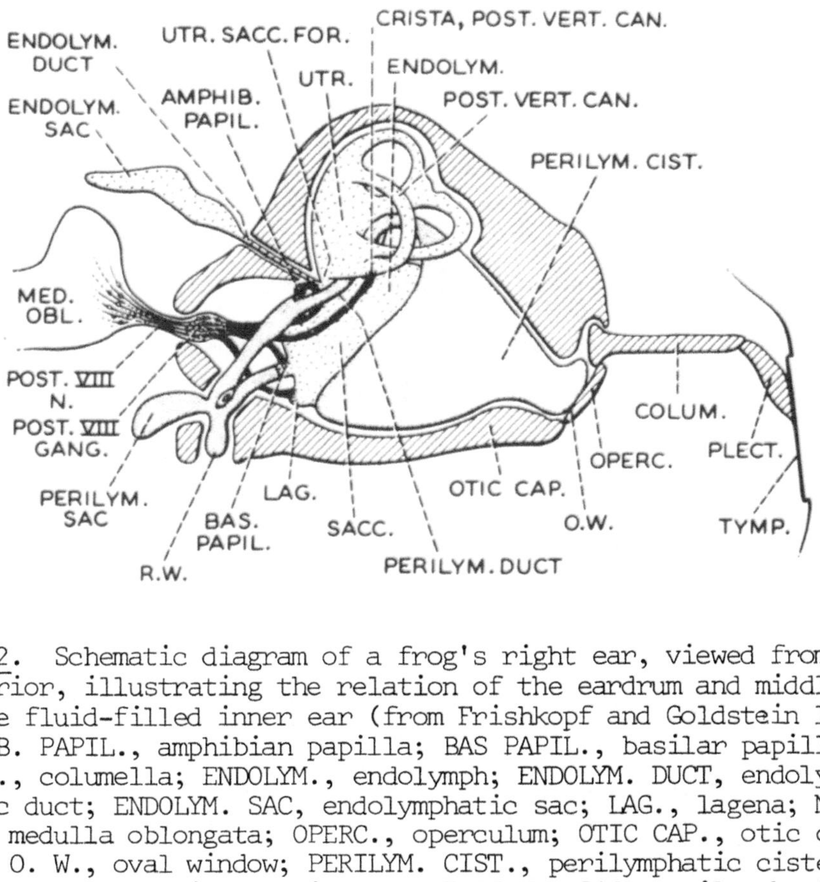

<u>Fig. 2</u>. Schematic diagram of a frog's right ear, viewed from the posterior, illustrating the relation of the eardrum and middle ear to the fluid-filled inner ear (from Frishkopf and Goldstein 1963). AMPHIB. PAPIL., amphibian papilla; BAS PAPIL., basilar papilla; COLUM., columella; ENDOLYM., endolymph; ENDOLYM. DUCT, endolymphatic duct; ENDOLYM. SAC, endolymphatic sac; LAG., lagena; MED. OBL., medulla oblongata; OPERC., operculum; OTIC CAP., otic capsule; O. W., oval window; PERILYM. CIST., perilymphatic cistern; PERILYM. DUCT, perilymphatic duct; PERILYM. SAC, perilymphatic sac; PLECT., plectrum; POST. VIII GANG., posterior eighth nerve ganglion; POST. VIII N., posterior eighth nerve; POST VERT. CAN., posterior vertical canal; R. W., round window; SACC., saccule; TYMP, tympanum; UTR., utricle; UTR. SACC. FOR., utricular-saccular foramen.

on the surface of the head is exposed directly to the air. Sound pressure fluctuations in an acoustic stimulus cause the membrane to vibrate and these vibrations are mechanically coupled to the fluid-filled inner ear by means of the middle-ear linkage: plectrum, columella and operculum. Two sensory organs, amphibian papilla and basilar papilla, are specialized to detect acoustical vibrational frequencies transmitted through the fluids. Notice that the two organs are separated anatomically. Each papilla possesses its own set of sensitive hair cells and a separate gelatinous tectorial membrane. Mechanical energy coupled through the fluids leads to vibration of the tectorial membranes of the two organs, which in turn causes bending of the ciliary processes of the hair cells. As in all auditory organs, bending of the ciliary processes excites the receptor cells and action potentials are initiated in the auditory nerve fibers that synapse on the basal end of the hair cells. The resultant pattern of neural excitation in the auditory nerve is transmitted to the central nervous system for further processing and decoding.

The peripheral innervation pattern of the eighth nerve is indicated in figure 3. The nerve, upon entering the wall of the otic capsule, splits into several branchlets. The amphibian papilla and the basilar papilla are innervated by branchlets within the posterior half of the nerve. Recently, Feng et al. (1975) managed to insert microelectrodes into the individual branchlets and then recorded the activity of single fibers in response to acoustic stimulation. Their study verified that there are three types of auditory nerve fibers in the bullfrog's eighth nerve. Two types of fibers derive their input from the amphibian papilla whereas the other type receives its input from the basilar papilla. To appreciate this difference, first consider the frequency sensitivity of an auditory nerve fiber to tones. In general, a fiber possesses a characteristic threshold so that tones must be sufficiently intense to excite it. The threshold varies with tonal frequency and such a plot represents the fiber's tuning curve. It gives a measure of the frequency selectivity of that fiber, namely its tuning sharpness. The tuning curve generally shows greatest sensitivity, i.e., possesses the lowest threshold, to a particular frequency called the fiber's best excitatory frequency.

Typical tuning curves for the three different fiber types in the bullfrog's eighth nerve are shown in figure 4. The low-frequency-sensitive fiber (unit 1) has a best excitatory frequency of 250 Hz; the best excitatory frequency of unit 2 is 600 Hz; and the high-frequency sensitive fiber 3 has a best excitatory frequency of 1400 Hz. Each fiber possesses a very distinct excitatory frequency and its threshold to tones either above or below this frequency increases monotonically.

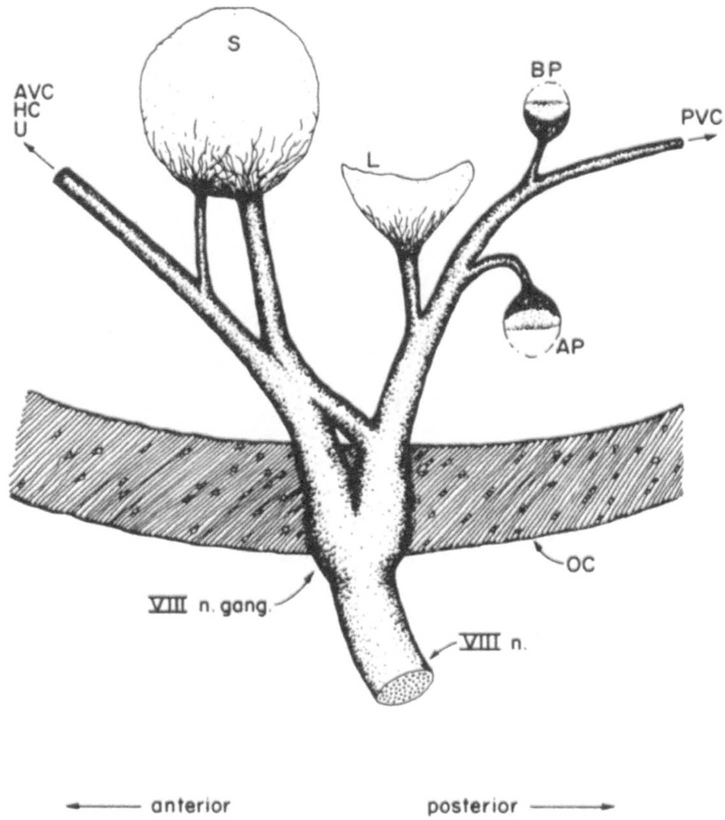

Fig. 3. Schematic of the peripheral innervation of the bullfrog's right ear by the VIIIth nerve, dorsal view. S, saccule; L, lagena; BP, basilar papilla; AP, amphibian papilla; PVC, posterior vertical canal; AVC, anterior vertical canal; HC, horizontal canal; U, utricle; VIII n., eighth cranial nerve; VIII n. gang., eighth nerve ganglion; OC, otic capsule.

Feng et al. (1975) recorded from a large number of different fibers in the bullfrog's eighth nerve. Figure 5 displays a histogram of the best excitatory frequencies of these fibers. There are three peaks in the histogram. One population is most sensitive to the low-frequency range 200-300 Hz. A second population of fibers is most sensitive to the mid-frequency range around 700-800 Hz. The best excitatory frequencies of the third population fall in the higher-frequency region around 1400 Hz. The low-frequency-sensitive fibers are distinct in that they show tone-on-tone suppression. That is, the response of a unit in this population to an excitatory tone could be totally suppressed by the

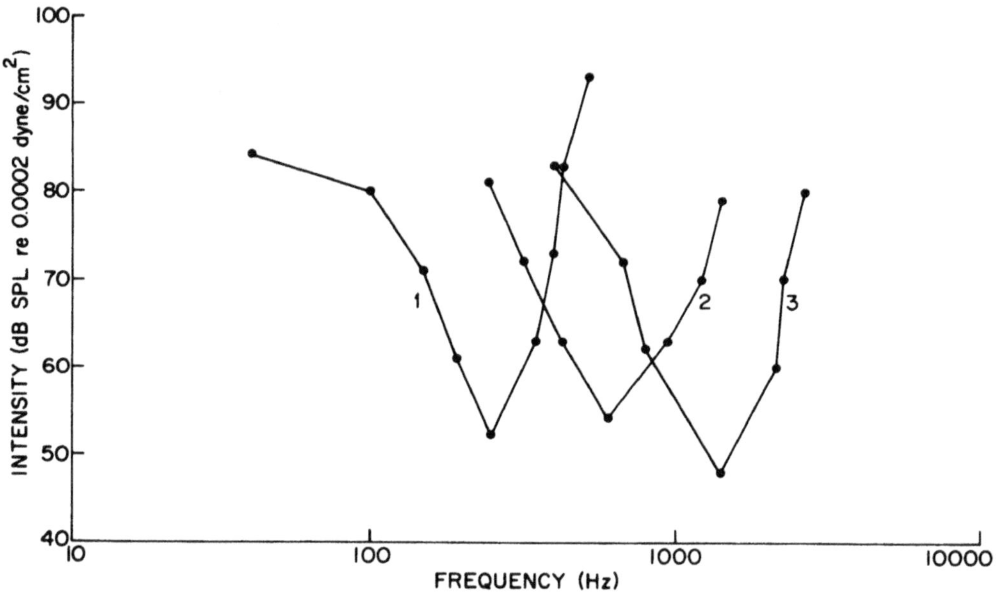

Fig. 4. Representative excitatory tuning curves for the three populations of auditory nerve fibers in the bullfrog (from Feng et al. 1975).

simultaneous addition of a second tone of higher frequency and appropriate intensity. In general the best inhibitory frequencies of these low-frequency fibers occur about 300 to 600 Hz above their best excitatory frequencies. The fibers which comprise the mid-frequency and high-frequency populations do not exhibit tonal inhibition.

By recording from the branchlets directly within the inner ear, Feng et al. (1975) verified unequivocally that the low-frequency sensitive and the mid-frequency sensitive fibers innervate the amphibian papilla while the high-frequency fibers innervate the basilar papilla. Thus the two organs are tuned to separate frequency ranges. These ranges of sensitivity bear a close relationship to the spectral energy in the signals in the bullfrog's vocal repertoire (Capranica 1968). In particular the low-frequency peak in the bullfrog's mating call excites the population of fibers from the amphibian papilla tuned to around 200 Hz while the high-frequency energy excites the fibers from the basilar papilla tuned to 1400 Hz. Other calls of the bullfrog excite different combinations of the three populations of fibers. Thus we find that the auditory organs in the bullfrog's inner ear are selective for the spectral features in his species-specific calls. Frishkopf and Goldstein (1963) showed that individual fibers in

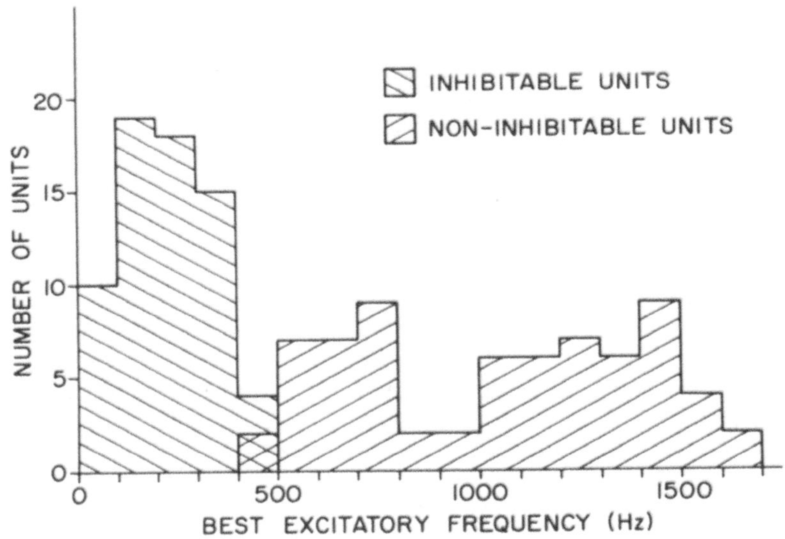

Fig. 5. Histogram of best excitatory frequencies of 135 units recorded in the eighth nerve of the bullfrog. Each of the units comprising the population with a peak at 200 Hz could be totally inhibited by a second tone of appropriate frequency and intensity. The units comprising the other two populations could not be inhibited by the addition of a second tone. The cross-hatched area indicates overlap between the low- and mid-frequency populations (from Feng et al. 1975).

the bullfrog's eighth nerve are also sensitive to the temporal features in the mating call. That is, auditory nerve fibers can fire synchronously to pulse trains having repetition rates of 100/sec. The bullfrog therefore represents an example of species-specificity in the spectral and temporal sensitivity of the peripheral auditory system to detect sounds of biological interest.

Acris crepitans

Cricket frogs are tiny anurans, weighing between one and two grams and averaging less than one inch in length. They get their name from the cricket-like sound of their mating call. In the United States there are two different species of cricket frogs: *Acris crepitans* and *Acris gryllus*. *A. crepitans* is found throughout most of the U. S. east of the Rockies while *A. gryllus* is found only in the Southeastern states; the two species overlap in the Southeast from Virginia to Louisiana. In their region of sympatry, both species may share common breeding sites and mate at the same time of year. Nevertheless, cross breeding must be rare since natural hybrids seldom are found.

A major basis for their reproductive isolation resides in the species-specific characteristics of the males' mating calls. Nevo (1969) and Capranica (1972) conducted behavioral studies in the field to verify the selectivity of females to calls of conspecific males. These experiments involved collection of females in amplexus and their subsequent release midway between two loudspeakers through which mating calls of each species were broadcast simultaneously. It was found that a female upon release would approach and make contact with a speaker broadcasting a call of a male of her own species but not of the other species. Furthermore, Capranica et al. (1973), in a study of A. crepitans, noted a geographic variation in the temporal and spectral characteristics of its mating calls. By employing natural and synthetic calls in playback experiments, they verified that females could discriminate between the recorded call of a local male and the call of a male from a different geographic region. These studies indicated that the mating call of A. crepitans is not only species-specific but that it is also geographically specific.

To determine the basis for this selectivity, Capranica et al. (1973) conducted an electrophysiological study of the auditory nervous system of cricket frogs from different geographic regions. Because these animals were so small, it was not possible to record directly from fibers in the eighth nerve. Instead they studied second-order cells in the medulla, which receive their input from the eighth nerve. In general two distinct types of cells were found in the medullary nucleus of cricket frogs. One type was sensitive to low-frequency tones and the other type was sensitive to high-frequency tones. Based on the results of Feng et al. (1975), it seems likely that the medullary cells tuned to lower frequencies derive their input from the auditory nerve fibers which innervate the amphibian papilla in the cricket frog's inner ear, while the cells tuned to higher frequencies derive their input from fibers from the basilar papilla.

Capranica et al. (1973) found geographic variation in the sensitivity of the medullary cells in these animals. For example, figure 6(A) shows representative tuning curves for the two types of cells in A. crepitans from New Jersey. The low-frequency-sensitive unit has its best excitatory frequency at 3550 Hz. The tuning curves do not overlap even at high sound intensities, emphasizing that the two auditory organs are tuned to separate frequency regions. Capranica et al. (1973) found that the best frequencies of the lower-frequency cells in the New Jersey animals were distributed over the range 200 to 1000 Hz while all the high frequency cells had their best frequencies narrowly clustered around 3500 to 3550 Hz. The lower-frequency cells possess moder-

(A)

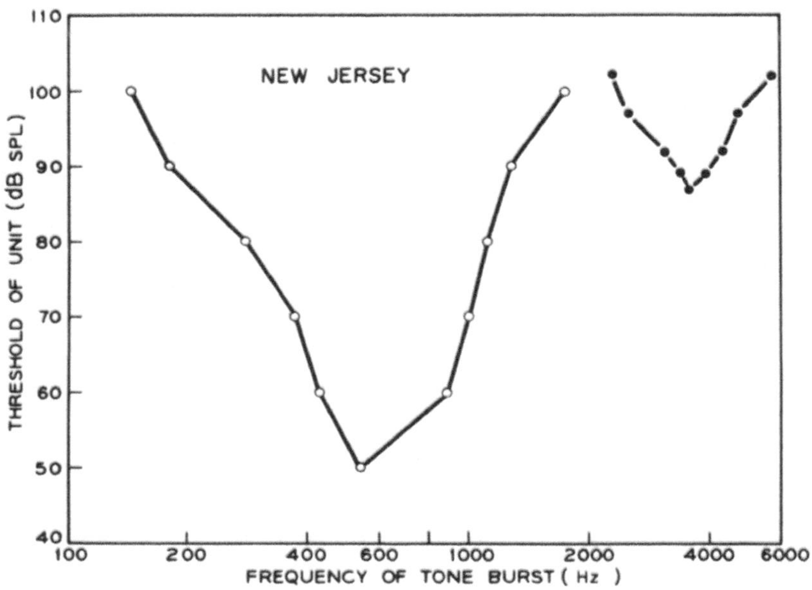

(B)

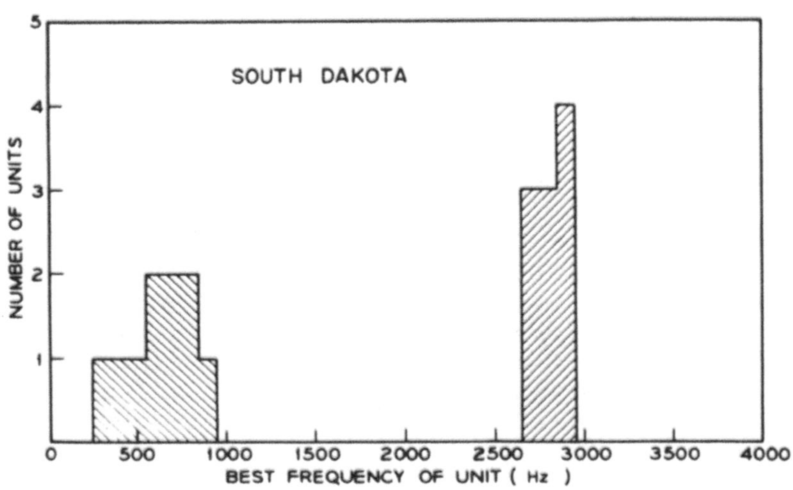

Fig. 6. (A) Tuning curves of a low-frequency unit (best frequency 550 Hz) and of a high-frequency unit (best frequency 3550 Hz) in *A. crepitans* from New Jersey. (B) Histogram of best frequencies for 20 low- and high-frequency units in *A. crepitans* from South Dakota (from Capranica *et al.* 1973).

ately sensitive thresholds of 35 to 60 dB SPL whereas all of the high-frequency cells have relatively poor thresholds around 75 to 85 dB SPL.

Suppose we consider a different geographic population of Acris crepitans. Figure 6(B) shows a histogram of best frequencies for medullary cells in cricket frogs from South Dakota. The histogram is clearly bimodal and emphasizes the disjoint frequency sensitivities of the two populations of cells. The units sensitive to lower frequencies, tuned to the range 200 to 1000 Hz, had their thresholds distributed between 35 and 60 dB SPL. These cells are very similar to the low-frequency cells in A. crepitans from New Jersey. On the other hand, the high-frequency-sensitive cells are all narrowly tuned, not to 3500 Hz as in New Jersey cricket frogs, but rather to 2900 Hz. All of these high-frequency units, as in the New Jersey animals, were fairly insensitive with thresholds at their best frequencies around 75 to 85 dB SPL. These results indicate that the frequency sensitivity of the basilar papilla varies with geographic origin, whereas the sensitivity of the amphibian papilla seems to remain constant throughout the geographic range.

In view of these results, let us now turn to the signal characteristics of the mating call in Acris crepitans. A typical call from an adult male recorded in New Jersey consists of a sequence of clicks having a stereotyped temporal pattern, as seen in figure 7(A). The energy in each click has a distinct spectral peak centered around 3500 Hz. This frequency is a characteristic feature of the mating calls of males in New Jersey. A comparison with the sensitivity of the high-frequency cells (Fig. 6A) indicates that these units are all selectively tuned to the spectral peak in the calls of males from New Jersey. Thus the auditory sensitivity in this geographic population is closely matched to the local mating call.

Suppose we consider the mating call of a typical male in South Dakota, shown in figure 7(B). The pattern of clicks is different from New Jersey and is an important feature in the dialect of this species. The energy peak in the South Dakota call occurs, not at 3550 Hz, but at 2900 Hz. This peak is closely matched to the frequency sensitivity of the high-frequency cells in the auditory system of South Dakota cricket frogs (Fig. 6B).

To summarize, we can conclude that the basilar papilla plays the principal role in detecting the mating calls of Acris crepitans. There is no lower frequency energy in the call that would excite the amphibian papilla. Furthermore the frequency sensitivity of the basilar papilla in this species varies within the geographic range to match the spectral peak in the local dialect. Thus, not only is the frequency sensitivity of this organ species-specific,

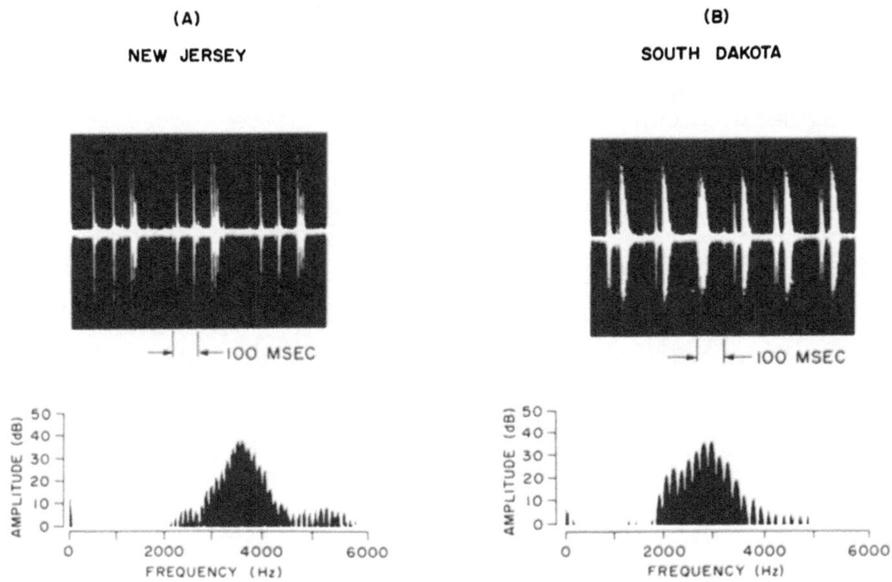

Fig. 7. (A) The mating call of an adult male A. crepitans from New Jersey consists of a repeated sequence of clicks. The spectographic section shows the distribution of energy in a single sequence; this distribution remains the same throughout the call. (B) Mating call of an adult male A. crepitans from South Dakota (from Capranica et al. 1973).

it is also geographically specific. The relatively poor threshold of the basilar papilla may seem puzzling at first, but these animals breed in very dense aggregations in which the distance between neighboring males is often only 1 to 2 meters. The intensity of each male's call is quite high, generally about 95 dB SPL or even greater. The relative insensitivity of the basilar papilla thus permits spectral peak detection, narrowly tuned to the dominant frequency in the local call but not stimulated by adjacent frequencies of lower amplitude.

The temporal pattern of the mating call in A. crepitans also varies geographically, but there was no evidence for selectivity of these temporal features in the response properties of auditory cells in the medulla. It would therefore seem that temporal pattern discrimination occurs at a higher center in the auditory system.

The lower-frequency sensitivity of the amphibian papilla does not show geographic variation. Apparently low-frequency sounds are important in the lives of cricket frogs but their significance is unknown. Perhaps the amphibian papilla in these tiny frogs serves to detect sounds made by larger predators (Capranica et al. 1973).

Eleutherodactylus coqui

The treefrog Eleutherodactylus coqui is found principally on the island of Puerto Rico. Each evening, beginning around sunset, males utter a distinct two-note call "Co-Qui." Both notes are stereotyped and have a tonal quality (Narins and Capranica 1977). Figure 8 shows a narrow-band sonagram of the call. The "Co" note is rather brief, about 100 msec in duration, and of relatively constant frequency around 1150 Hz. Following an interval of about 250-300 msec, the "Qui" note is given. It lasts for about 170-190 msec and sweeps upward in frequency from about 1800 Hz to about 2100 Hz. Because of their tonal quality, both notes can be heard over considerable distance.

A puzzling question arises. Why does the male produce a two-note call instead of simply a single note? One possibility is that this sequential pattern enhances the probability of detection by other members of his species in the noisy environment in which they live. That is, there might be less chance that both notes would be masked by background noise compared to just a single note. Another possibility is that a more complicated call can convey a greater gradation of information. Thus, if each note is graded,

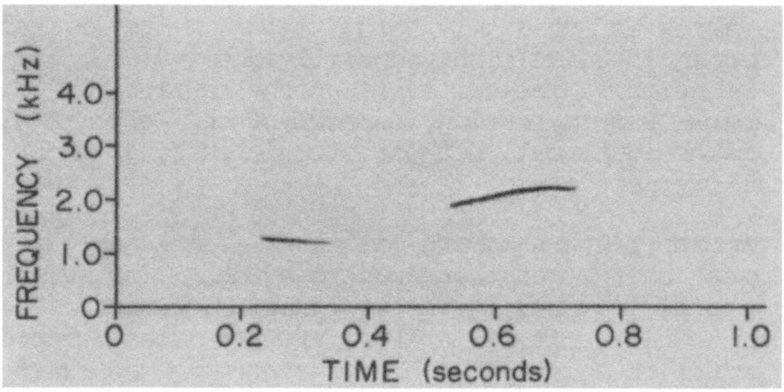

Fig. 8. Narrow-band (45 Hz) sonagram of the principal call of a male E. coqui in the El Verde rain forest in Puerto Rico (from Narins and Capranica 1977.

the variety of signals that can be transmitted increases as more notes are added. But since the call is so stereotyped, this second hypothesis seems improbable. A third possibility is that each note could convey a separate significance for the two sexes.

To test these alternatives, Narins and Capranica (1976) first conducted behavioral studies and then electrophysiological studies of the auditory sensitivity of E. coqui. Using an evoked calling paradigm, they found that males would answer a sufficiently intense playback of just the "Co" note with a similar "Co" note response, but the "Qui" note alone had no effect on the calling behavior of males. The situation is just the opposite for females. In discrimination studies involving release between two loudspeakers, females were attracted by the "Qui" note but paid little attention to the broadcast of the "Co" note. These studies therefore verify that the two notes signal the two sexes separately. The "Co" note is employed for male-male interaction and probably serves to maintain spacing between neighboring males, but does not attract a female. On the other hand, the "Qui" note serves to attract females but plays no role in male-male encounters.

In view of these behavioral results, suppose we consider the frequency sensitivity of the peripheral auditory system of this species. Upon recording from single fibers in the eighth nerve, Narins and Capranica (1977) identified three distinct populations of auditory units. One group has its best excitatory frequencies in a low-frequency region below 600 Hz; a second group has its best excitatory frequencies in a mid-frequency region between about 600 Hz and 1400 Hz; and the third group has its best excitatory frequencies distributed between 1800 and 3700 Hz. Each of the fibers in the low-frequency-sensitive group could be inhibited by the addition of a higher frequency tone; the fibers in the other two groups did not show two-tone inhibition. The low-frequency and mid-frequency fibers presumably derive their sensitivity from the amphibian papilla, whereas the high-frequency fibers derive their sensitivity from the basilar papilla.

An interesting finding in these electrophysiological studies is that best excitatory frequencies of the three populations are not the same in the two sexes. Figure 9 shows a histogram of the best excitatory frequencies based on a large number of fibers in each sex. The population of low-frequency-sensitive fibers is not significantly different in the two sexes. There is a significant ($P < .001$) difference between males and females in the mean best excitatory frequencies of both the mid- and high-frequency fibers. In males there is a large population of fibers with best excitatory frequencies in the range of the "Co" note but relatively few units in the range of the "Qui" note. On the other hand in females there are fewer units tuned to the "Co" note but a sizeable proportion

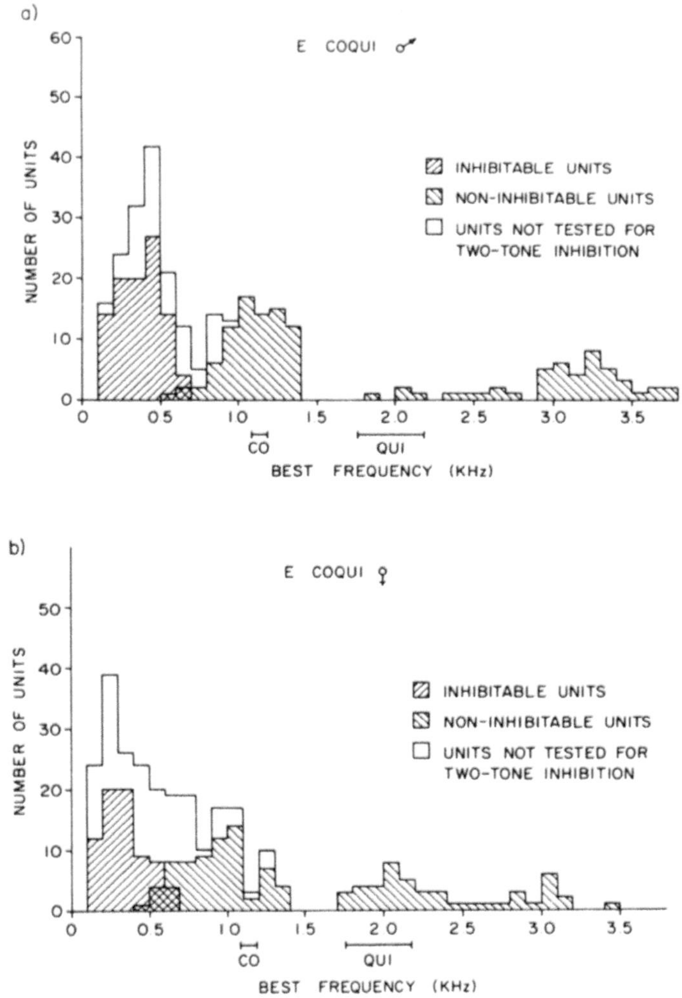

Fig. 9. Histograms of best excitatory frequencies for auditory nerve fibers in *E. coqui*. The striped bars within the histogram represent units tested for two-tone inhibition. The distribution of energy in each of the two call notes based on a sample of 16 males is indicated below the histogram. (A) Results for 284 auditory fibers from 20 male *E. coqui*. (B) Results for 279 single fibers recorded from 9 female *E. coqui* (from Narins and Capranica 1976).

is tuned to the "Qui" note. Thus the peripheral auditory sensitivity of males and females is tuned differently to the two notes in the male's call. These results indicate that in this species the sensitivity of their auditory organs is not only species-specific, but also sexually specific.

The role of the fibers tuned to the highest frequencies above 2500 Hz and extending to 3700 Hz is unknown. They may be involved in detection of occasional calls by male E. coqui that contain energy as high as 2700 Hz (Narins and Capranica 1976). The male's call lacks low-frequency energy that might excite the fibers from the amphibian papilla tuned below 500 Hz. As in cricket frogs, the significance of such low-frequency sounds in the lives of these animals is unknown.

CONCLUDING REMARKS

In this paper we have reviewed the neural basis for detection of the mating calls of three anuran species. The inner ear of Rana catesbeiana is selective for the species-specific features in the male's call. A further degree of selectivity is found in the cricket frog Acris crepitans. In this species the sensitivity of its peripheral auditory system is not only species-specific but also geographically specific. And finally the inner ear of Eleutherodactylus coqui shows a further degree of selectivity which provides a basis for sexual discrimination of the two notes in the male's call.

In the evolution of animal sound communication, anurans "adopted" a strategy of peripheral sensory filtering. This emphasis of peripheral processing plays a key role in maintaining reproductive isolation between species. Such specialization, however, poses a serious constraint. While the central nervous system of higher vertebrates may be "plastic" and readily alterable through experience, the peripheral auditory system of anurans is likely to be more static. To change the sensitivity of a sensory receptor, particularly the ear, which is a mechanical organ, would seem to require a very gradual evolutionary process. The inadvertent introduction of a foreign species or a rapid change in the sonic and breeding environment could be very disruptive. Anurans are very specialized animals. This is an important consideration for those of us who wish to preserve the integrity of existing anuran species around the world.

ACKNOWLEDGEMENTS

Most of the research reported in this chapter was supported by Research Grant NS-09244 from the National Institutes of Health

REFERENCES

Blair, W. F. 1958. Mating call in the speciation of anuran amphibians. Am. Natur. 92:27-51.

Blair, W. F., and M. J. Littlejohn. 1960. Stage of speciation of two allopatric populations of chorus frogs (Pseudacris). Evolution 14:82-87.

Bogert, C. M. 1960. The influence of sound on the behavior of amphibians and reptiles. IN: W. E. Lanyon and W. N. Tavolga (eds.), Animal sounds and communication. Am. Inst. Biol. Sci., Washington, D. C., pp. 137-320.

Capranica, R. R. 1965. The evoked vocal response of the bullfrog. Res. Mono. No. 33, M. I. T. Press, Cambridge, Mass.

Capranica, R. R. 1966. Vocal response of the bullfrog to natural and synthetic mating calls. J. Acoust. Soc. Am. 40:1131-1139.

Capranica, R. R. 1968. The vocal repertoire of the bullfrog Rana catesbeiana. Behaviour 31:302-325.

Capranica, R. R. 1972. Acoustic and behavioral assessment of sound. IN: F. G. Worden and R. Galambos (eds.), Auditory processing of biologically significant sounds. Neurosci. Res. Prog. Bull. 10:16-17, 65-66.

Capranica, R. R., L. S. Frishkopf and E. Nevo. 1973. Encoding of geographic dialects in the auditory system of the cricket frog. Science 182:1272-1275.

Cochran, D. M. 1961. Living amphibians of the world. Doubleday and Co., Garden City, New York.

Feng, A. S., P. M. Narins and R. R. Capranica. 1975. Three populations of primary auditory fibers in the bullfrog Rana catesbeiana: Their peripheral origins and frequency sensitivities. J. Comp. Physiol. 100:221-229.

Frishkopf, L. S., and M. H. Goldstein, Jr. 1963. Responses to acoustic stimuli from single units in the eighth nerve of the bullfrog. J. Acoust. Soc. Am. 35:1219-1228.

Gerhardt, H. C. 1973. Reproductive interactions between Hyla crucifer and Pseudacris ornata (Anura: Hylidae). Am. Midl. Natur. 89:81-88.

Gerhardt, H. C. 1975. Sound pressure levels and radiation patterns of the vocalizations of some North American frogs and toads. J. Comp. Physiol. 102:1-12.

Littlejohn, M. J., and T. C. Michaud. 1959. Mating call discrimination by females of Strecker's chorus frog *Pseudacris streckeri*. Texas J. Sci. 11:86-92.

Liu, C. C. 1935. Types of vocal sac in the Salientia. Proc. Boston Soc. Nat. Hist. 41:19-40.

Loftus-Hills, J. J., and M. J. Littlejohn. 1971. Mating-call sound intensities of anuran amphibians. J. Acoust. Soc. Am. 49:1327-1329.

Martin, W. F. 1971. Mechanics of sound production in toads of the genus *Bufo*: passive elements. J. Exp. Zool. 176:273-294.

Narins, P. M., and R. R. Capranica. 1976. Sexual difference in the auditory system of the treefrog *Eleutherodactylus coqui*. Science 192:378-380.

Narins, P. M., and R. R. Capranica. 1977. An automated technique for analysis of temporal features in animal vocalizations. Anim. Behav. (in press).

Nevo, E. 1969. Discussion on the systematic significance of isolating mechanisms. IN: Systematic Biology. Proc. Interm. Conf., Nat. Acad. Sci., Washington, D. C., pp. 485-489.

HORMONAL INFLUENCES ON CALLING AND REPRODUCTIVE BEHAVIOR IN ANURANS

Hans-Joachim Obert

Zoologisches Institut der Universität Bonn, Poppelsdorfer Schloss, D 5300 Bonn, Germany

All the Central European anurans have an annual reproductive cycle; calling, mating and egg-laying occur during a period beginning in the spring or early summer. Examination of the time course of this reproductive behavior has shown, however, that there are two distinct groups of anurans, which for the present I shall simply term Type I and Type II. In the following discussion I shall use mating-call production by male anurans as an indicator of the presence of reproductive activity.

Type I (Fig. 1a) is characterized by a reproductive season that includes only a single calling and spawning period. The duration of this period can vary among the different species, but in all cases it is a clearly defined time of more or less continuous activity. Reproductive periods of this type are found in the Central European members of the families Ranidae (Wahl 1969; Obert 1975, 1976), Bufonidae (Obert, 1971), Hylidae (Schneider 1967) and Pelobatidae (Obert unpublished data). Type II is represented in Central Europe by three species of the family Discoglossidae (Lörcher 1969; Heinzmann 1970; Obert 1973). The reproductive season of these toads is made up of several distinct periods of calling and spawning (Fig. 1b).

As an example of Type I, I shall describe the mating behavior of the common European Toad Bufo bufo. After they leave their winter quarters, the toads migrate to the spawning waters, where they are active only at night. This first part of the reproductive period is called the prespawning time. The onset of the next stage, the main spawning time, is marked by the fact that many animals extend their activity into the daylight hours. During the main

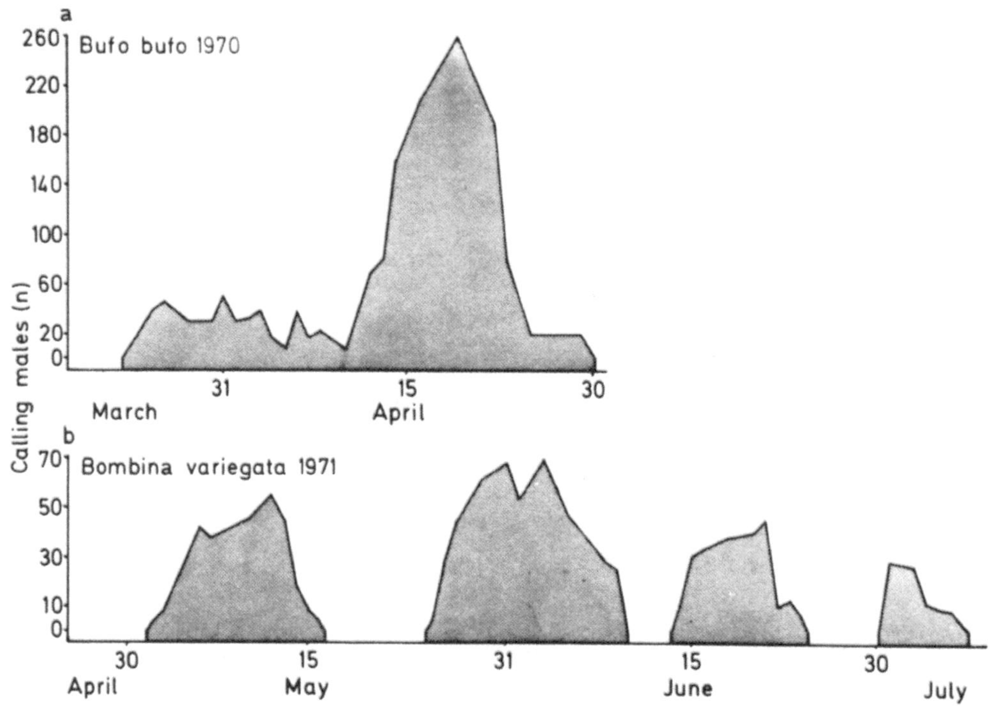

Fig. 1. The monophasic type of mating and calling activity in the common toad *Bufo bufo* (a) and the polyphasic type in the yellow-bellied toad *Bombina variegata* (b).

spawning time the males produce mating calls and the females lay eggs. The final stage, when all the females have left the spawning grounds and only unpaired males remain active there, is called the postspawning time.

The endocrine organs in Type I always change monotonically during the reproductive season. The only factors known to modify this linear development are those associated with the weather. During a period of cold weather the endocrine organs are less active than during a period of warmer weather. The central structure that regulates this hormonal control is the pituitary. Of the six different cell types there, only two or three are likely to be involved in control of reproductive behavior. These are the basophilic cells of Types 2 and 3 (B_2 and B_3), which possibly show luteinizing (LH) and follicle stimulating (FSH) activity, respectively, and perhaps also those acidophilic cells (A_1) possibly with luteotropic (LTH) activity. The proportions of these three types of pituitary cells change during the reproductive period (Fig. 2). The A_1 cells decrease slightly in number, whereas the number of B_2 cells is greatly reduced and the B_3 cells become much more numerous. The interstitial tissue of the testes, which produces

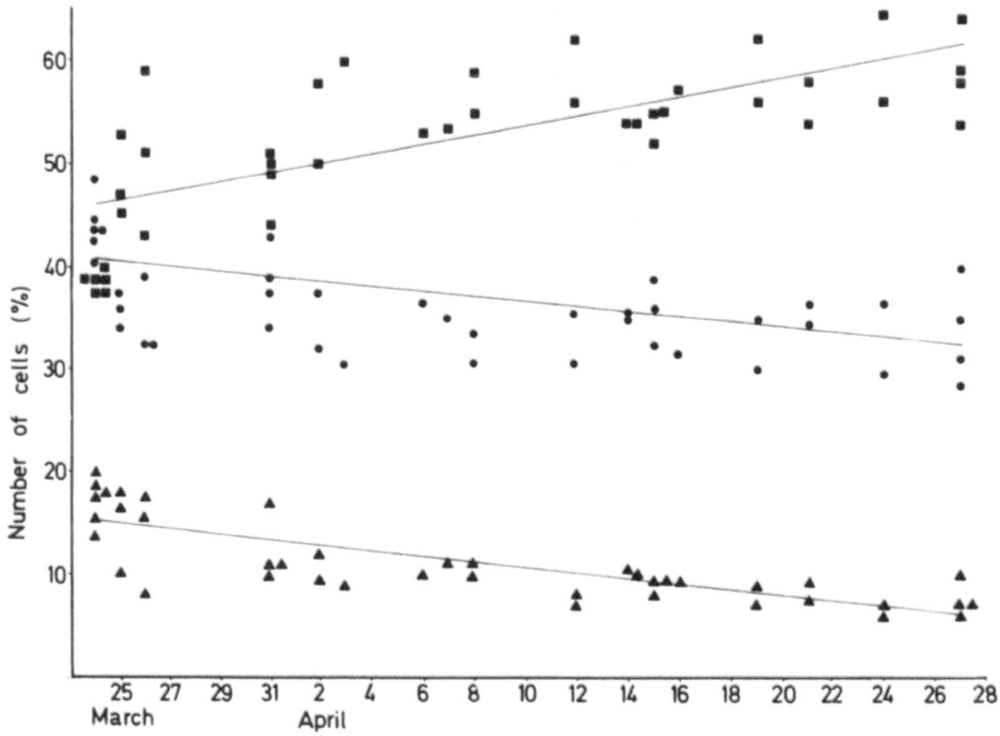

Fig. 2. Change of proportions of the B_2-(triangles), A_1-(circles) and B_3-(quadrangles) pituitary cells in Bufo bufo during the reproductive period.

androgenic hormones, also changes its activity. At the beginning of the reproductive period (Fig. 3a) the cells are large, with large round nuclei. There is a great deal of fatty material in the interstitial tissue (Fig. 4a). By contrast, at the end of the reproductive period the cells are small and their nuclei, now also small (Fig. 3b), are no longer rounded. Only a little fat remains in the testes (Fig. 4b). The decrease in androgen production, which is reflected in these changes, is also monotonic.

In the species of Type I, then, the hormone level is high at the beginning of the reproductive period and falls off steadily, so that stimulation is diminished and reproductive activity gradually comes to an end.

In the species with Type II reproductive behavior, endocrine activity follows a different pattern. As examples we shall take two discoglossids, Bombina bombina (L.) and Bombina v. variegata (L.). After toads of these species have arrived at the spawning ground they begin a calling period that lasts 10 to 12 days. During this time the first eggs are deposited in small clumps. Then

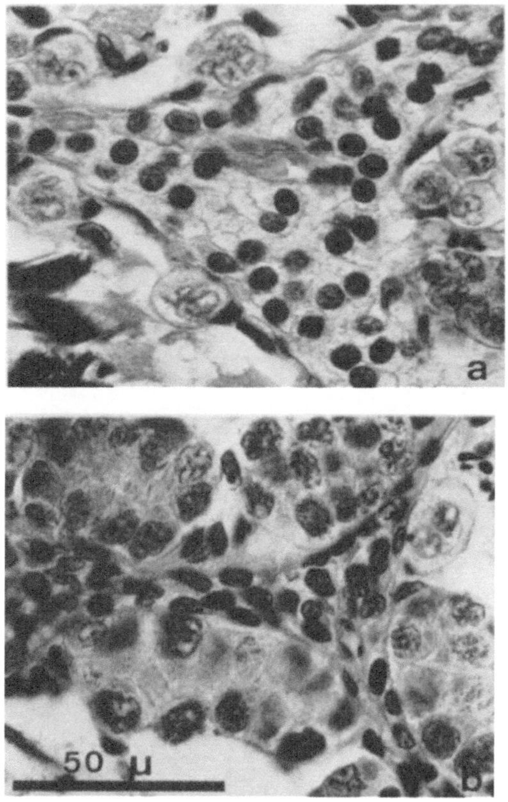

Fig. 3. The interstitial tissue of the testes in Bufo bufo at the beginning of the reproductive period (a) and at its end (b). (Hematoxylin-Erythrosin).

there is a pause of about 10 days, followed by a second calling period. Usually there is a third period, which follows another pause. Sometimes there is also a short fourth period. The females lay eggs again during each of the successive calling periods.

The finding of such intermittent behavior raises the questions: are there correlates in the endocrine activity; and what exogenous or endogenous factors trigger the individual periods of calling and spawning?

During the natural reproductive period, the numbers of cells of the different types in the anterior lobe of the pituitary do not change monotonically but follow functions of higher degree (Fig. 5). There are also not such steady changes in the development of the testicular interstitial tissue. Basically, the hormone-producing cells are highly active in males during a calling period, and less

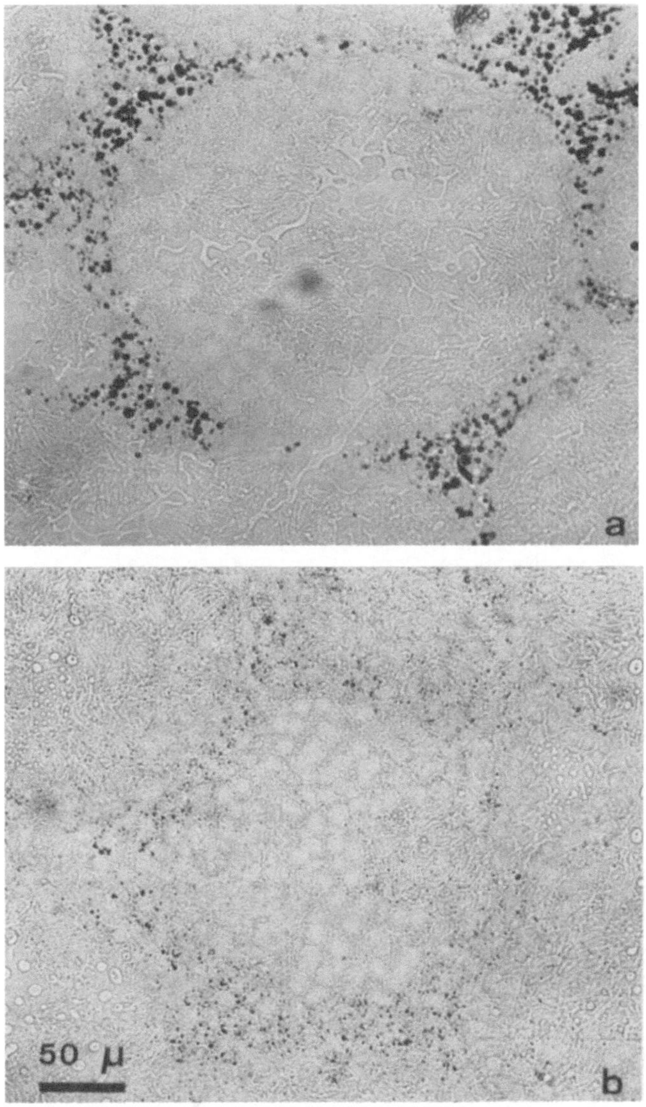

Fig. 4. Fatty material in the interstitial tissue of the testes in *Bufo bufo* at the beginning of the reproductive period (a) and at its end (b). (Sudan-black).

so during the pauses between periods.

A useful discovery, since it enables us to study calling in the laboratory as well as in the field, is that calling activity can be stimulated in all the Central European discoglossids by

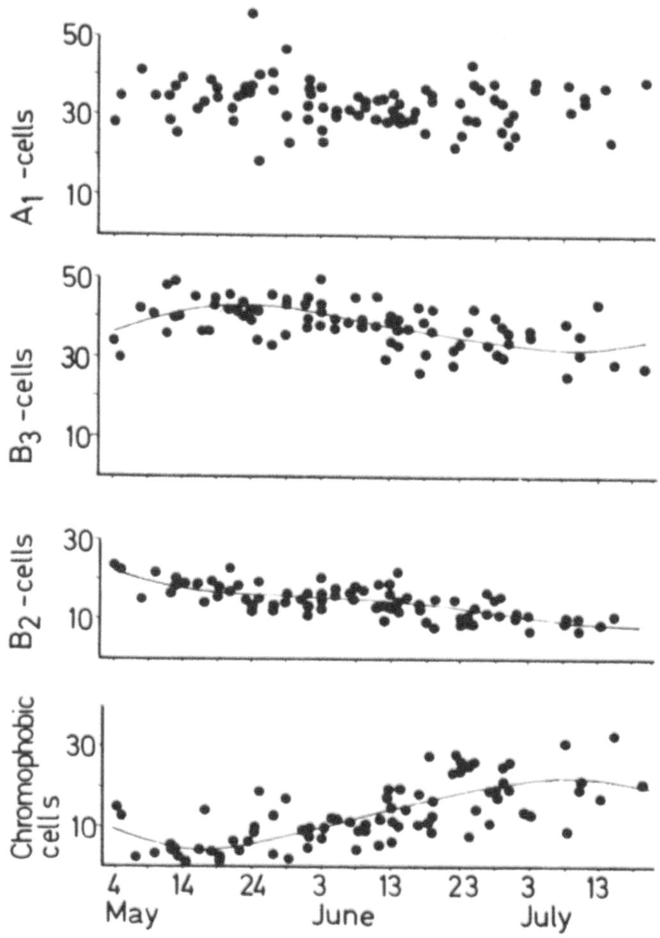

Fig. 5. Change of proportions of A_1-, B_3-, B_2- and chromophobic pituitary cells in *Bombina variegata* during the reproductive period.

injections of chorionic gonadotropins. The calling produced artificially in these Type II toads by a single such injection proceeds exactly like that observed in the natural habitat, even having a subdivision into distinct calling periods. This alternation between activity and inactivity in behavior is paralleled by an endocrinological alternation. During each calling period, the nuclei of the interstitial cells are large and the cells themselves present a characteristic active appearance (Fig. 6a). The activity of the 3β-hydroxysteroid dehydrogenases is high (Fig. 7a); this is the enzyme that splits off hydrogen at carbon 3 of the steroid molecule producing 3-oxysteroids. During the pauses, on the other hand, the cells are relatively quiescent (Fig. 6b); they are smaller, with small nuclei and slight 3β-hydroxysteroid dehydrogenase activity (Fig. 7b). It thus appears that the polyphasic occurrence of calling

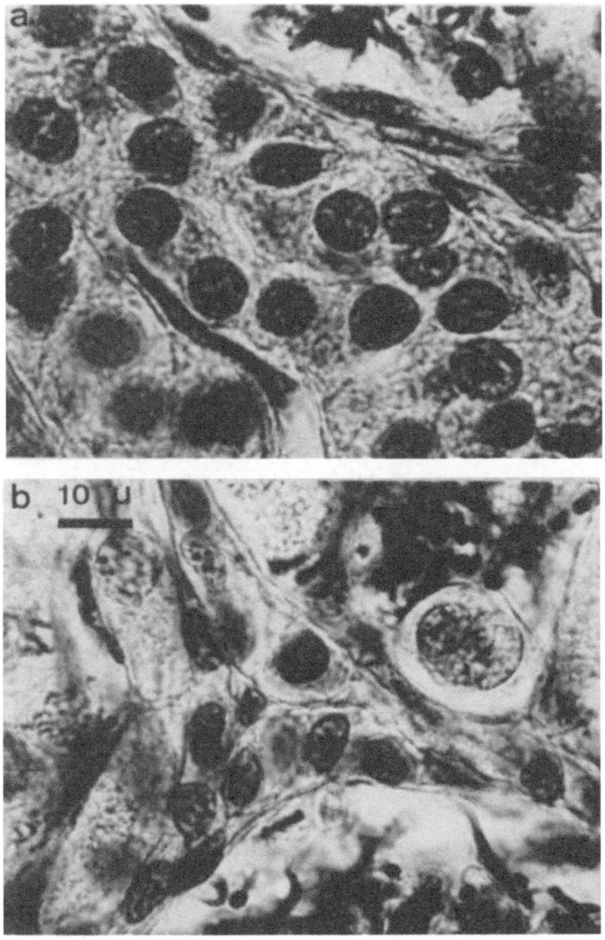

Fig. 6. Cells of the interstitial tissue of the testes in Bombina variegata during a calling period (a), and a calling pause (b). (Hematoxylin-Erythrosin).

is directly related to a fluctuation in the endocrinological function of the testes. It may be that reciprocal effects involving the androgens formed in the interstitial elements of the testes and the gonadotropic hormones elaborated by the pituitary are responsible for the polyphasic character of reproductive activity. I have demonstrated the existence of this sort of negative feedback in Bombina bombina (Obert 1973). The fact that polyphasic androgen production can be triggered experimentally by a single step increase in a gonadotropic hormone suggests that a similar mechanism may contribute to initiation of the calling cycle under natural conditions.

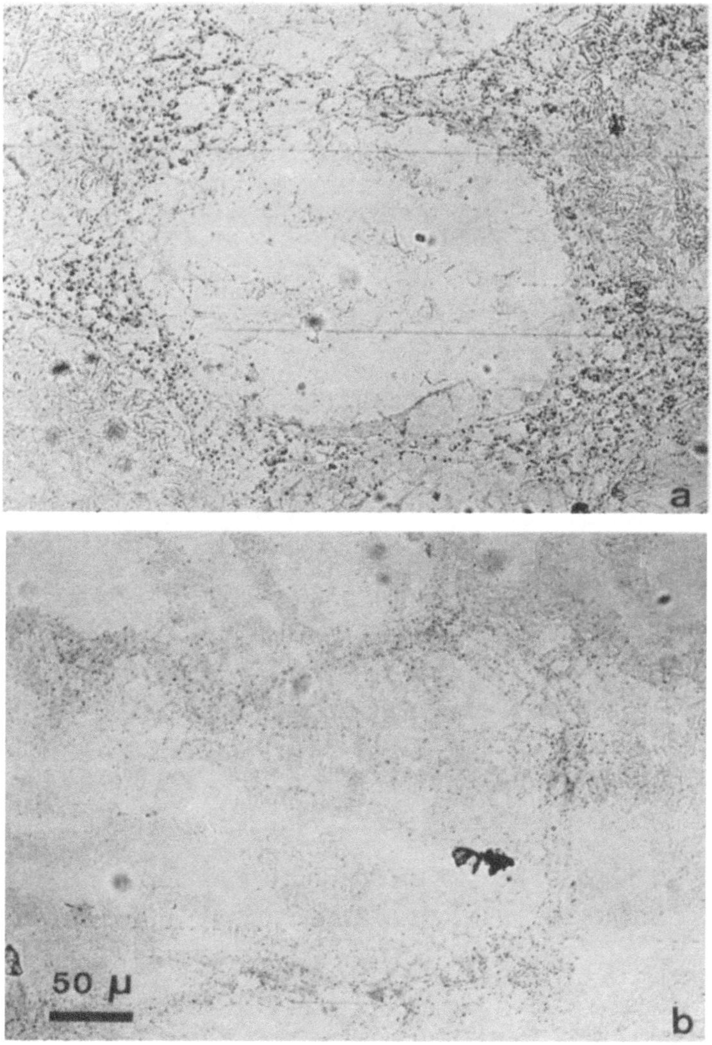

Fig. 7. Droplets of 3β-steroid dehydrogenase in the interstitial tissue of *Bombina bombina* during a calling period (a) and a calling pause (b).

As to the particular androgen most prominently involved, testosterone appeared a likely candidate. In fact it proved possible to stimulate calling activity in these anurans by injecting testosterone, but the calls produced by this method are atypical and resemble those given by animals in the field at the beginning of a calling period. Under natural circumstances these initial calls develop gradually, over several hours, into the typical mating calls. This result suggests to me that testosterone is not the only hormone responsible for control of calling activity, but that other steroids, precursors or breakdown products of testosterone, are also involved. Some of these hormones influence the secondary sexual characteristics in anurans (Müller et al. in press).

This much, then, is known about the endogenous factors that affect mating activity. It remains to be explained why many members of a given population in the field begin giving mating calls simultaneously. Observations have shown that the individual calling periods are often preceded by abundant rainfall. This suggests that rain may be an important exogenous factor. It is not the only synchronizer of calling activity, however. A second important factor is calling itself. If experimental animals not in a state of readiness to call are allowed to hear conspecific mating calls for a few days, subsequent examination of their interstitial tissue reveals that the cell content of the 3β-hydroxyster d dehydrogenase is considerably higher than in the control animals. Thus conspecific mating calls stimulate males that are at a low level of endocrine stimulation. The calls of the first males to become active can stimulate the others and thus tend to synchronize the population as a whole.

In summary we find two types of mating and calling activity in the Central European anurans. The first could be termed a "monophasic type" and the second a "polyphasic type." The mechanisms underlying these different forms of behavior involve differences in function of the endocrine organs, in particular the tests. The endogenously controlled behavior can be modified and synchronized by exogenous factors.

REFERENCES

Heinzmann, U. 1970. Untersuchungen zur Bio-Akustik und Ökologie der Geburtshelferkröte. Oecologia (Berl.) 5:19-53.

Lörcher, K. 1969. Vergleichende bio-akustische Untersuchungen an der Rot- und Gelbbauchunke, Bombina bombina (L.) und Bombina v. variegata (L.). Oecologia (Berl.) 3:84-124.

Müller, B., H.-J. Obert and H. Schneider. Untersuchungen zur hormonalen Steuerung der Ausbildung der Brunftschwielen bei der Riesen-Erdkröte Bufo bufo spinosus Dandin 1803 (Bufonidae, Anura). Zool. Jb., in press.

Obert, H.-J. 1971. Morphologie von Hoden und Hypophyse der Erdkröte (Bufo b. bufo) während der Fortpflanzungsperiode. Zool. Jb. Physiol. 76:185-197.

Obert, H.-J. 1973. Untersuchungen zur hormonalen Steuerung der Ruf- und Paarungsaktivität bei der Rot- und Gelbbauchunke *Bombina bombina* (L.) und *Bombina v. variegata* (L.). *Zool. Jb. 77*: 166-198.

Obert, H.-J. 1975. The dependence of calling activity in *Rana esculenta* Linné 1758 and *Rana ridibunda* Pallas 1771 upon exogenous factors (Ranidae, Anura). *Oecologia* (Berl.) *18*:317-328.

Obert, H.-J. 1965. Some effects of external factors upon the reproductive behavior of the grass frog *Rana t. temporaria* L. (Ranidae, Anura). *Oecologia* (Berl.) *24*:43-55.

Schneider, H. 1967. Rufe und Rufverhalten des Laubfrosches *Hyla arborea arborea* (L.). *Z. vergl. Physiol. 57*:174-189.

Wahl, M. 1969. Untersuchungen zur Bioakustik des Wasserfrosches *Rana esculenta* (L.). *Oecologia* (Berl.) *3*:14-55.

SYSTEMATICS AND HYBRIDIZATION IN THE *RANA ESCULENTA* COMPLEX

Leszek Berger

Ecological Institute, Polish Academy of Sciences
Poznań, Poland

Water frogs inhabit nearly all of Europe, northern Africa and western Asia to Pakistan. The European forms were named *Rana esculenta* by Linnaeus in 1758. Since that time many forms have been described, but the systematic status of the central European forms has been much disputed.

Three morphological forms occur in Europe: the pool frog (*Rana lessonae* Camerano), the lake frog (*Rana ridibunda* Pallas) and the edible frog (*Rana esculenta* Linnaeus). While extensive morphological variation has been widely recognized, there has been little agreement among zoologists about the systematics of this group. In the last 10 to 15 years at least seven hypotheses have been published on the affinities of these frogs; my hypothesis (Berger 1967), that *R. lessonae* and *R. ridibunda* were separate species and that *R. esculenta* was a hybrid between them, raised the most objections. For some years, this hypothesis was not accepted, but its independent confirmation by the Swiss zoologists Blankenhorn *et al.* (1971) marked a turning point. Since that time the European water frogs have inspired much interest in biologists of various specialties.

We have at present numerous observations dealing with special biological problems. We have selected for discussion data that highlight the systematics of the water frogs. Few of these observations have been published. Many of them have been made by my colleagues, above all by R. Günther (Berlin), H. Roguski (Krakow), H. G. Tunner (Vienna) and T. Uzzell (Philadelphia), to each of whom I am much obliged for their manuscripts or information.

DESCRIPTION OF FORMS

The two most useful morphometrical characters for distinguishing the three European forms of water frogs are the ratio of the length of first toe (digitus primus) to the length of the inner metatarsal tubercle (callus internus), and the length of the tibia relative to the length of the femur (Fig. 1). In these two features, *Rana esculenta* is intermediate between *Rana lessonae* and *Rana ridibunda*, but in the rest of its characteristics, it is similar to either *R. ridibunda* or *R. lessonae* (Table 1).

R. lessonae is a small form. During the spring and summer it lives in small bodies of water; it leaves these in September and hibernates on land (Berger 1970). It is a relatively terrestrial, fossorial species and is sexually mature after the first hibernation. *R. ridibunda*, in contrast, is the largest European species of amphibian. It lives in rivers and lakes and hibernates in water.

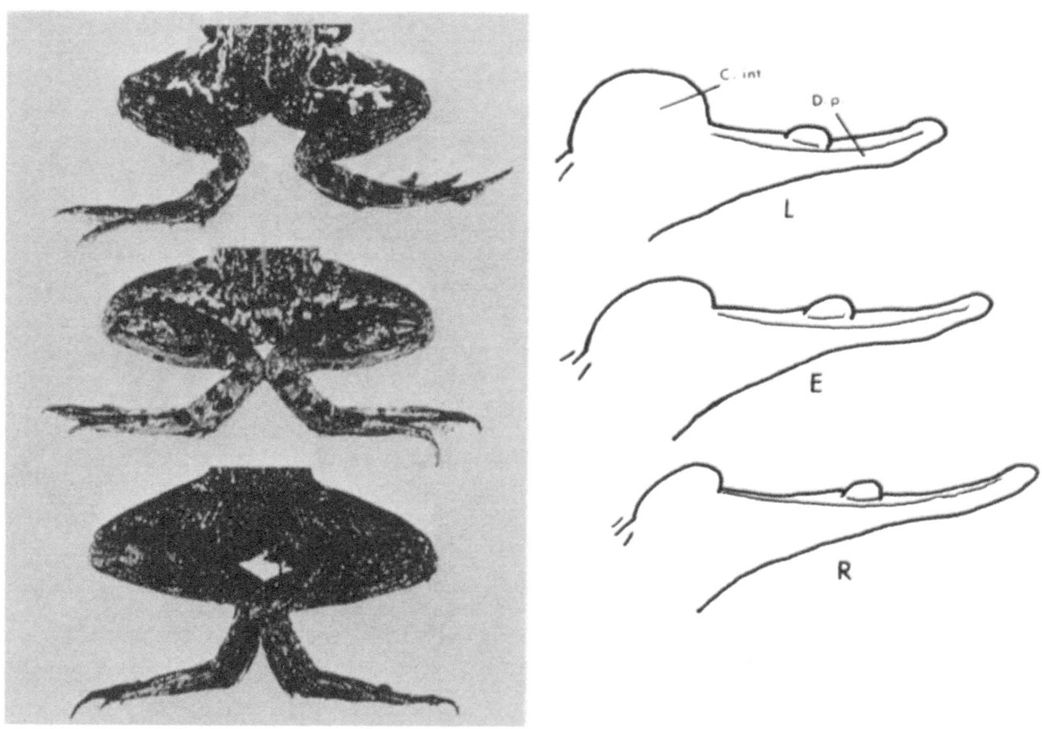

Fig. 1. A comparison of morphological features of three females, all of 75 mm snout-vent length, belonging to the three forms of water frogs. The *Rana lessonae* (L) is an adult full of eggs; the *Rana esculenta* (E) and the *Rana ridibunda* (R) are subadults. D.p., first toe; C. int., metatarsal tubercle (after Berger 1973a).

Table 1. Some characteristics of water frogs from populations from near Poznań, Poland (after Berger 1973c, Berger and Uzzell 1977)

Phenotypes	Rana lessonae	Rana ridibunda	Rana esculenta from field	Rana esculenta from laboratory
		Females		
Number of frogs	142	90	99	97
Body length in mm	42-83	72.5-107	54-115	54-83
Mean	58.3	86.7	81.0	66.3
Body weight in g	7.3-49.3	32.7-145.4	17.0-150.0	-
Mean	22.2	66.2	56.3	-
Number of eggs	480-4465	1230-12520	1085-11334	670-5484
Mean	1656	5469	5325	2680
		Males		
Number of frogs	106	30	31	229
Body length in mm	42.5-65.0	65-99	51-86	42-72
Mean	52.0	78.0	70.0	56.4
Body weight in g	6.6-24.0	21.3-88.0	15.0-54.2	-
Mean	18.0	50.7	30.7	-

It is an aquatic species and produces gametes after the second hibernation. During the breeding season it sometimes leaves its native bodies of water and migrates to ponds situated near rivers and lakes. These ponds, which are often occupied by R. lessonae, are the sites of hybridization of these forms in Poland.

R. esculenta shows no ecological specialization. It lives in all types of bodies of water, usually with individuals of one or the other of its parental species. When individuals of R. esculenta live in small bodies of water, they hibernate on land, but in larger bodies of water they hibernate under water. Tadpoles that live in poor conditions (for example, in aquaria) are small and reach much smaller adult body sizes than those grown in normal conditions (Table 1). The former are very similar in length to individuals of R. lessonae (Berger 1970). Frogs that live in good trophic conditions as tadpoles have body sizes almost as large as individuals of R. ridibunda. Males of R. esculenta are sexually mature after their first hibernation as are both sexes of R. lessonae, but the females mature sexually after the second hibernation, as do both sexes of R. ridibunda (Berger 1970).

The coloration of the body of individuals of R. esculenta is quite similar to that of individuals of R. lessonae. The dorsal side of both forms is usually grass-green, covered with small black spots of a distinct shape. Frogs of these two forms are very rarely brown. During the breeding season the males of both forms, and especially of R. lessonae, turn yellow. The similarity of their coloration led herpetologists in Europe to believe that R. lessonae was a synonym of R. esculenta. Individuals of R. ridibunda are colored quite differently. The upper part of their body is brown-green or dirty-green, never grass-green. On this background are scattered dark spots of indistinct shape. The males do not assume a yellow breeding color.

HYBRIDIZATION

During 1963-1975, we crossed more than 400 individuals belonging to all three phenotypes and to different generations. From nine possible phenotypic combinations we obtained nearly 8000 froglets; about 1500 were saved for further rearing (Berger 1968a, 1970, 1971a, 1973b, 1976).

Our discussion is based primarily on the offspring that were obtained from three series of crosses (Fig. 2), but additional data are included wherever possible. In these series we simultaneously crossed three females and three males of each form. The offspring of all nine combinations completed metamorphosis successfully and reached maturity, but their relative vitality and their patterns of inheritance were surprising. The embryos from two R. esculenta parents had the lowest vitality. Almost all zygotes from 34 pairs perished in development, although many of the tadpoles from one pair reached metamorphosis (Berger 1967); some of them even reached maturity (Berger 1971a). Froglets from this combination often developed from large eggs, which we discuss later. If the same individuals of R. esculenta were crossed with R. lessonae or R. ridibunda, we obtained offspring, many of which reached metamorphosis. The progeny of R. ridibunda and R. lessonae parents and their hybrids were always characterized by high vitality (Berger 1967).

The inheritance of morphological and color features was as follows (Fig. 2, F_1). The offspring of R. ridibunda and R. lessonae were like their parents, while the reciprocal hybrids (RL and LR) were generally intermediate; they had esculenta phenotypes and included both sexes. The inheritance of features in the offspring of the five remaining combinations was remarkable. Progeny from crosses between R. lessonae and R. esculenta had esculenta phenotypes irrespective of the origin of the R. esculenta parent (Berger 1968a, 1971a, 1973b, 1976, Günther 1973, Blankenhorn et al. 1971).

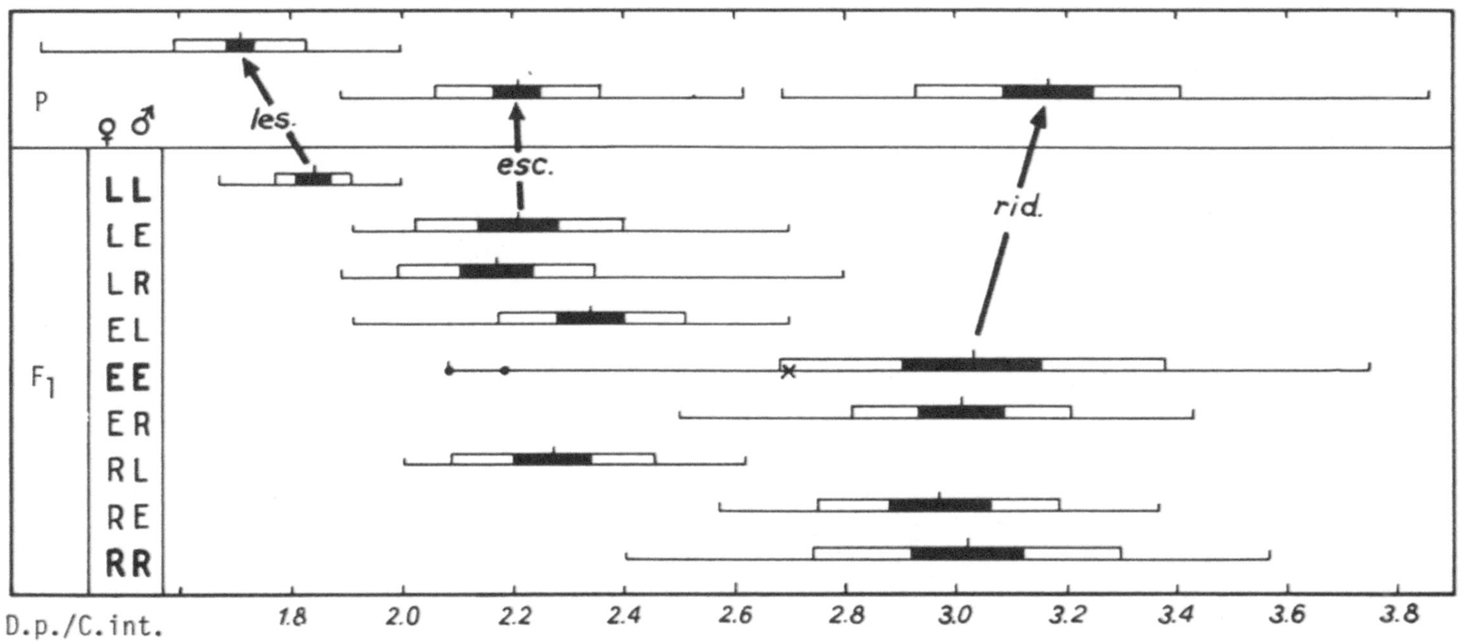

Fig. 2. Inheritance of morphological features in water frogs: Rana lessonae = les, R. esculenta = esc, R. ridibunda = rid. Progeny of R. esculenta parents belong to two phenotypes: esculenta (two EEe specimens marked by dots) and ridibunda (the others, EEr specimens, their lowest index value marked by a cross; after Berger 1970).

If the parents had esculenta or ridibunda phenotypes, the progeny always had ridibunda phenotypes.

Progeny fathered by R. lessonae or R. ridibunda always included both sexes irrespective of the phenotype of the mother (Table 2), but those fathered by R. esculenta usually were either all females or, more rarely, all males; both sexes almost never appeared in a single sibship (Berger 1971b, 1973b, 1976). If the father was a hybrid between an R. ridibunda female and an R. lessonae male (RL), all the offspring were daughters, but if it was hybrid from the reciprocal cross (LR), the offspring were all males (Table 3).

HYBRIDOGENESIS

After analyzing the pattern of inheritance of different features, and especially of electrophoretic phenotypes of serum albumins of the frogs of different phenotypes and generations, Tunner (1973, 1974) concluded that individuals of R. esculenta usually reproduce hybridogenetically. This means that they discard the genome received from one of the parental species during meiosis. This phenomenon was first observed by Schultz (1969) in unisexual fishes of the genus Poeciliopsis.

Diploid Rana esculenta forms two systems (Uzzell and Berger 1975). The individuals from the first (L-E) system almost always transmit a ridibunda genome to progeny and discard the lessonae

Table 2. Sex ratio of the F_1 progeny of water frogs (after Berger 1971b).

Parents: females x males	Symbols	F_1 Progeny: N	Sex ratio: ♀♀	♂♂	?	♀:♂
lessonae x lessonae	LL	453	188	265		0.71:1
" x esculenta	LE	302	263	39		6.75:1
" x ridibunda	LR	353	160	193		0.83:1
esculenta x lessonae	EL	179	80	98	1	0.82:1
" x esculenta	EE	256	249	2	5	124.50:1
" x ridibunda	ER	131	68	55	8	1.24:1
ridibunda x lessonae	RL	172	55	112	5	0.49:1
" x esculenta	RE	194	190	3	1	63.33:1
" x ridibunda	RR	239	115	119	5	0.97:1
TOTAL		2279	1369	885	25	1.55:1

Table 3. Vitality, sex ratio and phenotype of backcrosses from four females that were crossed with different males (after Berger 1973b).

Parent				Progeny (F_2 and B_1)					
	cross			First	Alive after		Sex		Pheno-
No.	♀♀ F_1	♂♂ F_1	P	clea-vage	stage 25 (in percent)	N	♀♀	♂♂	type
11a	RL[1]	LL		475	99	68	31	37	esc
b	"	LE		502	74	61	61	-	rid
c	"	LR		113	10	11	-	11	rid
d	"	LR		423	79	58	-	58	rid
e	"	EL		378	67	58	58	-	rid
f	"	RL		70	0	-	-	-	-
g	"	RL		5	100	5	5	-	rid
29a	EL	EL		287	0				
b	"	LE		109	0				
c	"	LR		123	0				
d	"		les	115	64	12	7	5	esc
e	"		rid	110	42	11	6	5	rid
36a	EEe	LE		100	54	28	28	-	rid
b	"	LL		171	97	65	31	34	esc
c	"	LL		155	94	39	22	17	esc
d	"		esc	87	63	24	24	-	rid
e	"		rid	87	74	39	20	19	rid
41a	EEr	LE		431	73	48	48	-	rid
b	"	LR		328	60	42	-	42	rid
c	"	EL		303	62	37	37	-	rid

[1]The first letter indicates phenotype of mother, the second, of the father; L = R. lessonae, E = R. esculenta, R = R. ridibunda; EEr, EEe = respectively froglets with ridibunda and esculenta phenotypes derived from crosses between two R. esculenta.

genome. Populations that belong to this system are composed, as a rule, of individuals of Rana lessonae and Rana esculenta. Such populations occur in Poland (Berger 1968a), Germany (Günther 1973) Austria (Tunner 1974, Tunner and Dobrowsky 1976) and Switzerland (Blankenhorn et al. 1971). Individuals of Rana esculenta from the second (R-E) system in which R. ridibunda males and females are also present, transmit lessonae chromosome sets to the progeny two thirds of the time on the average and ridibunda chromosome sets one third of the time (Uzzell et al. 1977). Such populations, in which males of Rana esculenta often predominate, are best known from Germany (Günther 1973, 1974).

The male R. esculenta of both systems are heterogametic, but the genetic material of their gametes differs. The progeny fathered by R. esculenta from the L-E system are, as a rule, daughters. When crossed to female Rana esculenta from the L-E system, 96.5% of them had ridibunda phenotypes (Berger 1968a); all were females (Berger 1971a). Esculenta phenotypes appeared in 3.5% of them; both males and females were present. In progeny fathered by R. esculenta from the R-E system, sons received lessonae genomes from the father, whereas daughters received ridibunda sets (Uzzell et al. 1977).

PLOIDY

An interesting phenomenon in water frogs is the production by R. esculenta of gametes that vary strikingly in size (Table 4; Berger 1967, Günther 1970). On the basis of 134 females we found (Berger and Uzzell 1977) that in populations from near Poznań, Poland, 74% laid eggs that form two (36% of females), three (32%), or even four (7%) size classes. The remaining females (26%) laid eggs in only one size class. The froglets raised from large eggs were usually very vital (Berger and Uzzell 1976). On the basis of their morphology, electrophoresis of some enzymes, and cell sizes, we found that they were generally triploids (Uzzell et al. 1975). If they were fathered by R. lessonae, their genotypes were composed of two sets of lessonae and one set of ridibunda chromosomes (LLR); but if the fathers had esculenta or ridibunda phenotypes, the froglets had one set of lessonae and two sets of ridibunda chromosomes (LRR).

The young from the medium-sized eggs were also very vital, but those from small eggs usually perished in development. The froglets from these two groups of eggs were diploids; they were very similar because they had the same genotypes. If they were fathered by R. lessonae, their genotypes were RL and if by R. ridibunda or R. esculenta, their genotypes were RR.

These data suggest that the large eggs were diploid with one ridibunda and one lessonae genome; the small and medium-sized eggs were haploid with one ridibunda genome.

The coloration of RL, LLR and LRR hybrids was similar to that of R. esculenta individuals from nature; the morphology of LLR hybrids was intermediate between that of R. lessonae and of LR individuals, whereas LRR hybrids are intermediate between LR individuals and R. ridibunda (Uzzell et al. 1975). Günther (1975a) found

Table 4. Examples of differentiated eggs laid by Rana esculenta females (after Berger and Uzzell 1977, Berger and Roguski unpublished). The ploidy of the progeny derived from larger eggs (the sizes outlined by boxes) are indicated. Females 1-5 were reared in the laboratory; 6-17 were caught in nature. One lab-reared female laid eggs in two successive years (1/73 and 26/74). The eggs of females 15-17 were not differentiated.

Females			Diameter of Eggs (mm)														Ploidy of Progeny
No.	N/y[1]	Length (mm)	1.0	1.1	1.2	1.3	1.4	1.5	1.6	1.7	1.8	1.9	2.0	2.1	2.2	2.3	
1	11/76	60			5	11	30	25	23	34	24	12	16				3n
2	5/71	78					6	38	83	63	31	14	8	2			2n + 3n
3	7/76	61	12	1	55	46	1	14	7	2	7	3					2n + 3n
4	1/73	78		3	4	11	98	82	11		4	3	3				3n
5	26/74	82				2	7	8	42	83	27		1	3	2	1	3n
6	14/75	54	5	16	31	2	3	3	1								3n
7	15/75	73		2		5	42	3	3	5							3n
8	13/74	79			2	3		34	15	2	1	5	1				3n
9	14/74	79			2	3	8	26	20	6		4					2n + 3n
10	2/76	80				9	1		30	8		4					3n
11	16/75	83		7	26		28	23		2	4	1					2n + 3n
12	5/76	86				10	2		30	5	1	4	4				3n
13	6/76	91					3	10	1	15	30	3	2	4	2		3n
14	7/75	91				1	2		7	35	25	3	5	8	4		2n + 3n
15	25/68	85						23	27								2n
16	20/74	86			7	26	15	2									2n
17	8/74	91						9	36	5							2n

[1]N = The number of a female in a particular year; y = the year in which the cross was made.

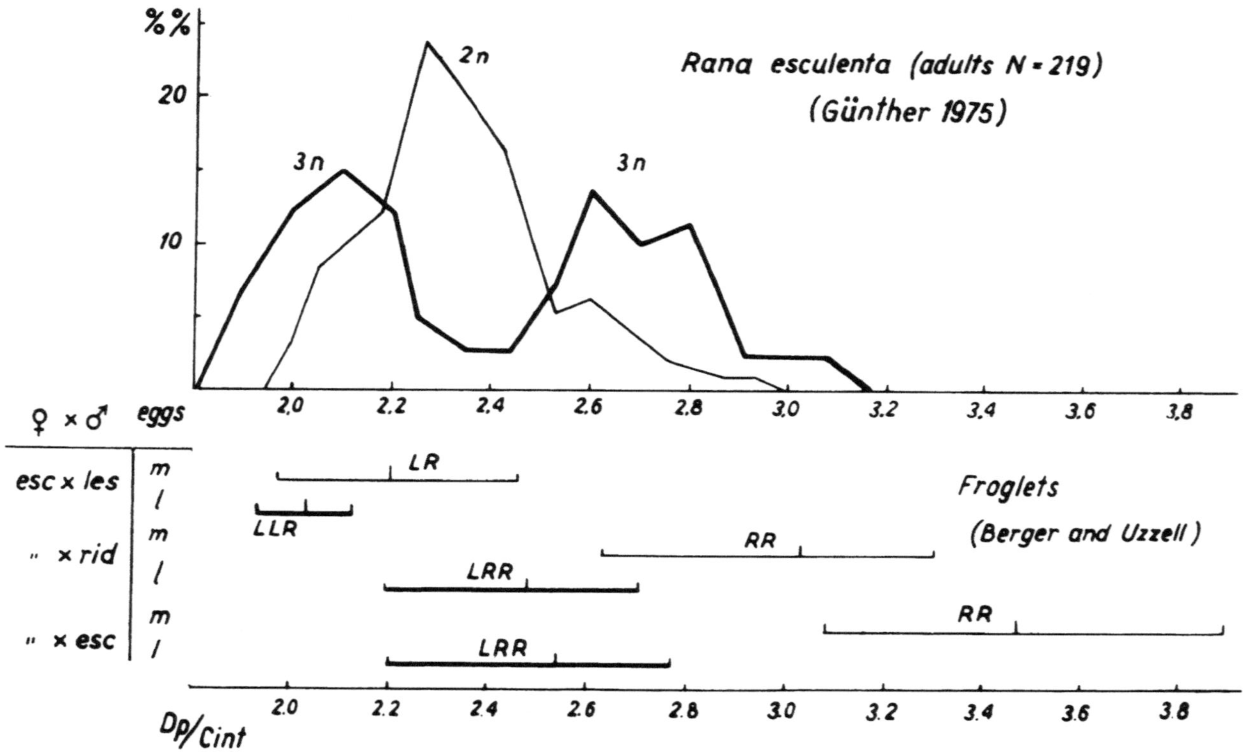

Fig. 3. Above, ploidy and morphological differentiation of adult Rana esculanta caught in nature (after Günther 1975a).

Below, ploidy and morphology of R. esculenta (LR, LLR and LRR) and R. ridibunda (RR) froglets derived from medium sized (m) and large (l) eggs in laboratory (after Berger and Uzzell 1976).

two morphological classes of triploids (Fig. 3) and surmised that they are represented by these two classes.

The mean diameter within particular classes of eggs was often arranged in such a way that their volumes were as 1:2:4. In these cases it was easy to divide the eggs into separate size classes (Fig. 4). It is, however, impossible to establish the ploidy of the progeny if we know only the diameter of the eggs (Uzzell et al. 1975, Roguski, unpublished, Berger and Roguski, unpublished; cf. Table 4).

The differentiated eggs are not evenly distributed in the size classes (Table 5). The medium-sized class contains 87.9-99.7% (mean 96.22%) of all eggs laid, the small-size class 0.09-11.84% (mean 3.28%), and the large-size class 0.12-3.02% (mean 0.83%).

HETEROSIS

Heterosis or hybrid vigor is widely believed to occur in hybrids in many animal groups. Possible heterosis in esculenta hybrids includes their rapid development rate. R. esculenta froglets, especially those resulting from crosses between R. ridibunda and R. lessonae, show a surprising tendency to migrate not manifested by other phenotypes (Berger 1970, 1973a). Because of this, in Poland R. esculenta always appears before the two other species in all new artificial bodies of water situated within a few km of natural bodies of water (Berger, unpublished). Tadpoles and froglets seldom appear in these new bodies of water but, when they do, they have ridibunda phenotypes (Table 6, No. 5).

Hybrid tadpoles with esculenta phenotypes have unusual characteristics. If they live with R. lessonae tadpoles in a common aquarium they inhibit the development of their aquarium-mates; in overcrowded aquaria, some of the organs in R. lessonae individuals are even deformed (Berger 1968b).

The superiority of hybrids is especially distinct in adult R. esculenta females. For body lengths below 85 mm, R. esculenta females are heavier than the R. ridibunda of the same size (Berger and Uzzell 1977). At 80-85 mm, the difference is about 12%; above 90 mm, R. ridibunda females are heavier than R. esculenta females (Fig. 5). Their superiority is emphasized in a special way by two factors: 1) Most R. esculenta females lay eggs at about 75-85 mm body length while R. ridibunda females first lay at 85-90 mm; 2) at this body length R. esculenta females lay, on the average, 1700 more eggs than R. ridibunda females; with larger body size this difference diminishes but never averages less than 500 more eggs. This is reflected by the fertility index, which gives the

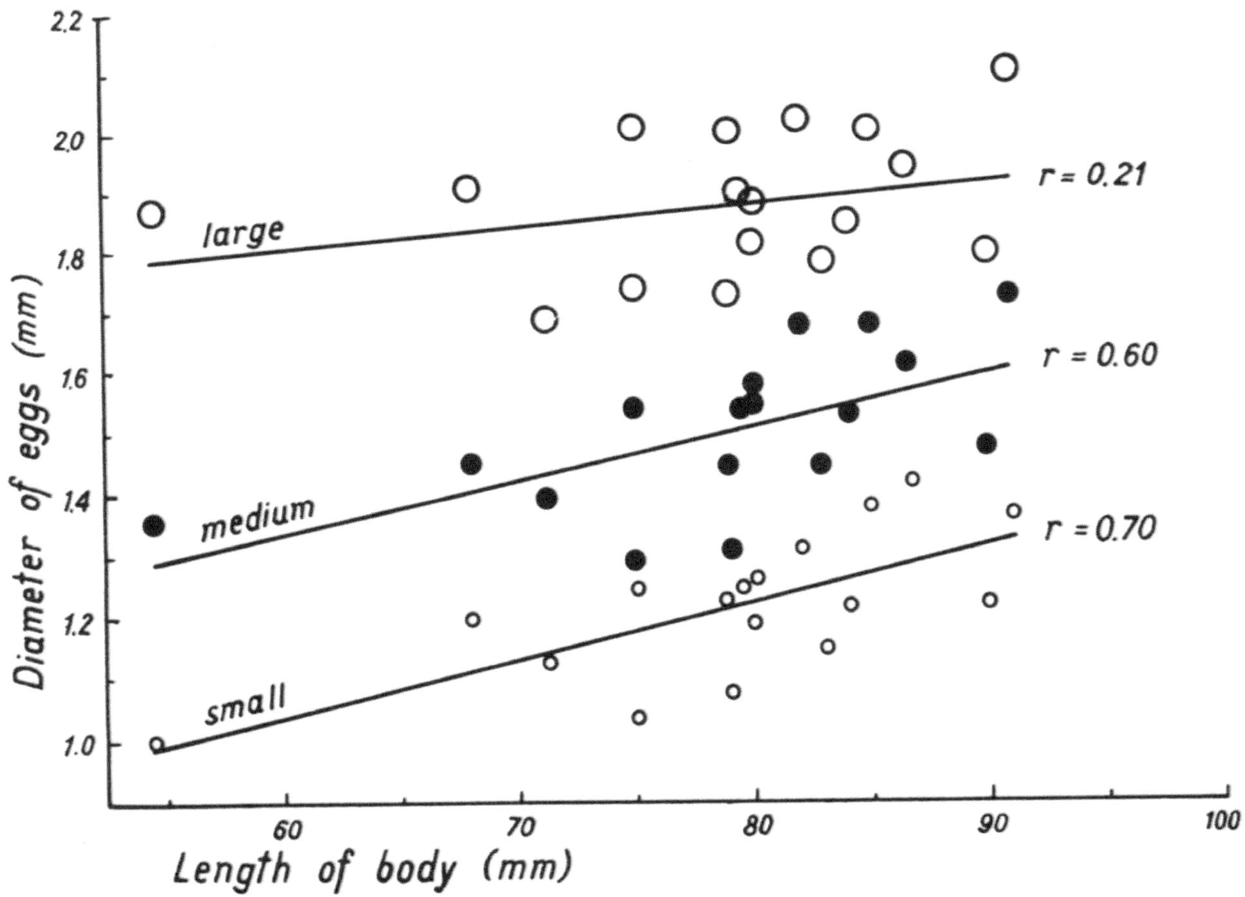

Fig. 4. The eggs of 17 *Rana esculenta* females having ova that were differentiated into three size classes. Mean values of each class are marked by separate points (small, medium and large), r = correlation coefficient (after Berger and Uzzell 1977).

Table 5. The eggs of Rana esculenta females divided on the basis of size. Female No. 4 (7/76) laid two classes of large eggs that are grouped together (cf. Table 4, no. 3; after Berger and Uzzell 1977).

Females			Eggs						
		Length		Number			Percent		
No	N/y[1]	mm	Total	s	m	l	s	m	l
			From laboratory						
1	1/73	78	2285	?	?	29	?	?	1.27
2	26/74	82	5484	156	5289	39	2.84	96.44	.71
3	31/74	78	3993	30	3963	-	.75	99.25	-
4	7/76	61	1358	37	1280	41	2.72	94.26	3.02
5	13/76	60	836	15	808	13	1.79	96.65	1.56
			From nature						
1	7/74	70	1657	-	1652	5	-	99.70	.30
2	15/74	72	1087	7	1073	7	.64	98.71	.64
3	11/75	78	5579	244	5306	29	4.37	95.11	.52
4	10/75	82	2972	183	2776	14	6.12	93.41	.47
5	16/75	83	3335	216	3103	16	6.48	93.04	.48
6	2/72	85	4305	8	4292	5	.18	99.70	.12
7	5/76	86	3800	450	3341	9	11.84	87.92	.24
8	7/75	91	5723	5	5673	45	1.09	99.12	.79
9	6/76	91	5837	241	5579	17	4.13	95.58	.29
10	3/76	100	8599	58	8840	101	.67	98.15	1.18
	Mean	79.8	3897	127	3783	26	3.28	96.22	.83

[1]N = the number of the female in a particular year;
y = the year in which the cross was made.

number of eggs per gram of body weight. In R. esculenta females this averages 102, while in R. lessonae and R. ridibunda females it is 75 and 66, respectively.

In R. esculenta males we can see the reverse phenomenon; their fertility is low as a rule; many of them are quite sterile (Berger 1970, 1971a, 1973c, Günther 1973).

Table 6. Phenotypes of froglets from several populations near Poznań in 1963 (Berger, unpublished)

Phenotypes of Adults	Phenotype and Sex of Progeny								
	lessonae ♀♀	lessonae ♂♂	ridibunda ♀♀	ridibunda ♂♂	esculenta ♀♀	esculenta ♂♂	lessonae ♀ : ♂	ridibunda ♀ : ♂	esculenta ♀ : ♂
1. _R. lessonae_ + _R. esculenta_	255	256	87	9	264	205	1 : 1	9.67 : 1	1.29 : 1
2. _R. lessonae_ + _R. esculenta_	3	1			109	57	3 : 1		1.91 : 1
3. _R. ridibunda_ + _R. esculenta_			135	59	4	3		2.29 : 1	1.33 : 1
4. pure _R. ridibunda_			236	254				0.93 : 1	
5. pure _R. esculenta_			58	14				4.14 : 1	

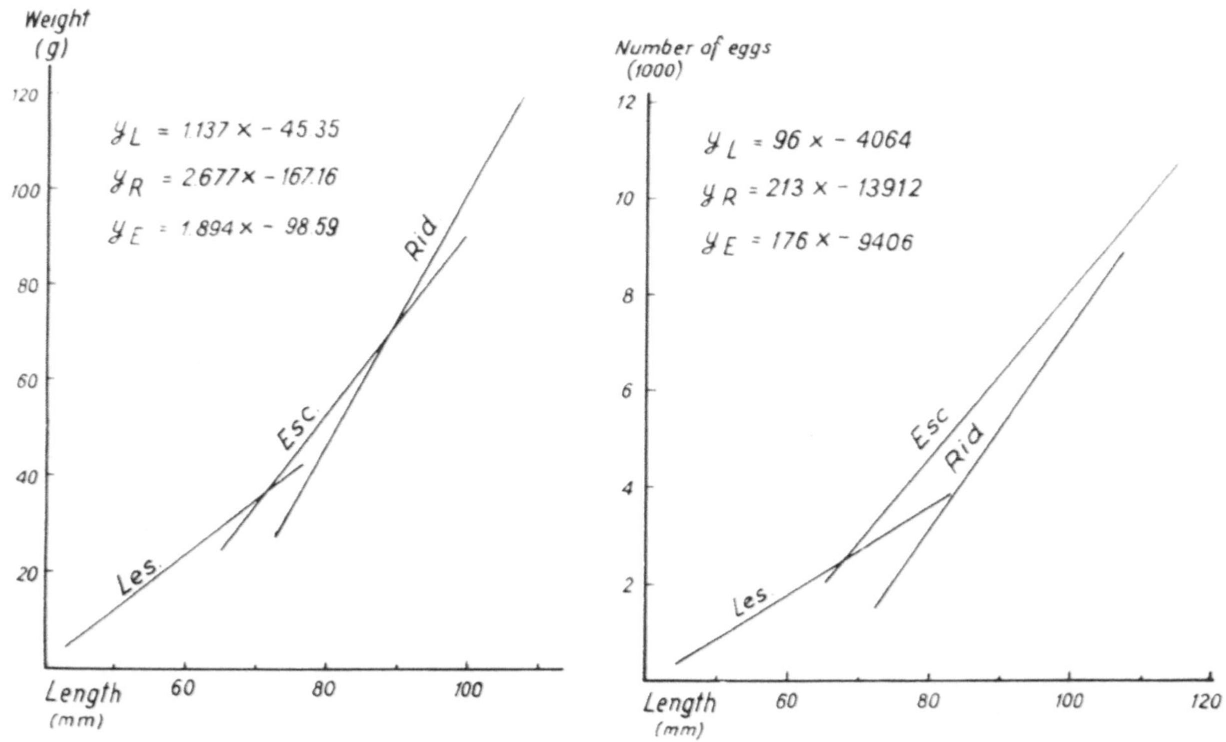

Fig. 5. The relation of body length of Rana lessonae, R. esculenta and R. ridibunda females to body weight in g and to the number of eggs in thousands (after Berger and Uzzell 1977). The regression lines were fit by least squares.

POPULATIONS

Distribution of the forms of water frogs in Europe and especially near the borders of their ranges is poorly known. Our best data are from central Europe. Individuals with esculenta phenotype are in general distributed throughout the common range of R. lessonae and R. ridibunda. Most often R. esculenta occurs with R. lessonae, rarely with R. ridibunda. In mixed R. lessonae-R. esculenta populations, the R. esculenta are generally more numerous than the R. lessonae (Berger unpublished, Blankenhorn et al. 1971, Halfmann and Müller 1972, Blankenhorn 1973, Tunner 1974, Günther 1975a, Tunner and Dobrowsky 1976). Some parts of Germany and Poland, where pure populations of R. esculenta live, are exceptions.

Pure populations of R. esculenta that live in the northern parts of Germany and Poland are among the more interesting ones. Often more than 75% of their members are triploid (compare Fig. 3; Uzzell and Berger 1975, Günther 1975a). From crosses among them Günther (1973) received offspring belonging to all three phenotypes. In northern Poland, we caught in one pond (Uzzell and Berger 1975) 36 adults of which 27 were triploids. The meiosis of triploid individuals will be discussed separately (Günther et al. 1977).

I have examined frogs from central Italy that in morphology resemble R. esculenta of central Europe. I have no other information on them that would permit us to draw any conclusions on their relationship. It is not unlikely that these frogs form a separate taxon as Hotz (1974) supposed. To solve this problem, special field and laboratory investigations should be undertaken.

In the L-E system in Austria, interesting mixed populations were described (Tunner 1974, Tunner and Dobrowsky 1976) that consist entirely of female R. esculenta with rare males and females of R. lessonae. We do not yet know the mechanism that controls the inheritance within these populations, but Tunner's data suggest that the lessonae males are homogametic rather than heterogametic, in contrast to Polish and Swiss populations (Berger 1966, Blankenhorn et al. 1971). If further investigations confirm this, the populations discussed would belong to the ZW type and would be an exception not only among water frogs, but also for the genus Rana in general (Kawamura and Nishioka, this volume).

The composition of adult frog populations near Poznań, Poland, depends entirely on the type of biotope. The mixed R. lessonae-R. esculenta and R. ridibunda-R. esculenta populations are the most numerous; pure R. ridibunda and pure R. esculenta populations are very rare; we have not yet found, near Poznań or anywhere else in Poland, pure R. lessonae populations. We know little about the ploidy of R. esculenta individuals in Poland (Uzzell and Berger 1975) but Günther's (1975a, 1975b) research shows that throughout the German Democratic Republic triploid individuals having esculenta phenotypes are very common.

In many cases it has been difficult to explain the phenotypic structure of mixed R. lessonae-R. esculenta populations near Poznań. Of many such populations we will discuss only two, which inhabit shallow ponds situated in the forests. Each population consists of about 2,000 adults. In the first population about 95% of the frogs are R. lessonae and 5% R. esculenta (Table 6, population 1); in the second, the ratio is reversed (about 5% R. lessonae, 95% R. esculenta; Table 6, population 2). These ratios have remained about the same for over ten years. The phenotypes of the progeny in these populations for 1963 suggest that there is much differential survival of phenotypes (Table 6).

Froglets of all three phenotypes appeared among the offspring from the first population, but those of R. lessonae and R. esculenta emerged almost every year, whereas froglets with R. ridibunda phenotypes were caught only in some years. The completely metamorphosed froglets that appeared first (July 10-20) always had esculenta phenotypes; several days later froglets with lessonae phenotypes emerged. Froglets with ridibunda phenotypes appeared 5-6 weeks later at the beginning of September. The sex ratio within the lessonae phenotype is 1:1; in the esculenta phenotype there are a few more females than males, but in the ridibunda phenotype most are females. In the second population, female R. esculenta froglets are nearly twice as numerous as the males. The preponderance of female R. esculenta in such populations suggests that some of them had lessonae mothers and esculenta fathers; generally only females appear from such crosses (compare Table 3 and Berger 1971b).

In mixed R. ridibunda-R. esculenta populations near Poznań (Table 6, population 3) the froglets have, as a rule, a ridibunda phenotype; most of them are females. The few specimens among them having esculenta phenotypes appear to be LRR triploids on the basis of morphological features; cell size and electrophoretic data are lacking. The R. esculenta parents in these populations therefore appear to be migrants from an L-E system population rather than members of an R-E system population. In pure R. ridibunda populations only froglets with ridibunda phenotypes appeared; the sex ratio is 1:1. In pure R. esculenta populations R. ridibunda froglets appeared; most of them were females.

The data discussed, which were difficult to explain in 1973, can be easily elucidated on the basis of Tunner's (1973, 1974) hybridogenesis hypothesis. Some problems are not clear as yet. We can not explain, for example, why the progeny fathered by EL or LE hybrid males are almost all females (Table 3). According to Tunner's hypothesis, in the first cross all progeny should be females, in the second, males only.

CONCLUSIONS

Although many problems remain to be solved, that *R. esculenta* represents a hybrid between *R. ridibunda* and *R. lessonae* has been shown by experimental crosses (Berger 1968a, 1973b, 1976, Blankenhorn *et al*. 1971, Günther 1973) and by electrophoretic examination of serum albumin (Tunner 1970, 1972, 1974, Engelmann 1972, 1973) and several enzymes (Engelmann 1974, Uzzell and Berger 1975, Günther and Hahnel 1976, Tunner and Dobrowsky 1976). In this situation the question arises: what is the systematic status of the parental forms (*Rana ridibunda* and *Rana lessonae*)?

We know that the parental forms are characterized by many distinct features; in spite of that, they easily produce hybrids that not only display higher vitality but that also have more advantageous biological features than their parental forms. From that we might draw the conclusion that these forms are closely related to each other, and that *R. ridibunda* and *R. lessonae*, which seem like species with respect to their morphological and ecological features, should be considered as taxa (subspecies or semispecies) that have a common gene pool and that can freely exchange their genes. Analysis of the features of successive generations, however, does not confirm this view. It appears rather that *R. lessonae* and *R. ridibunda* rarely exchange alleles. If gene exchange takes place, as some results suggest (Günther and Hahnel 1976, Uzzell *et al*. 1977), its biological and genetical impact is rather slight. Thus, there are no intermediate individuals other than triploids among the three phenotypes of frogs in nature; introgressant alleles might, of course, contribute adaptively valuable variation to either parental species.

The morphological analysis of F_1 hybrids (Berger 1968a, 1970, Blankenhorn *et al*. 1971, Günther 1973) and of hybrids of later generations obtained from different lines and crossing directions (Berger 1971b, 1973b, 1976) allows a statement that, regardless of the origin of the parents, the inheritance of characters within the L-E system always follows the same scheme:

1. Frogs with lessonae and ridibunda phenotypes give progeny that always resemble their parents, while their reciprocal hybrids (RL and LR) always have esculenta phenotypes and intermediate characteristics.

2. Progeny of crosses between *R. lessonae* and *R. esculenta* (LE and EL) always have esculenta phenotypes, and are morphologically indistinguishable from hybrids of the F_1 generation (RL, LR).

3. Progeny of crosses between R. esculenta and R. ridibunda (RE and ER) and of R. esculenta with R. esculenta (EE) almost always have ridbunda phenotypes.

4. Disturbances of the sex ratio depend totally on the origin of males having an esculenta phenotype; the progeny of fathers that arise from RL crosses are always females; those from the reciprocal (LR) crosses are always males.

The above scheme permits us to conclude that, in European water frogs, the phenotype of the progeny is predictable provided that the phenotype of the parents is known; further, the progeny's sex is predictable if the exact origin of a father having an esculenta phenotype can be traced back and if we know the ploidy of the gametes produced.

It is thus clear that, while the individuals having esculenta phenotype are vessels in which alleles of both forms are present, some genetic difference serves as a barrier that prevents exchange between their chromosomes during meiosis. That is to say, independent assortment of chromosomes does not take place in specimens of the esculenta phenotype in central Europe.

REFERENCES

Berger, L. 1966. Biometrical studies on the population of green frogs from the environs of Poznan. Annales Zool. 23:303-324.

Berger, L. 1967. Embrional and larval development of F_1 generation of green frogs different combinations. Acta Zool. Cracoviensia 12:123-160.

Berger, L. 1968a. Morphology of the F_1 generation of various crosses within Rana esculenta-complex. Acta Zool. Cracoviensia 13:301-324.

Berger, L. 1968b. The effect of inhibitory agents in the development of green-frog tadpoles. Zool. Poloniae 18:381-390.

Berger, L. 1970. Some characteristics of the crosses within Rana esculenta complex in postlarval development. Annales Zool. 27: 373-416.

Berger, L. 1971a. Viability, sex and morphology of F_2 generation within forms of Rana esculenta-complex. Zool. Poloniae 21:345-393.

Berger, L. 1971b. Sex ratio in the F_1 progeny within forms of Rana esculenta complex. Genetica Polonica 12:87-101.

Berger, L. 1973a. Systematics and hybridization in European green frogs of Rana esculenta complex. J. Herp. 7:1-10.

Berger, L. 1973b. Some characteristics of backcrosses within forms of Rana esculenta complex. Genetica Polonica 14:413-430.

Berger, L. 1973c. Sexual maturity of males within forms of Rana esculenta complex. Zool. Poloniae 22:177-188.

Berger, L. 1976. Hybrids of B_2 generations of European water frogs (Rana esculenta complex). Annales Zool. 33:201-214.

Berger, L. and T. Uzzell. 1977. Vitality and growth of progeny from different egg size classes of Rana esculenta L. (Amphibia, Salientia). Zool. Poloniae (in press).

Berger, L. and T. Uzzell. 1977. The eggs of European water frogs and their hybrids (Rana esculenta complex). (in preparation)

Blankenhorn, H. J. 1973. Zum Stand der Forschung über die Verbreitung der Grünfrösche im Kanton Zürich. Revue Suisse Zool. 80:656-661.

Blankenhorn, H. J., H. Heusser and P. Vogel. 1971. Drei Phänotypen von Grünfröschen aus dem Rana esculenta-Komplex in der Schweiz. Revue Suisse Zool. 78:1243-1247.

Engelmann, W. E. 1972. Disk-Elektrophorese der Serumproteine von Wasserfröschen. Ein Beitrag zur Diskussion über den Hybridcharakter von Rana esculenta L. Acta biol. med. germ. 29:431-435.

Engelmann, W. E. 1973. Zur Frage der verwandtschaftlichen Beziehungen europäischer Grünfrösche (Gattung Rana). Eine vergleichende elektrophoretische Untersuchung der Serumproteine. Zool. J. Syst. 100:183-196.

Engelmann, W. E. 1974. Vergleichende Enzymuntersuchungen der Serum mitteleuropäischer Ranidae (Amphibia, Anura). Experientia 30:870-873.

Günther, R. 1970. Der Karyotyp von Rana ridibunda Pall. und das Vorkomen von Triploidie bei Rana esculenta L. (Anura, Amphibia). Biol. Zbl. 89:327-342.

Günther, R. 1973. Über die verwandtschaftlichen Beziehungen zwischen den europäischen Grünfröschen und den Bastardcharakter von Rana esculenta L. (Anura). Zool. Anz. 190:250-285.

Günther, R. 1974. Neue Daten zur Verbreitung und Ökologie der Grünfrösche (Anura, Ranidae) in der DDR. Mitt. Zool. Mus. Berlin 50:287-298.

Günther, R. 1975a. Zum naturlichen Vorkommen und zur Morphologie triploider Teichfrösche, "Rana esculenta" L., in der DDR (Anura, Ranidae). Mitt. Zool. Mus. Berlin 51:145-158.

Günther, R. 1975b. Untersuchungen der Meiose bei Männchen von Rana ridibunda Pall, Rana lessonae Cam. und der Bastardform "Rana esculenta" L. (Anura). Biol. Zbl. 94:277-294.

Günther, R. and S. Hähnel. 1976. Untersuchungen über den Genfluss zwischen Rana ridibunda und Rana lessonae sowie die Rekombinationsrate bei der Bastardform Rana "esculenta" (Anura, Ranidae). Zool. Anz. 197:23-38.

Günther, R., T. Uzzell and L. Berger. 1977. The meiotic pattern of triploid Rana esculenta (Amphibia, Salientia). (in prep.)

Halfman, H. and P. Müller. 1972. Populationsuntersuchungen an Grünfröschen im Saar-Mosel-Raum. Salamandra 8:112-116.

Hotz, H. 1974. Ein Problem aus vielen Fragen - europäische Grünfrösche (Rana esculenta-Komplex) und ihre Verbreitung. Natur u. Museum 104:262-272.

Schultz, R. J. 1969. Hybridization, unisexuality, and polyploidy in the teleost Poeciliopsis (Poeciliidae) and other vertebrates. Am. Natur. 103:605-619.

Tunner, H. G. 1970. Das Serumeiweissbild einheimischer Wasserfrösche und der Hybridcharakter von Rana esculenta. Verh. Dtsch. Zool. Ges. 23:352-358.

Tunner, H. G. 1972. Serologische und morphologische Untersuchungen zur Frage der Artabgrenzung bei Wasserfröschen aus der Umgebung von Mainz (Rhein-Main-Gebiet). Z. Zool. Syst. Evol.-Frosch. 10:127-132.

Tunner, H. G. 1973. Das Albumin und andere Bluteiweisse bei Rana ridibunda Pallas, Rana lessonae Camerano, Rana esculenta Linné und deren Hybriden. Z. Zool. Syst. Evol.-Forsch. 11:219-233.

Tunner, H. G. 1974. Die klonale Struktur einer Wasserfröschpopulation. Z. Zool. Syst. Evol.-Forsch. 12:309-314.

Tunner, H. G. and T. Dobrowsky. 1976. Zur morphologischen, serologischen und enzymologischen Differenzierung von Rana lessonae und der hybridogenetischen *Rana esculenta* aus dem Seewinkel und dem Neusiedlersee (Österreich, Burgenland). *Zool. Anz.* *197*:6-22.

Uzzell, T. and L. Berger. 1975. Electrophoretic phenotypes of *Rana ridibunda*, *Rana lessonae*, and their hybridogenetic associate, *Rana esculenta*. *Proc. Acad. Nat. Sci. Phila.* *127*:13-24.

Uzzell, T., L. Berger and R. Günther. 1975. Diploid and triploid progeny from a diploid female of *Rana esculenta* (Amphibia, Salientia). *Proc. Acad. Nat. Sci. Phila.* *127*:81-91.

Uzzell, T., R. Günther and L. Berger. 1977. *Rana ridibunda* and *Rana esculenta*: a leaky hybridogenetic system. *Proc. Acad. Nat. Sci. Phila.* *128*: 147-171.

REPRODUCTION AND MATING BEHAVIOR IN *RANA LESSONAE* - *RANA ESCULENTA* MIXED POPULATIONS

H.J. Blankenhorn
Ethologie und Wildforschung
University of Zurich, Switzerland

From the results of Berger's studies on European green frogs in Poland, it is known that *Rana esculenta* Linnaeus is a hybrid between *Rana lessonae* Camerano and *Rana ridibunda* Pallas (Berger 1964, 1966, 1967, 1968a, 1968b, 1970). When the current study was begun, only *Rana esculenta* had been reported for the greater part of Switzerland. Actually, three types of green frogs, *Rana esculenta*, *Rana lessonae* and *Rana ridibunda* occur in this region (Blankenhorn *et al.* 1971). The natural ratio of the three phenotypes is peculiar. Out of 24 frog populations explored in the course of the study, only three were pure ridibunda populations, whereas 21 populations were mixed populations of *R. lessonae* and *R. esculenta* (Blankenhorn 1973). The *R. ridibunda* populations are separated geographically from the *R. lessonae* and *R. esculenta* region, and there is strong evidence that *R. ridibunda* was introduced between 1950 and 1970 by medical institutes using *R. ridibunda* from eastern Europe for pregnancy tests (Blankenhorn 1973).

Since there are no other *R. ridibunda* populations known in the region of eastern Switzerland, I presumed that the natural green frog populations in this area are of the *R. lessonae*-*R. esculenta* type.

The emphasis of my paper lies, therefore, on some causal and functional aspects of breeding behavior in *R. lessonae*-*R. esculenta* populations, on the presentation of ethological experiments on reproductive behavior of *Rana lessonae* and *Rana esculenta* and on speculations about the adaptivity of this "species-hybrid" system. I will develop this in three sections, presenting data from a field study, an experimental approach to sexual behavior and a synthesis in form of a discussion.

The main study area was a gravel pit near Zurich (Fig. 1), which contained an estimated 220 individuals, 147 of which were individually marked (Fig. 2), 131 of esculenta-phenotype and 16 of lessonae-phenotype. The ratio between R. lessonae and R. esculenta individuals was estimated at 1 to 10, R. lessonae being caught more easily than R. esculenta individuals (Blankenhorn 1974).

REPRODUCTION IN A NATURAL R. lessonae-R. esculenta POPULATION

In the breeding season, the males gather together on a small traditional site in the pond and show increased activity and calling. High densities (25 males per m^2) are common. The females come singly to this arena, copulate, lay their eggs in the center of the arena and leave it again. The location of the arena was the same in three successive years.

Answers to the following questions were sought: 1) How are the different types and age classes of the males (Fig. 3) distributed

Fig. 1. The gravel pit in Gossau near Zurich. The grid is 1 x 1 m.

Fig. 2. An individually marked male R. esculenta.

in the arena? 2) Do the different males show differences in behavior, especially in sexual and aggressive behavior? 3) What are the mating rates for the males?

Four colleagues helped me in collecting the data. They were first trained in the use of the R. esculenta-R. lessonae ethogram (Blankenhorn 1974). The investigation started only when each observer's observations agreed with mine over 80% of the time. Observation conditions were excellent because the flight distances of the frogs are reduced to between 1 and 2 m during the breeding period. On each of three days, between 0800 and 1200 and between 1400 and 1800, data were collected during two hour intervals each time giving a total of 48 hours of investigation. In each two hour session, four marked animals of each age and type class were observed for 10 minutes. The remaining observations (eight at 10 minutes each) were on randomly chosen marked animals. Each hour, marked and unmarked animals were mapped.

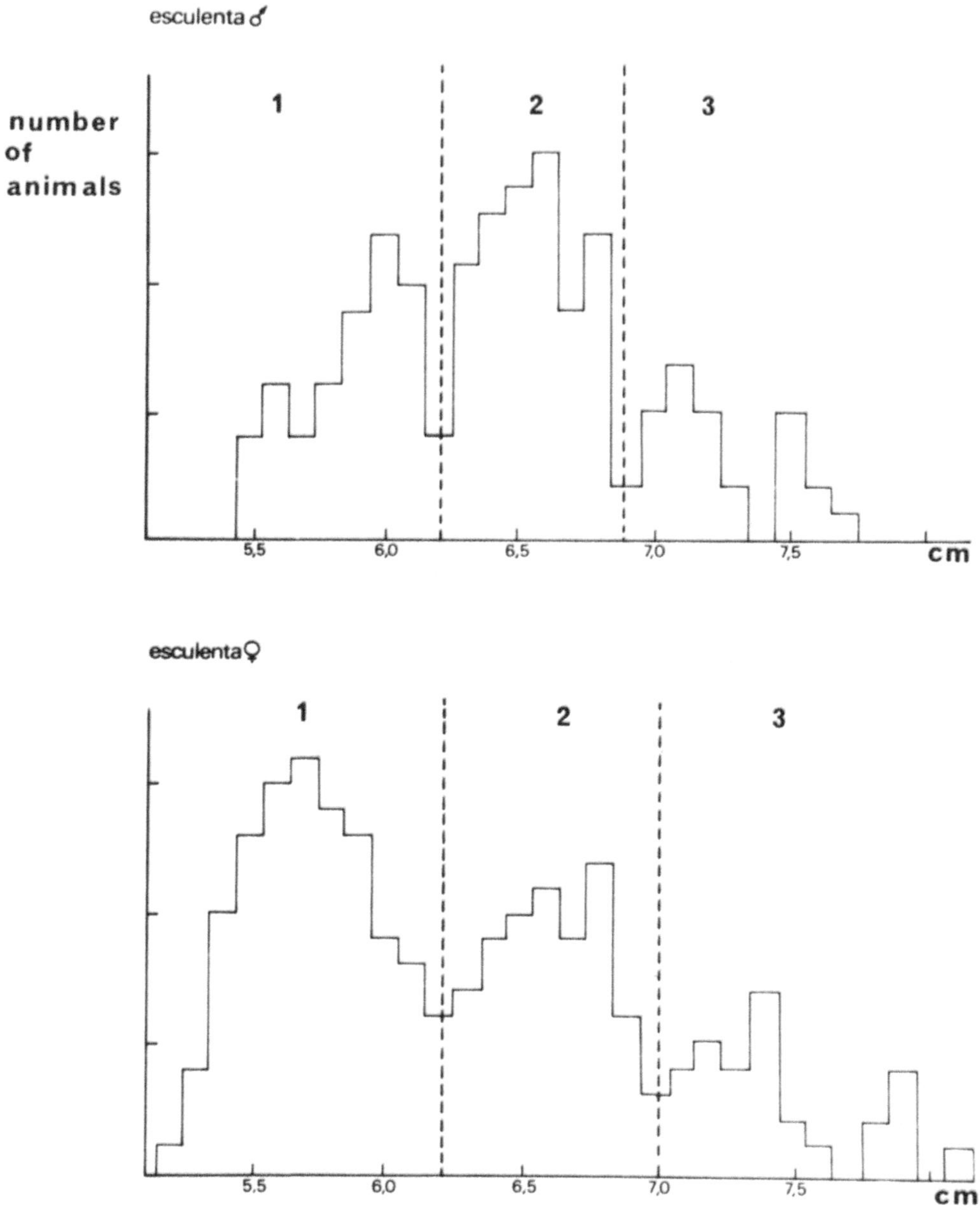

Fig. 3. Snout-vent length histograms for individual *R. esculenta* caught and measured in 1970 and 1971.

1 = first year of sexual maturity = small *R. esculenta*.
2 = second year of sexual maturity = medium *R. esculenta*.
3 = third year of sexual maturity = large *R. esculenta*.

The distribution of males in the arena was not random. Figure 4 shows that, in general, the R. lessonae males and the small (young) R. esculenta males prefer to stay at the edge of the arena, whereas middle-sized and large (old) R. esculenta males stay in the central parts of the arena. When the females arrive, this picture becomes disturbed during the course of the day, but the general outline remains the same. If we divide the breeding place arbitrarily into three parts, the arena zone in the center, an intermediate zone and an edge zone (Fig. 4), and count the animals of the different classes at different times of day, it can be shown that the distribution of the males is significantly nonrandom (Tables 1 and 2).

The next question was whether or not these differences in distribution represent functional differences in the behavior of the males. For this purpose, I counted sexual and aggressive behavioral

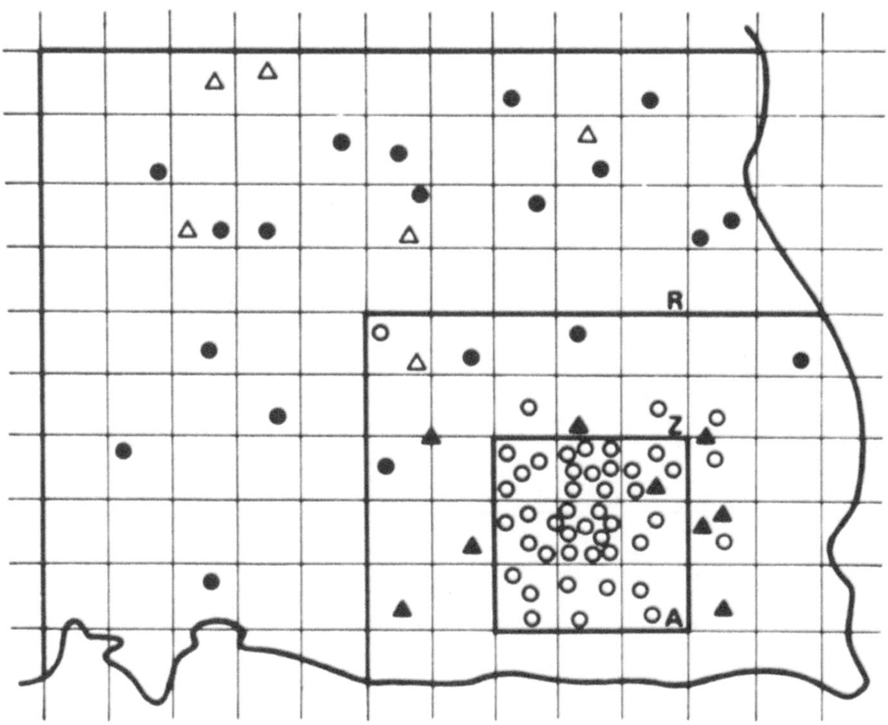

A = arena zone
Z = intermediate zone
R = edge zone

Fig. 4. Distribution of marked males in the arena on 4 June 1971. Observations from 0930 and 1030 combined. Black triangles - large R. esculenta; open triangles - medium-sized R. esculenta; black circles - small R. esculenta; open triangles = R. lessonae.

Table 1. Observed distributions of males in the arena at 0730 in the morning on 4-6 June 1971. Expected numbers in parentheses.

	R. esculenta Small	*R. esculenta* Medium	*R. esculenta* Large	*R. lessonae*	T
Central zone	2 (16.6)	55 (28.8)	1 (7.4)	0 (5.2)	58
Intermediate zone	7 (9.7)	9 (16.9)	16 (4.4)	2 (3.1)	34
Edge zone	29 (11.7)	2 (20.4)	0 (5.2)	10 (3.6)	41
Total	38	66	17	12	133

Contingency-table $X^2_{6\ d.f.} = 141.34$***

Table 2. Observed distributions of males in the arena at 2000 in the evening on 4-6 June 1971. Expected numbers in parentheses.

	R. esculenta Small	*R. esculenta* Medium	*R. esculenta* Large	*R. lessonae*	T
Central zone	35 (32.0)	64 (56.2)	11 (14.8)	4 (10.9)	114
Intermediate zone	0 (0.6)	1 (1.0)	0 (0.2)	1 (0.2)	2
Edge zone	6 (8.4)	7 (14.8)	8 (3.9)	9 (2.9)	30
Total	41	72	19	14	146

Contingency-table $X^2_{6\ d.f.} = 33.14$***

sequences (Blankenhorn 1974). First I looked at animals located in their preferred zone: *R. lessonae* males and young *R. esculenta* males in the edge zone, old *R. esculenta* males in the intermediate zone and middle-aged *R. esculenta* males in the central zone. Table 3 shows that the middle-aged *R. esculenta* males were sexually and aggressively the most active of the males; the old *R. esculenta* males showed much aggressive behavior and little sexual behavior; *R. lessonae* and young *R. esculenta* showed very little activity with

Table 3. Comparison of sexual and aggressive sequences of the four male classes, each male class in its preferred zone. Per class four males pooled. Expected numbers in parentheses.

	R. esculenta			R. lessonae	T
	Small	Medium	Large		
Sexual sequences	31 (19.26)	54 (56.43)	13 (31.35)	18 (8.69)	116
Aggressive sequences	12 (23.74)	72 (69.56)	57 (38.65)	2 (11.04)	143
Total	43	126	70	20	259

Contingency-table $X^2_{3\ d.f.} = 49.13$***

sexual behavior dominating. In general, we found high activity in the center of the arena and low activity at the edge of the arena, much aggressive behavior in the center and more sexual behavior at the edge. Animals not located in their preferred zone showed the same pattern of behavior regarding the ratio of sexual and aggressive behavior but a respectively lower or higher overall activity. The behavior pattern is thus type-specific, the activity pattern zone-specific.

At the end of the breeding period, there is a general tendency towards more aggressive behavior, the distances between the animals become greater, and the arena is dissolved. This process can last several days. The longest lasting arena was 10 days in 1970, the shortest one only three days in 1971. In all three years observed, mating occurred during the first three days and then the number of males present in the arena gradually decreased.

The mating success of the classes of males seems to be correlated with their place in the arena (Table 4). In more than two-thirds of the pairs with marked males, either R. lessonae males or young R. esculenta males were involved. Middled-aged R. esculenta males and especially old R. esculenta males showed very low pairing rates. The males at the edge of the arena have great success in mating, the males in the center do not. There are two possible explanations for this fact. First we have the behavior of the females, which come from the bank to the arena and are then "caught" by males at or in the edge zone. I have no evidence of females selecting males; on the contrary, females always flee when a male approaches. During the breeding season, they just do not flee as far. There is always some chasing before copulating. The males definitely have the active role just before copulation. The second reason may be found in the behavior of the males. R. lessonae males and young

Table 4. Mating success of the four male classes.

	R. esculenta Small	*R. esculenta* Medium	*R. esculenta* Large	*R. lessonae*	Total
Seen in arena	15	24	7	6	52
Seen in copula	12	6	1	13	32
Expected	9.23	14.76	4.30	3.69	32
Rate of mating	0.80	0.25	0.14	2.16	

Contingency-table $X^2_{3\ d.f.} = 32.05$***

R. esculenta males show sexual behavior almost exclusively and this very specifically when a female is in sight. Middle-aged and old *R. esculenta* males show sexual behavior in the presence of other males, of pieces of wood, of shoes lying in the water, of fishes and of other non-specific objects, as well as when other females are present. They can also be very aggressive when a female is present. It is surprising that these males have any success in copulating.

The successful pattern of the *R. lessonae* males is, in light of these findings, composed of being in the right place in the arena and reacting with the appropriate behavior in the presence of a female. For *R. esculenta* males, as hybrids, it is not surprising that they show some defects in this respect.

The question of why the animals show this special spatial and behavioral organization remains unanswered. It looks, indeed, as if the *R. esculenta* males do the "dirty" work of attracting the females while the *R. lessonae* males then copulate with the females. First I thought that this arena behavior was a specialty only of mixed *R. lessonae*-*R. esculenta* populations with a high ratio of *R. esculenta* individuals providing the absolutely necessary *R. lessonae* ♂-*R. lessonae* ♀ and *R. lessonae* ♂-*R. esculenta* ♀ pairings. But populations with high proportions of *R. lessonae* also have arenas, as do *Rana temporaria*, *Bufo bufo*, and *Rana ridibunda* populations. All of these species live in permanent ponds of some size, produce large numbers of eggs per female and form arenas during their short breeding periods. Species living in periodic ponds with shallow water, such as *Bombina variegata*, *Hyla arborea* and *Bufo calamita* produce fewer eggs per female, have a long breeding time and do not form arenas. We can deduce that it must be very important for species living in large ponds to concentrate their breeding behavior in space and time. What, then, are the selective forces behind this phenomenon? Let me take the green frog as an example. In early spring, Rana temporaria and Bufo bufo lay their eggs in the same pond before returning to their terrestrial habitats. Experiments by Heusser (1971) have shown that their larvae prefer to feed on spawn. The larvae of Bufo bufo, especially, which move through the pond in big schools, would be a serious danger for spawn being laid too early or singly throughout the pond. These larvae finish their metamorphosis at the end of May. On the other hand, there are the fishes occurring in the same pond. Most important here is Tinca tinca, the young of which are very numerous from July to August and also feed on spawn. The development of the larvae of the green frogs lasts two and a half months and metamorphosis has to be finished before the cold season begins. I believe that mainly these are the ecological factors responsible for the timing and concentration in green frog breeding. It seems that R. esculenta individuals maintain the possibility of taking part in arena behavior, but show some defects in sexual behavior. The combination of these characteristics may be responsible for their strange but still, for the population, adaptive behavior.

EXPERIMENTAL ANALYSIS OF THE MATING BEHAVIOR OF R. lessonae AND R. esculenta MALES

It is perhaps wise to remember that in mixed R. lessonae-R. esculenta populations, only the pairings ♂ R. lessonae x ♀ R. lessonae and ♂ R. lessonae x ♀ R. esculenta give vital offspring. The pairing of R. esculenta ♂ x R. lessonae ♀ was seen only twice in the last six years and R. esculenta ♂ x R. esculenta ♀ results in offspring dying between the fourth and eighth larval weeks.

The results of the mating success studies presented in the last section, and theoretical considerations showing that with a proportion of one R. lessonae individual to 10 R. esculenta individuals in a population of 220 animals mating by chance could not result in sufficient offspring of the essential R. lessonae-R. lessonae type, were the basis for further experiments. The questions to be answered were: 1) Are there intra- and interspecific pairing preferences? 2) Which type-specific behavior patterns are correlated with such preferences? 3) Which causal factors influence these preferences?

Maintaining the animals presented a major problem in these experiments. Animals brought in fresh from the field and placed in aquaria showed only flight behavior; their body colors became darker and darker. It seemed that these animals would not exhibit normal behavior. After trying with interruptions for two years, I finally found a solution. First I had to make the animals hibernate artificially in a refrigerator at 4° C for at least two months before putting them into an aquarium. There, they were brought gradually from 10° C to 20° C and from a day length of 10 hours to a 16 hour day. The aquaria were furnished with one-way screens and the animals were kept as quiet as possible. This resulted in quite normal behavior; copulations were frequent and they resulted in viable offspring.

The experimental design consisted of all combinations, balanced and crossed with the two types, R. lessonae and R. esculenta, the two age classes, small and large animals, and the two sex classes as independent variables (Table 5). Thirty-six males and 36 females were tested in three series in 12 aquaria. The responses of the males in the presence of a female were recorded for 10 minutes by using an ethogram and taken as the dependent variable. As I had previously performed a sequence analysis of the behavior of the green frogs (Blankenhorn 1974), it was possible to give each behavioral element, corresponding to its place in the sequence, a point value allowing the quantification of the behavior. The results are presented in table 6. Each cell represents six replicates; each replicate is formed by the addition of the points of elements belonging to sexual motivation and divided by the total number of elements occurring during the testing time of 10 minutes. The analysis of the data was carried out by a two-way analysis of variance. The two factors were the females, with the four levels, small R. lessonae, large R. lessonae, small R. esculenta and large R. esculenta, and the males, with the four levels, small R. lessonae, large R. lessonae, small R. esculenta and large R. esculenta. The

Table 5. Experimental design for the analysis of pairing preferences. Numbers are the sequence of tests on the first day. On the second day the order was inverted.

♂ ♀	Small R. lessonae	Large R. esculenta	Large R. lessonae	Small R. esculenta
Small R. esculenta	1	12	16	5
Large R. lessonae	9	4	6	13
Large R. esculenta	11	2	8	15
Small R. lessonae	3	10	14	7

Table 6. Results from the experiments on pairing preferences.

	Small male R. lessonae	Large male R. lessonae	Small male R. esculenta	Large male R. esculenta
Small female R. lessonae	1.5192	0.3016	0.0000	0.0000
	0.0000	0.1892	0.0000	0.0000
	1.0122	0.3415	0.0000	0.0000
	0.7969	0.0000	0.0000	0.0000
	0.0000	0.2911	0.0000	0.0000
	0.4828	0.6628	0.0000	0.0000
Large female R. lessonae	0.5085	1.7424	0.0000	0.0000
	0.6500	1.1446	0.0000	0.0000
	3.4717	0.1412	0.0000	0.0000
	1.3814	2.2877	0.0000	0.0000
	3.4717	2.0667	0.1351	0.0000
	0.4535	3.4667	0.1132	0.0000
Small female R. esculenta	0.0000	0.4333	0.0000	0.0000
	0.0000	0.6188	0.0000	0.0000
	0.0000	1.3286	0.0000	0.0000
	0.0000	0.8462	0.0000	0.0000
	0.5342	0.3030	0.0000	0.0000
	0.7297	0.2326	0.0000	0.0000
Large female R. esculenta	0.0000	1.6429	0.5484	0.2963
	0.0000	2.3729	0.4844	0.1923
	0.0000	3.4615	0.7273	0.7473
	0.0000	2.6866	0.6543	0.4478
	0.0000	4.3654	0.4073	0.4127
	0.0000	3.9800	0.0000	0.4615

analysis showed that both female and male factors have significant effects (Table 7).

Let us first look at the effect the females have. It is composed of the differences between the four female classes. There are two groups of females; the big females of both types are much preferred over the small females of both types (Scheffé's multiple comparison, significant). In the males, R. lessonae males react with much higher intensity in the presence of a female than R. esculenta males. The big R. lessonae males show a greater activity than the small R. lessonae males. There is no significant difference within the R. esculenta males.

Table 7. Analysis of variance.

Source	Sum of squares	d.f.	MS
Female factors	2.1960	3	0.7320
Male factors	1.8528	2	0.9264
Female-male factor interaction	0.4086	6	0.0681
Error	0.5481	36	0.0152
Total about grand mean	5.0055	47	

F_1 3/36 = 48.16***
F_2 2/36 = 60.95***
F_3 6/36 = 4.48***

The female-male factor interactions are significant too. Some of the male-female combinations show effects independent of the single male or female class expressed in the effects of the female or male factor. These effects are of course especially interesting, because they give a direct measure of the pairing preference in each male-female combination, independent of the more general effects of the female and male classes. In this respect, we find very clear preferences in pairings of small *R. lessonae* ♂ x small *R. lessonae* ♀, small *R. lessonae* ♂ x large *R. lessonae* ♀ and large *R. lessonae* ♂ x large *R. esculenta* ♀. Avoided are the combinations of small *R. lessonae* ♂ x large *R. esculenta* ♀, large *R. lessonae* ♂ x small *R. lessonae* ♀, large *R. lessonae* ♂ x small *R. esculenta* ♀, small *R. esculenta* ♂ x large *R. lessonae* ♀, and large *R. esculenta* ♂ x large *R. lessonae* ♀.

If we look at the behavior of the males, we find a picture similar to that in the field: the *R. lessonae* males react very strongly to the presence of preferred females and indifferently toward non-preferred ones. The *R. esculenta* males show the same strange division in sexual and aggressive behavior with aggressiveness being stronger when non-preferred females are present.

What conclusions can we draw from these experiments? Looking at figure 5 we can say that there is, first of all, a general tendency for males to prefer females that are larger than themselves. This is the case for the small *R. lessonae* males preferring big *R.*

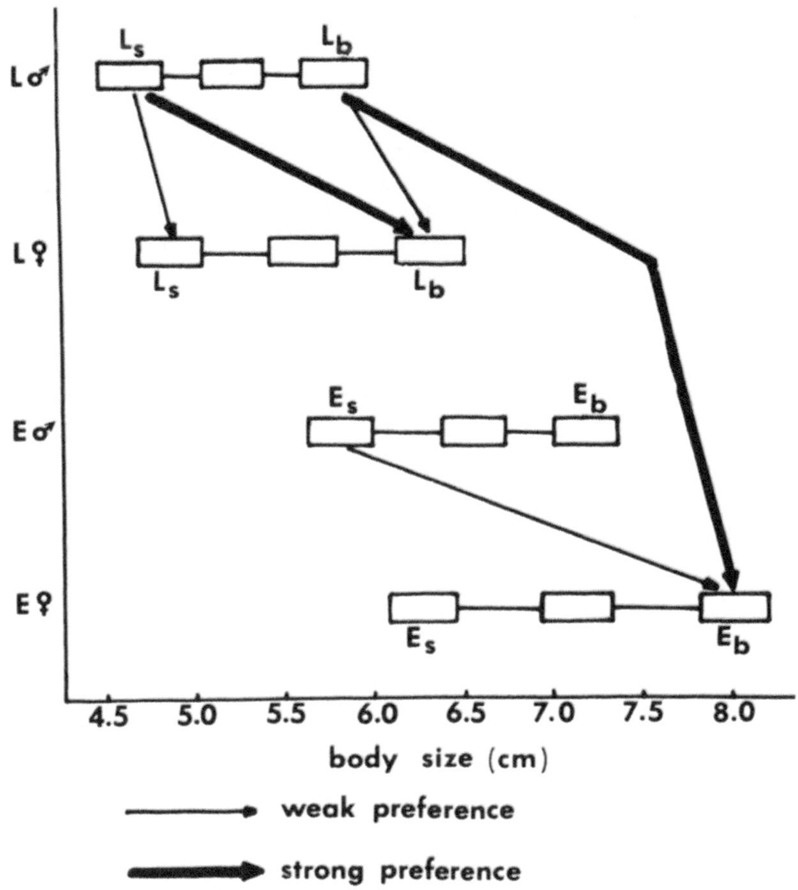

Fig. 5. Body size and pairing preferences. L_s = small R. lessonae, L_b = large R. lessonae, E_s = small R. esculenta, E_b = large R. esculenta.

lessonae females over small ones, and for small R. esculenta males preferring big R. esculenta females rather than small R. esculenta females. The big R. lessonae males prefer the larger R. esculenta females over big R. lessonae and small R. esculenta females. Both small R. lessonae and large R. lessonae males nevertheless prefer females of their own type rather than R. esculenta females of the same size. We can assume, therefore, that in natural populations the small (young) R. lessonae males produce the vital R. lessonae-R. lessonae pairings, whereas the big (old) R. lessonae males are mainly concerned with the R. lessonae-R. esculenta pairings that give R. esculenta offspring. R. esculenta males seem to perform, almost exclusively, pairings with females of their own type. The larvae die between the fourth and eighth weeks. These effects are reinforced by the spatial arrangement of the males in the arena described in the last section.

The experiments did not give any evidence of the nature of the preference of homotypic pairings, but there are cases known where the pairing call plays an important part in bringing together males and females of the same species (Littlejohn 1959). Indeed, there are specific differences in the calling of *R. esculenta* and *R. lessonae* males. It would be surprising if *R. lessonae* females had no such mechanisms at their disposal allowing them to actively select males of their own type. For *R. esculenta* females, as hybrids, I suspect that there is no such mechanism.

THE *R. lessonae*-*R. esculenta* SYSTEM: A DISCUSSION

The crucial question arising from the last two sections is certainly: What are the selective forces bringing about such a strange species-hybrid system, with its several specific adaptations and modifications? This question will be discussed now on a broader base, including genetic, faunistic and ecological results and considerations.

Genetics

Berger (1964-70) has demonstrated that, of the three phenotypes of European green frogs, *Rana lessonae* and *Rana ridibunda* behave genetically like true species giving *inter se* offspring of their own type. *Rana esculenta* is formed either by hybridization of *R. lessonae* with *R. ridibunda* or by back-crossing of *R. esculenta* with *R. lessonae*. Much evidence for the hybrid character of *Rana esculenta* has accumulated (Berger 1964, 1966, 1967, 1968a, 1968b, 1970, 1971a, 1971b, 1971c, 1973a, 1973b, Blankenhorn 1973, 1974, Blankenhorn *et al*. 1971, 1973, Günther 1968, 1969, 1970, 1973, Hemmer 1973, Heusser and Blankenhorn 1973, Hotz 1974, Tunner 1970, 1972, 1973a, 1973b).

In Poland (Berger 1970) the breeding of *R. esculenta* results in a very high larval mortality; some individuals reach metamorphosis and even sexual maturity, however. The offspring of Swiss *R. esculenta* show, on the other hand, complete mortality between the fourth and eighth larval weeks (Heusser and Blankenhorn 1973), and no specimens reach metamorphosis.

Faunistics

As described above, there is strong evidence that only *R. esculenta* and *R. lessonae* occur autochthonically in eastern Switzerland, whereas *R. ridibunda* has been introduced secondarily in this region. In fact, in the undisturbed region, only mixed R. lessonae-R. esculenta populations could be found; there were no pure R. lessonae or R. esculenta populations. For a broader view of the distribution of European green frogs refer to Hotz's (1974) excellent publication.

Breeding behavior

The ethological studies presented in the two preceding sections show tha R. lessonae males have a much higher sexual activity, choose better places for copulating at the breeding site and seem to prefer, to some extent, females of their own type over females of the R. esculenta type. On the whole, the organization at the breeding place and the differences in behavior of the two types seem perfectly apt in maintaining populations with even such extreme R. lessonae to R. esculenta ratios as one to 10.

Social organization

With the exception of the mating season, these populations show a territorial organization. Before the mating period, the males occupy a small portion of the bank and defend it against other males, although they tolerate females within it. The females usually form small groups at particular locations where exposure to sunlight is longest. The females do not show any aggressive behavior at this time. After the breeding season, the females also become very aggressive and establish territories. During this period, which lasts about two weeks, one third of the population leaves the breeding pond and settles in adjacent ponds as far as 2 km away. Young R. esculenta individuals emigrate most frequently (Blankenhorn 1974).

Biotope type and R. lessonae to R. esculenta ratio

A quantitative analysis of 10 populations in different ponds showed a clear correlation between biotope type and the ratio of R. lessonae to R. esculenta (Fig. 6). In gravel pits, R. esculenta is much more frequent than R. lessonae and ratios of one R. lessonae individual to 10 R. esculenta individuals can be reached. In ponds within marshes and forests, the ratios are quite the opposite, with many more R. lessonae individuals than R. esculenta, R. lessonae occurring, in some cases, 15 times as frequently as R. esculenta. In small lakes, the ratio is around one to one.

Behavior of the larvae

Qualitative observations during the raising of different types of larvae showed evidence for differences in swimming behavior. The lethal R. esculenta by R. esculenta larvae show very conspicuous, spiral swimming movements; the hybrid R. lessonae by R. esculenta

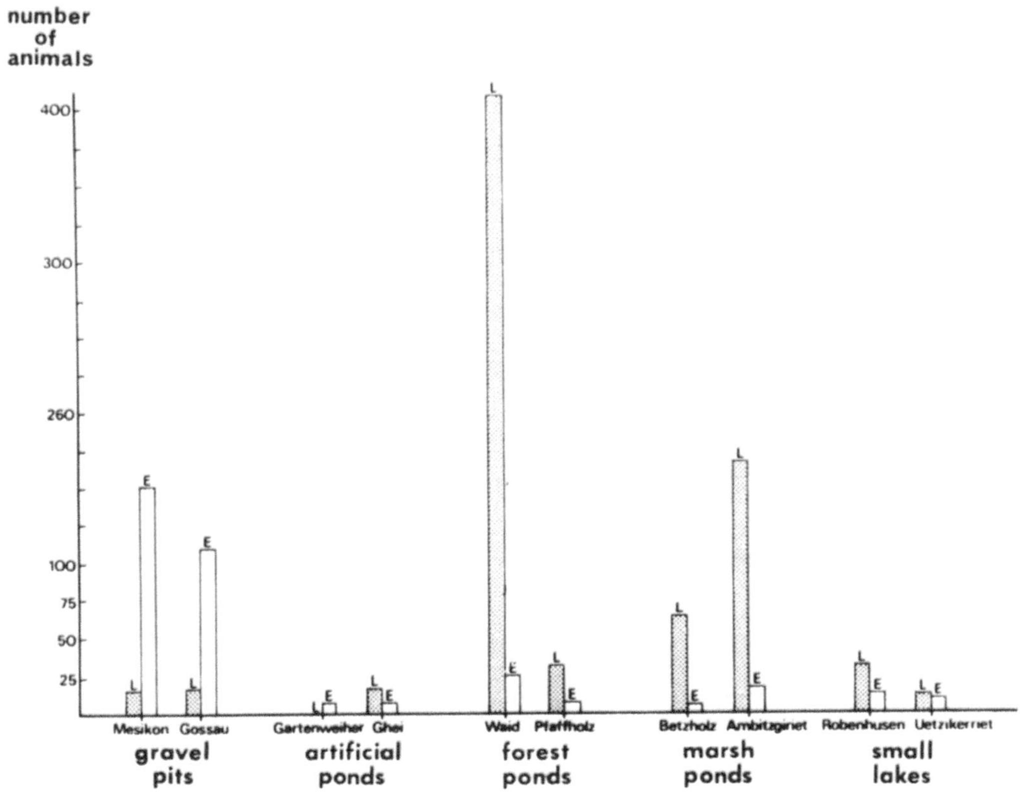

Fig. 6. *R. lessonae* to *R. esculenta* ratios in 10 biotopes of eastern Switzerland. L = number of *R. lessonae* individuals. E = number of *R. esculenta* individuals.

larvae move in very straight lines with long resting times at the bottom; *R. lessonae* larvae could be seen very often in free water with movements less straight than those of *R. lessonae* by *R. esculenta* larvae.

Discussion

These findings confront us with several problems. First of all, we must remember that these populations are formed by individuals belonging to one species (*Rana lessonae*) and by hybrids (*Rana esculenta*) between this species and another (*Rana ridibunda*) which does not occur in the populations in this geographical region. There is no doubt that R. esculenta is the product of back-crossing R. esculenta with R. lessonae in these populations. This does not correspond to the well-known genetic instability, lowered fertility and vitality of F_2 hybrids (Mayr 1963). One would rather expect to see a dying-out of the hybrid without the constant supply of new hybrid offspring from R. lessonae by R. ridibunda crossings. There is some evidence that R. esculenta in mixed R. lessonae-R. esculenta populations is not as genetically unstable as thought. Larval mortality seems to be stabilized. In Poland, Berger (1970) found that the mortality of R. esculenta by R. esculenta offspring occurred from the blastula stage to sexual maturity with a peak at metamorphosis. Some individuals even reached sexual maturity. In our ponds, R. esculenta by R. esculenta offspring are completely inviable. After nine weeks, all larvae were dead. Obviously there is a difference in genetic stabilization between Poland and Switzerland. In fact, Berger's findings correspond rather well with the situation one would expect to find in a zone of secondary or tertiary intergradation. Genetic mechanisms like hybridogenesis (Schultz 1969) might have been involved in stabilizing the genetics of the R. esculenta phenotype. The question remains: Why did evolution take this path to maintain the hybrids in the R. lessonae-R. esculenta system? Certainly, the genetic stabilization must be regarded as one of the important causal conditions for the survival of mixed R. lessonae-R. esculenta populations.

Another causal factor, this one behavioral, must exist to maintain both phenotypes in the populations. The study of male sexual behavior and selectivity shows clearly that there is no problem in producing sufficient R. esculenta offspring in these populations. The probability of forming R. esculenta female-R. lessonae male pairs seems to be guaranteed through the high pairing rates of R. lessonae males and the preference of older R. lessonae males for older R. esculenta females. On the other hand, the sufficient supply of R. lessonae offspring poses some problems, especially in populations with high R. esculenta proportions. Ratios of one R. lessonae to 10 R. esculenta seem to work; I doubt that enough R. lessonae by R. lessonae pairings will occur in a population with higher R. esculenta ratios. If these pairings are not formed, there will be only R. esculenta offspring, and the population is doomed. Certainly the sexual behavior of the R. lessonae males and the not very skillful behavior of the R. esculenta males are important causal factors for the survival of the mixed populations, but why take the risk of extinction?

The ecology of these mixed populations poses another problem. R. lessonae-R. esculenta populations are, in fact, found in almost any type of pond. The ratio of R. lessonae to R. esculenta individuals varies with the biotope type. Is this variation adaptive?

My hypothesis is, of course, that the changing R. lessonae-R. esculenta ratios are adaptive. I think that the number and the ratio of the three larval types are important. We know that in gravel pits, as the extreme example, a considerable proportion of the larvae are of the lethal R. esculenta by R. esculenta type. These larvae show strange spiral swimming movements. Presumably this type of swimming is more conspicuous to possible predators than the normal swimming of the vital R. esculenta by R. lessonae and R. lessonae by R. lessonae larvae. It is interesting that ponds in gravel pits usually have rather few possible hiding places and cover for larvae, compared to other biotopes. It may well be that predators form a search image for the slow and conspicuously swimming R. esculenta by R. esculenta larvae. In this case, the feeding pressure on the vital larvae would be lowered, a fact that could be especially important in ponds with little cover. There is, to date, one fact that suggests this. The peak of mortality of R. esculenta by R. esculenta larvae occurs just when the vital larvae have finished their first growth spurt and are very difficult to catch (Heusser and Blankenhorn 1973). Why do we have such small proportions of R. esculenta in small forest and wood ponds? Here, I think, Berger's (1970) findings on overwintering may offer a clue. He found that R. ridibunda and most R. esculenta choose wintering places under water; R. lessonae, on the other hand, hibernates on land. The shallow water of these ponds often freezes down to the bottom, and the animals there will die. The fact remains, however, that R. lessonae-R. esculenta populations occur in a wide variety of biotopes and that there is a correlation between biotope type and the R. lessonae to R. esculenta ratio.

It is generally acknowledged that R. lessonae and R. ridibunda are ecologically quite well separated, R. ridibunda occurring more in larger ponds, lakes and rivers, R. lessonae in small ponds and marshland. Mixed R. lessonae-R. esculenta populations occur in either type. After the last Ice Age, certainly both R. ridibunda and R. lessonae already existed and almost certainly R. esculenta did, too. With the glaciers gradually receding, many new biotopes were colonized by the green frogs. We might well assume a competitive situation between different population types: R. ridibunda populations, R. lessonae populations, mixed R. ridibunda and R. esculenta populations, R. lessonae-R. esculenta populations, and perhaps R. esculenta populations. Certainly, the distribution of the biotopes to be colonized would not have been even. Marshlands, small ponds, lakes and rivers would have been available for colonization by the frogs at different times and different places. It seems clear to me that the wide ecological range of mixed R. lessonae-R. esculenta populations, as opposed to the ecological specificity of pure populations, was an advantage in colonizing the new biotopes. Of course, there would then have been a high selective pressure on stabilizing the R. lessonae-R. esculenta system genetically, ethologically and ecologically. Mixed R. lessonae-R. esculenta populations should therefore be found in areas that could only be colonized by the frogs relatively late, after the last Ice Age. The known recent distribution corresponds with this assumption. I must admit, however, that the knowledge of the distribution of the European green frogs and their ecology is not complete and new results would be a great help for the further understanding of this quite unusual and interesting system.

SUMMARY

From the results of Berger's studies on European green frogs in Poland, it is known that Rana esculenta Linnaeus is a hybrid between Rana lessonae Camerano and Rana ridibunda Pallas (Berger 1964-70). The natural ratio of these three phenotypes in Swiss populations is peculiar. Out of 24 frog populations explored in the course of the work, only three were pure R. ridibunda populations, whereas 21 populations were mixed R. lessonae-R. esculenta populations. The mating behavior and success of males in these mixed populations was investigated and it was shown that R. lessonae males were on the average more than twice as often in copulation with females than were R. esculenta males. Experiments in the laboratory showed that young R. lessonae males preferred females of their own type and older R. lessonae males preferred R. esculenta females. R. esculenta males copulate almost exclusively with females of their own type; their offspring are inviable. These findings are discussed in light of ethological, ecological, faunistic and genetic considerations.

ACKNOWLEDGEMENTS

This study was supported by the Swiss National Foundation for Scientific Research, Grant No. 3.256.69. I am indebted to my advisor, Prof. Dr. H. Kummer, to Dr. H. Heusser for financial and scientific support, to Prof. H. Hampel for statistical advice and to Dr. D.C. Turner for his help in stylistic matters.

REFERENCES

Berger, L. 1964. Is Rana esculenta lessonae a distinct species? Ann. Zool. Warszawa 22:245-261.

Berger, L. 1966. Biometrical studies on the population of green frogs from the environs of Poznan. Ann. Zool. Warszawa 23:303-324.

Berger, L. 1967. Embrional and larval development of F_1 generation of green frogs of different combinations. Acta zool. cracov., Krakov 12:123-160.

Berger, L. 1968a. Morphology of the F_1-Generation of various crosses within Rana esculenta complex. Acta zool. cracov., Krakov 13:301-324.

Berger, L. 1968b. The effect of inhibitory agents in the development of green frog tadpoles. Zool. Poloniae 18:381-390.

Berger, L. 1970. Some characteristics of the crosses within Rana esculenta complex in post larval development. Ann. Zool. Warszawa 27:373-416.

Berger, L. 1971a. Sex ratio in the F_1 progeny within forms of Rana esculenta complex. Genet. Polon. 12:87-101.

Berger, L. 1971b. Viability, sex and morphology of F_2-generation within forms of Rana esculenta complex. Zool. Poloniae 21:345-393.

Berger, L. 1971c. Inheritance of sex and phenotype in F_1 and F_2 crosses within Rana esculenta complex. Genet. Polon. 12:517-521.

Berger, L. 1973a. Systematics and hybridization in European green frogs of Rana esculenta complex. J. Herp. 7:1-10.

Berger, L. 1973b. Sexual maturity of males within forms of Rana esculenta complex. Zool. Poloniae 22:177-188.

Blankenhorn, H. J. 1973. Zum Stand der Forschung über die Verbreitung ger Grünfrösche im Kt. Zürich. Rev. suisse zool. 80: 655-662.

Blankenhorn, H. J. 1974. Soziale Organisation einer Mischpopulation von Rana lessonae Camerano und Rana esculenta Linnaeus. Unpubl. Dissertation, Univ. Zurich.

Blankenhorn, H. J., H. Heusser and P. Vogel. 1971. Drei Phänotypen von Grünfröschen aus dem Rana esculenta-Komplex in der Schweiz. Rev. suisse zool. 78:1242-1247.

Blankenhorn, H. J., H. Heusser and P. Notter. 1973. Zur Verbreitung von Rana lessonae und Rana esculenta im Kanton Zürich. Rev. suisse zool. 80:662-666.

Günther, R. 1968. Morphologische und ökologische Untersuchungen zur Unterscheidung von Rana esculenta L. und Rana ridibunda Pall. Zool. Jb. Syst. 95:229-264.

Günther, R. 1969. Paarungsrufe und reproduktive Isolationsmechanismen bei europäischen Anuren der Gattung Rana (Amphibia). Forma Functio 1:263-284.

Günther, R. 1970. Der Karyotyp von Rana ridibunda Pall. und das Vorkommen von Triplodidie bei Rana esculenta L. Biol. Zentralbl. 89:327-342.

Günther, R. 1973. Ueber die verwandtschaftlichen Beziehunger zwischen den europäischen Grünfröschen und dem Bastardcharakter von Rana esculenta L. Zool. Anz. 190:250-285.

Hemmer, H. 1973. Das Serumeiweissbild von Rana ridibunda perezi perezi im Rahmen des Rana esculenta-lessonae-ridibunda-Komplexes. Salamandra 9:168-172.

Heusser, H. 1971. Laichräubern und -Kannibalismus bei sympatrischen Anuren-Kaulquappen. Experientia 27:474.

Heusser, H. and H. J. Blankenhorn. 1973. Crowding Experimente mit Kaulquappen aus homo- und heterotypischen Kreuzungen der Phänotypen esculenta, lessonae und ridibunda (Rana esculenta Komplex, Anura, Amphibia). Rev. suisse zool. 80:543-569.

Hotz, H. 1974. Ein Problem aus vielen Fragen - europäische Grünfrösche (Rana esculenta Komplex) und ihre Verbreitung. Nat. u. Mus. Frankfurt a.M. 104:262-272.

Littlejohn, M. J. 1959. Call differentiation in a complex of seven species of Crinia. Evolution 13:452-471.

Mayr, E. 1963. Animal species and evolution. The Belknap Press, Cambridge, Massachusetts.

Schultz, R. J. 1969. Hybridization, unisexuality and polyploidy in the teleost Poeciliopsis (Peociliidae) and other vertebrates. Am. Natur. 103:605-619.

Tunner, H. G. 1970. Das Serumeiweissbild einheimischer Wasserfrösche und der Hybridcharakter von Rana esculenta. Verh. der deutsch. zool. Ges. 64:352-358.

Tunner, H. G. 1972. Serologische und morphologische Untersuchungen zur Frage der Artabgrenzung bei Wasserfröschen aus der Umgebung von Mainz. Z. zool. Syst. Evol. forsch. 10:127-132.

Tunner, H. G. 1973a. Demonstration of the hybrid origin of the common green frog *Rana esculenta* L. *Naturwissensch.* *60*:481-482.

Tunner, H. G. 1973b. Das Albumin und andere Bluteiweisse bei *Rana ridibunda*, *Rana lessonae* und *Rana esculenta* und deren Hybriden. *Z. zool. Syst. Evol. forsch.* *11*:219-233.

CYTOTAXONOMY AND CYTOGENETICS IN EUROPEAN NEWT SPECIES

Giorgio Mancino, Matilde Ragghianti and Stefania Bucci-Innocenti

Institute of Histology and Embryology, University of Pisa, Via A. Volta 4, Pisa

The study of the mechanisms of speciation and of evolution in the animal world is a subject of wide biological interest which in recent years has become the object of an increasingly vast literature. Protein, chromosome and anatomical evolution as well as changes in the patterns of gene expression represent some of today's most extensively studied themes which require strict interdisciplinary approaches. In each animal group, therefore, any conclusion regarding "phyletic relationship" can be drawn only on the basis of all the chromosomal and genetic, as well as biochemical and biometrical, evidence (White 1973a).

In this connection, the urodele amphibians are a particularly interesting vertebrate group, primarily in view of the present convergence of chromosomal, biochemical and genetical studies (cf. Hedgecock and Ayala 1974; Ayala 1975) with those on systematics and anatomy (cf. Özeti 1967; Özeti and Wake 1969; Wake and Özeti 1969; Thireau 1975a, 1975b) as well as on ecology and ethology (cf. Ahrenfeldt 1960; Salthe 1967; Avery 1968, 1971; Bell 1974; Halliday 1974a, 1974b, 1975; Bell and Lawton 1975). Moreover, urodeles are important because they have high DNA values and possess giant ovarian chromosomes, the so-called lampbrush chromosomes, which have resolution power approximately equal to that of the polytene chromosomes in *Drosophila* larvae. Thus, it is possible: 1) to perform morphological and structural studies which permit us to individualize, not only extensive (cf. Mancino *et al*. 1970), but also minute chromosome rearrangements; 2) to investigate structure and organization of chromosomal DNA to verify the validity of the interpretations about the evolutionary or adaptive value of the high nuclear DNA content in urodeles; 3) to analyze more completely organizational structure, as well as genetic and functional significance of euchromatin and heterochromatin; 4) to inquire into the gene arrangement on the chromosomes of closely related species and subspecies. These studies are fundamental for describing the cytogenetic and molecular events involved in speciation and in phylogeny.

THE EUROPEAN NEWT SPECIES

Our research with the European newts and, more specifically, the genus Triturus, benefits from sound taxonomic studies of the family Salamandridae (Thorn 1968; Tortonese and Lanza 1968; Steward 1969), recently improved by an analysis of the functional morphology of their feeding mechanisms (Wake and Özeti 1969).

According to Lantz (1947), the genus Triturus can be divided into three groups of related species:

1) The cristatus group can be considered as an Artenkreis because it contains two polytypic species only partially overlapping in the west central part of France (cf. Vallée 1959): T. cristatus has four geographical races, T. c. cristatus (Laurenti 1768), T. c. carnifex (Laurenti 1768), T. c. dobrogicus (Kiritzescu 1903 = danubialis Wolterstorff 1923) and T. c. karelinii (Strauch 1870); T. marmoratus has two subspecies, T. m. marmoratus (Latreille 1800) and T. m. pygmaeus (Wolterstorff 1905).

2) The vulgaris group contains six species; some of them have additional subspecies not included in Lantz's (1947) original scheme: T. vulgaris includes T. v. vulgaris (L. 1758), T. v. meridionalis (Boulenger 1882) and seven other subspecies according to Thorn (1968); T. helveticus (Razoumowsky 1789), T. italicus (Peracca 1898), T. montandoni (Boulenger 1880), T. boscai (Lataste 1879) and T. vittatus (Gray 1835) are the other five species. This group has also been considered as an Artenkreis, but the use of this term has been criticized by Spurway and Callan (1960) because the ranges of T. v. vulgaris and T. h. helveticus overlap widely.

3) The alpestris group includes only the polytypic species T. alpestris with the subspecies T. a. alpestris (Laurenti 1768), T. a. apuanus (Bonaparte 1839) and seven others (cf. Thorn 1968). This group could be considered a Rassenkreis if this term were to maintain its original meaning, a polytypic species (Rensch 1929).

CYTOTAXONOMIC AND CYTOGENETIC OBSERVATIONS

The different families of urodeles are characterized by a taxonomic divergence which parallels karyotypic divergence involving the number and shape of chromosomes (Morescalchi 1970; Léon and Kezer 1974). Within each of the more derived families, the situation is not as easily defined through cytotaxonomic investigations, because the different genera generally share the same chromosome number. Within the family Salamandridae, the only relevant variation is between the Nearctic and Palearctic genera, which have n = 11 or 12, 2n = 22 or 24, respectively. The 24-chromosome genera show karyotypic uniformity and stability although anatomic evolution within the single genera is considered to have reached different levels. Thus, the principle at the base of the so-called "karyotype systematics," that is the immediate relationship between taxonomic affinity and correspondence in karyotype, is not immediately applicable.

Using sophisticated cytological techniques such as C-staining, fluorescence, or *in situ* molecular hybridization applied to mitotic and giant lampbrush chromosomes, our studies offer the possibility of showing the distribution and localization of euchromatin and heterochromatin and the arrangement of certain genes along the chromosomes. Applied to giant lampbrush chromosomes, these methods permit establishment or rejection of homosequentiality, such as examination of the polytene chromosomes demonstrated to occur in complexes of closely related species of Hawaiian *Drosophila* (Carson *et al*. 1967) and to be absent in chironomid, sciarid and simulid midges.

To support and to improve the cytotaxonomic analysis, we also have studied interspecific hybrids, which permits us to verify the rate of chromosome differentiation between the genomes of the two parent species. The cytological analysis on hybrids can avert the criticism that the two karyotyping factors (length and centromere index) are insufficient to establish chromosomal homology or for deductions regarding speciation and karyological evolution (Atchley 1972).

The *cristatus* Group

Mitotic chromosomes. The karyotypes of *T. c. carnifex* and *T. c. dobrogicus* (Callan and Lloyd 1960a; our unpublished results) and of *T. m. marmoratus* (Nardi and Mancino 1971) include twelve chromosome pairs of decreasing length which can be divided into three groups: A, B and C. Group A consists of the largest elements in the two complements. Pair I, submetacentric in both species, is more heterobrachial in *T. marmoratus*; this pair is made up of two homologs of different length. Chromosomes II, III and IV are metacentric in both species. Group B includes the four medium-sized elements: chromosomes V, VII and VIII are metacentric, chromosome VI is submetacentric. Group C consists of the smallest elements in the two complements and includes two metacentric (IX and XI) and two submetacentric (X and XII) chromosomes. In *T. cristatus*, however, the centromere of chromosome X is localized at the limit between the submetacentric and metacentric positions. Therefore, the karyotypes of the two species appear similar in size and shape, apart from the morphology of chromosome I (Fig. 1). The most evident differentiations are represented by secondary constrictions which, on the basis of the structures identified on lampbrush chromosomes (Nardi and Mancino 1971; Mancino *et al.* 1972a), can be considered as the nucleolus organizer regions (NORs). These are two in *T. c. carnifex* (on the short arm of chromosomes VI and IX in subterminal and intercalary positions, respectively) but only one in *T. marmoratus* (subterminally on the long arm of chromosome X) (Fig. 1).

When the chromosomes of the two species are treated by Arrighi and Hsu's (1971) C-staining method, they are found to be morphologically and structurally differentiated in number, distribution pattern and in thickness of the C-bands. Such differentiation appears to be species-specific since the banding pattern of *T. c. carnifex* (Mancino *et al.* 1973) seems to be identical to those of *T. c. dobrogicus* (unpublished results) and *T. c. cristatus* (Rudak and Callan 1976), but different from that of *T. marmoratus* (Figs. 2 and 3).

The centromere (centromeric heterochromatin) appears as a dark granule which is smaller in *T. cristatus* than in *T. marmoratus*. In both species, there are thin and simple pericentric bands which are close to the centromeres but not continuous with the centromeric heterochromatin. These bands are clearly visible on both arms of the smallest chromosomes. Some discrepancies between the karyotypes

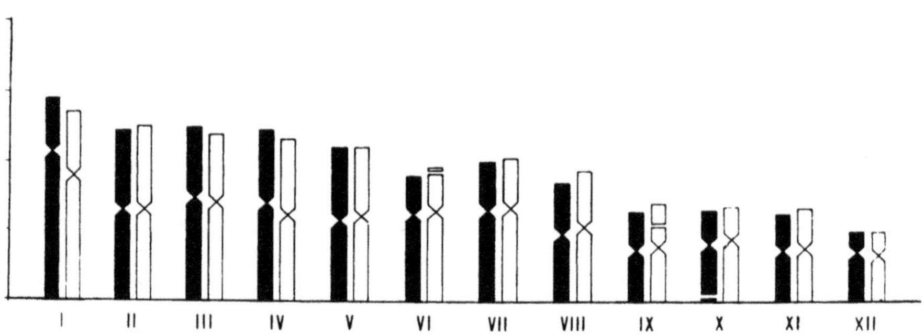

Fig. 1. Idiogram of the 12 mitotic chromosomes of the haploid sets of *T. m. marmoratus* and *T. c. carnifex*.

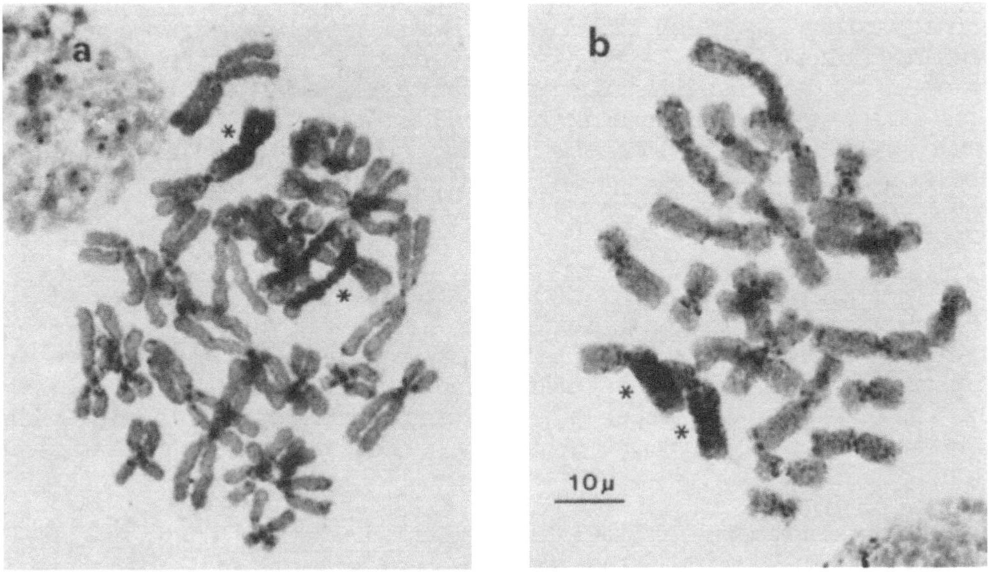

Fig. 2. The 24 C-stained mitotic chromosomes from two spermatogonia of T. c. dobrogicus. Chromosomes I are indicated by *.

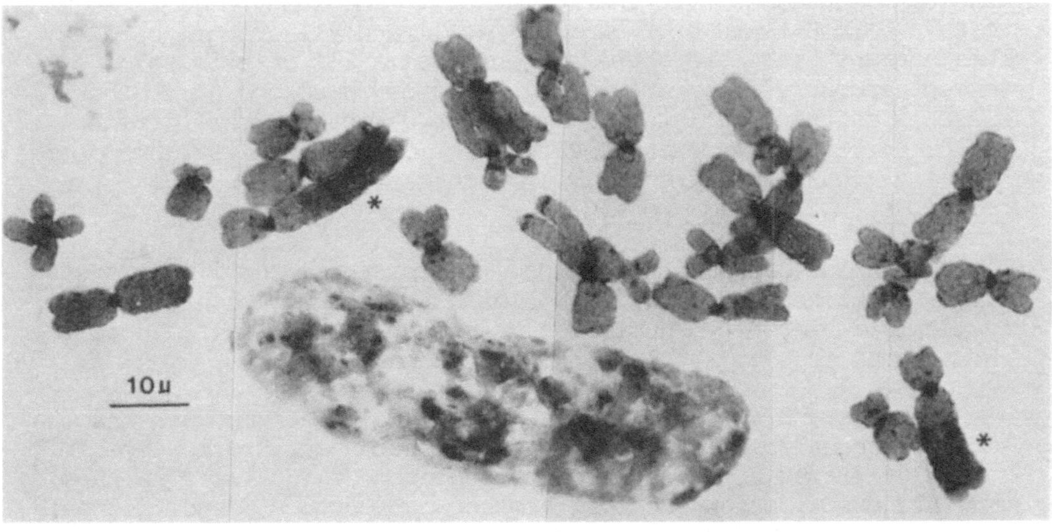

Fig. 3. A spermatogonial metaphase of T. marmoratus. Chromosomes of pair I are indicated by *. The chromosomes are C-stained.

of the two species concern the number of the subterminal C-bands. For instance, only chromosomes V and VIII of T. cristatus show tiny subterminal bands which can be related to the sphere sites on the corresponding lampbrush chromosomes. The appearance of the nucleolus organizers after C-staining is variable. They appear as granules or dark bands in some preparations, while in others they look like regions having slightly stained edges. Finally, an extended segment of the long arm of pair I is darkly stained by Giemsa in both *T. cristatus* and *T. marmoratus* (Figs. 2 and 3); its significance will be discussed in the section regarding sex chromosomes.

Lampbrush chromosomes. The maps of the four subspecies of *T. cristatus* (Callan and Lloyd 1960a, 1975) and of *T. m. marmoratus* (Nardi *et al*. 1972a) are available for cytotaxonomic studies. Microevolutive and microsystematic deductions can be drawn from the comparison of the maps at inter- and intraspecific levels which help us to analyze chromosome homologies and rate of differentiation.

Centromeres are particularly evident *in vitro* only in *T. c. karelinii*. In *T. marmoratus*, centromeres become easily identifiable only after C-staining (Batistoni *et al*. 1974).

At present, the comparative analysis of loops must be based only on those loops which are recognizable morphologically and thus can be used as cytogenetic markers. The lumpy objects seem to be a common feature of chromosomes II and VII of both species. The "currant buns" are shared by chromosome III of both species (except probably *T. c. cristatus*), but they also occur on other elements of the *T. cristatus* complement. Finally, the loops "GGL" and "GFL" are remarkable features of the smallest chromosomes; they are localized differently in each geographical race of *T. cristatus* and in *T. marmoratus*.

Nucleolus organizing regions. *T. cristatus* and *T. marmoratus* differ in number and localization of NOR markers, which are clearly identifiable when a nucleolus is still attached to the specific chromosome sites on the lampbrush chromosomes.

Spheres are characteristic landmarks of the lampbrush chromosomes of urodeles. Their number and position are species-specific. In the *cristatus* group, spheres have been individualized in subterminal positions on the long arms of chromosomes V and VIII only in the four subspecies of *T. cristatus*. Callan and Lloyd (1975) pointed out, however, that in *T. c. karelinii* the spheres are not reliable landmarks because they may or may not be present, but when present, they occupy the standard sites on chromosomes V and VIII. These sites correspond to C-bands on the mitotic chromosomes. Spheres carried by chromosomes, or free in the nucleoplasm, have not been found in *T. marmoratus*.

Axial granules are present, at similar sites, on chromosomes III, IV and VI of both species which give rise to reflected fusions.

As a common feature, the four subspecies of *T. cristatus* and *T. marmoratus* demonstrate a double-axis region in the right end of chromosome XII and, often, another one just to the left of the centromere (see also Callan 1955).

The *vulgaris* Group

The mitotic chromosomes of T. vulgaris. Two subspecies *T. v. vulgaris* (Callan 1942; Wickbom 1945; Ullerich 1970) and *T. v. meridionalis* (Barsacchi *et al*. 1970; Nardi *et al*. 1973; and our unpublished results) have been analyzed. In both subspecies the twelve chromosome pairs of the complement can be recognized and divided into the three usual groups A, B and C. Group A includes the four large-sized metacentric pairs (I, II, III and IV). Group B has two medium-sized metacentric (V and VIII) and two medium-sized submetacentric (VI and VII) chromosomes; chromosome VII is more heterobrachial than chromosome VI and often is subacrocentric. Group C has three very small submetacentric pairs (IX, X and XI) characterized by different values of centromeric index; one of them is often subacrocentric. Chromosome XII is metacentric.

The mitotic chromosomes of T. italicus. Notwithstanding some discrepancies regarding the smallest chromosomes, in particular chromosome XII, the karyotype of *T. italicus* seems to correspond to that of *T. vulgaris*. Group A includes the four longest pairs of metacentric chromosomes (I, II, III and IV). Group B has two medium-sized metacentric (V and VIII) and two medium-sized submetacentric (VI and VII) chromosomes, of which VII is more heterobrachial. Group C includes one small submetacentric (X) and two small metacentric (IX and XI) chromosomes. Chromosome XII is always subacrocentric (Mancino and Barsacchi 1969; Nardi *et al*. 1973).

The mitotic chromosomes of T. helveticus. As in *T. vulgaris* and *T. italicus*, the chromosomes can be arranged in three groups. Group A includes the four longest pairs of metacentric chromosomes (I, II, III and IV). Group B consists of four medium-sized chromosomes, of which two are metacentric (V and VIII) and two submetacentric (VI and VII, the latter being more heterobrachial). Group C has one small metacentric (IX) and three small submetacentric (X, XI and XII) chromosomes. Chromosome XII is the most heterobrachial (Mancino and Barsacchi 1966a; Namur and Signoret 1967).

The mitotic chromosomes of T. montandoni. The karyotype seems to correspond to that of *T. vulgaris*, except for some peculiarities (Fig. 4); in particular, chromosomes VI and IX appear metacentric

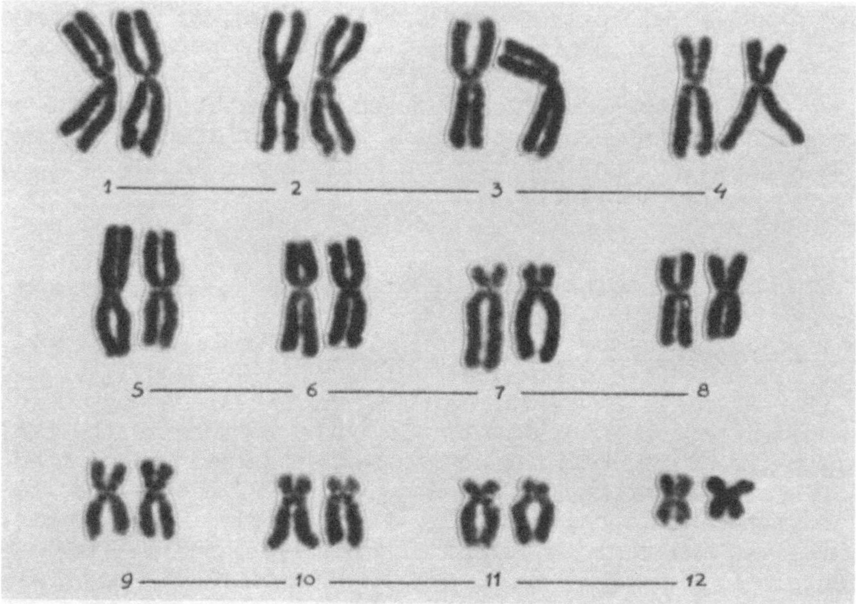

Fig. 4. The mitotic karyotype of T. montandoni. Courtesy of Dr. Cornelia Geormaneanu, Bucharest.

(Geormaneanu 1975) rather than submetacentric.

In conclusion, the mitotic chromosomes of the species of the T. vulgaris group examined here appear uniform when judged according to traditional morphometric criteria; the evolutionary significance of the previously indicated interspecific differences, for instance that concerning chromosome XII, which is metacentric in T. vulgaris and T. montandoni and heterobrachial in T. helveticus and T. italicus, must yet be ascertained.

By contrast, when the mitotic chromosomes of T. v. meridionalis, T. italicus and T. helveticus are treated by C-staining (Nardi et al. 1973; unpublished results), the results are distinctive.

T. v. meridionalis. The centromeres appear as dark granules (centromeric heterochromatin), while bands, usually double, are found on both sides of the centromere region (pericentric bands). There are also some subterminal regions, such as the one present on the long arm of chromosome XI, which correspond to the NOR (Fig. 5).

T. italicus. The chromosomes usually present unstained centromeric regions, except for chromosome XII, which consistently shows

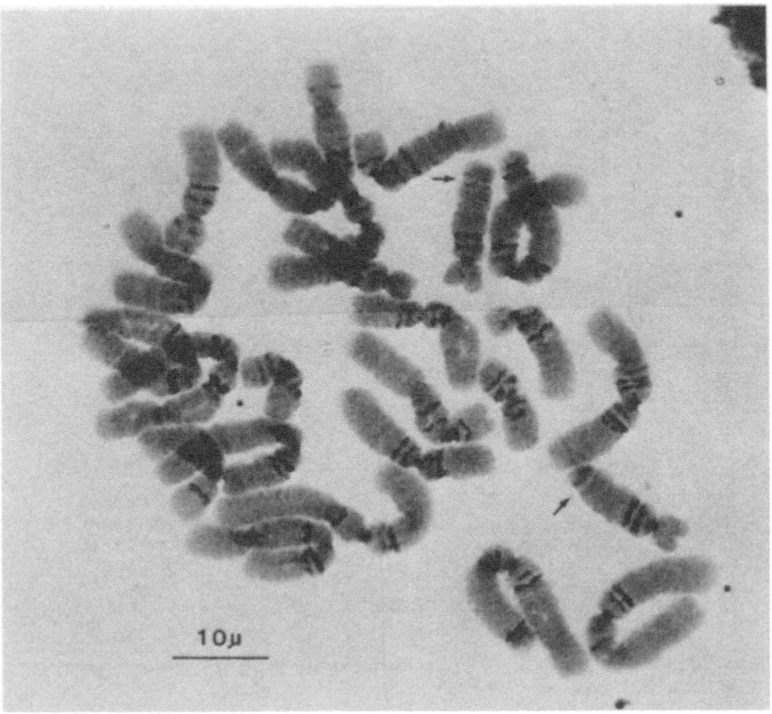

Fig. 5. The 24 C-stained mitotic chromosomes from a larva of *T. v. meridionalis*. NORs are indicated by arrows.

a dark centromeric granule. Pericentric bands, generally single, are present on all the elements of the complement, while subterminal regions are induced only on some chromosomes. Pair II shows a large dark terminal cap which is heterozygous in the male line (Fig. 6).

T. helveticus. The centromere regions of all the chromosomes become evident after C-staining, although with a variable intensity in different preparations. Pericentric bands are present on both arms of most chromosomes close to the centromere. They are particularly evident only on the short arm of chromosomes VI and VII, while in the smallest elements they resolve into granules. Finally, subterminal bands are present on a few elements of the complement, for example, on the short arm of chromosome II and on the long arm of chromosome VIII (Fig. 7).

The distribution of the pericentric and subterminal bands seems to correspond to that of spontaneous or cold-induced secondary constrictions described on mitotic chromosomes by various authors (Callan 1942; Namur and Signoret 1967; Namur 1969; Mancino and Barsacchi 1969; Barsacchi *et al*. 1970). In *T. vulgaris*, for example,

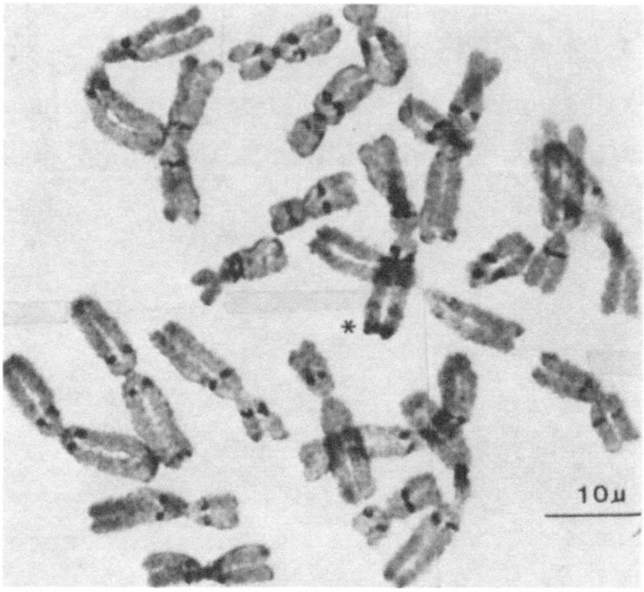

Fig. 6. The 24 C-stained mitotic chromosomes from a spermatogonium of *T. italicus*. The dark cap on one chromosome II is indicated by *.

the double pericentric C-bands appear to have the same localization as the cold-induced double secondary constrictions.

In conclusion, the interspecific differentiation in banding pattern does not seem to confirm the widespread opinion that the species of this group are karyologically identical. In fact, the karyotype of each species seems to differ in the quantity and in the distribution of the constitutive heterochromatin.

The lampbrush chromosomes. Lampbrush maps are available for *T. v. meridionalis*, *T. italicus* and *T. h. helveticus* (Mancino and Barsacchi 1966a, 1969; Barsacchi *et al.* 1970). In *T. italicus* centromeres are particularly evident and look like large loopless chromomeres. In the other two species, the centromere is evident only after C-staining. The three species lack "GGL" and "GFL" loops having landmark value. Nevertheless, some chromosomes can be identified because they have a few large loops with a dense or granular matrix. Spheres are present in all the species; two of them are found in *T. v. meridionalis* and *T. italicus*, while only one probably occurs in *T. helveticus*. In *T. v. meridionalis*, the smaller sphere-carrying chromosome also carries the nucleolus organizer.

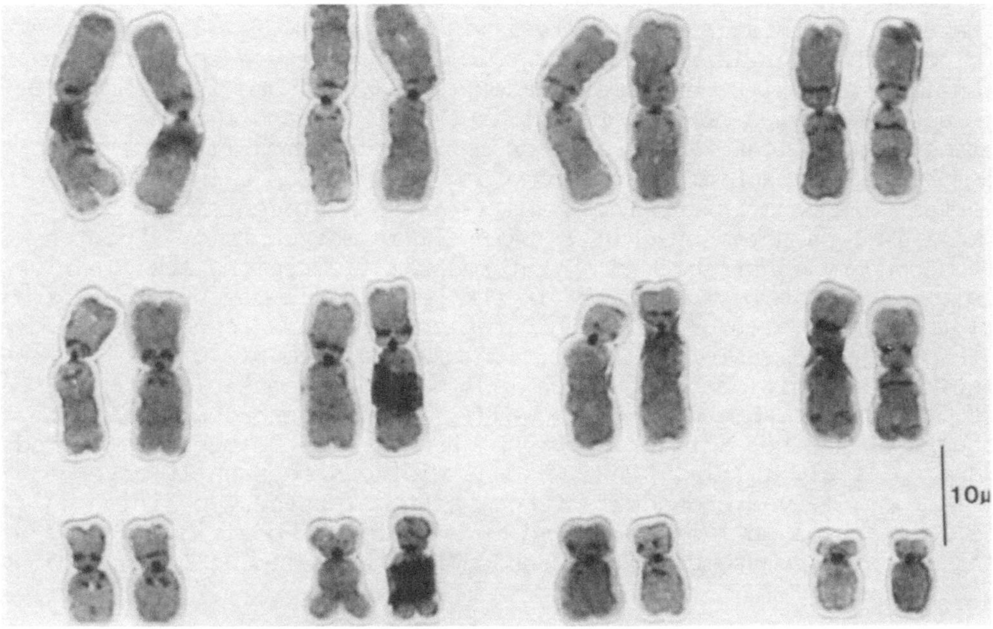

Fig. 7. The karyotype of T. h. helveticus mitotic chromosomes stained by the C-staining method.

The alpestris Group

The mitotic chromosomes. Karyotypes have been constructed only for T. a. alpestris and T. a. apuanus.

In the former, notwithstanding some difficulties arising from chromosomal analysis performed in different developmental stages and with different techniques (Namur and Signoret 1967; Ullerich 1970; Schmid and Krone 1976), the karyotype can be arranged in the usual three groups. Group A includes the four largest chromosome pairs, which are metacentric (I, II, III and IV). Group B includes the medium-sized chromosomes, of which two pairs are metacentric and two submetacentric. Group C includes three pairs of small submetacentric (IX, X and XI) and one pair of metacentric (XII) chromosomes. Secondary constrictions can be induced by cold treatment in the pericentric region of nearly all the chromosomes and in some subterminal regions (Namur and Signoret 1967).

Similar studies have also been carried out in T. a. apuanus (Nardi et al. 1972b; unpublished results). Among the twelve pairs of the diploid complement, some often have identical lengths; therefore, the alignment given in the karyotype is susceptible to changes if the statistical analysis is enlarged (Fig. 8). Group A includes the four largest metacentric chromosomes, each pair being identifiable by its centromere index (I, II, III and IV). Group B includes two pairs of submetacentric (V and VI) and two pairs of metacentric (VII and VIII) chromosomes; chromosome VIII appears to be more heterobrachial than chromosome VII. Group C includes the shortest elements; pairs IX, X and XI are submetacentric (chromosome X is less heterobrachial); pair XII is metacentric. Spontaneous secondary constrictions or heterochromatic regions are clearly visible in the embryonic cells in the pericentric region of nearly all the chromosomes. A few other regions are also found in the intercalary and subterminal position of some elements of the complement. The significance of these last secondary constrictions is not clear; for some of them, however, the correspondence with the spheres of the respective lampbrush chromosomes has been postulated (Mancino et al. 1972b). The two constrictions constantly observed in embryos, larvae and adults (present in intercalary position on the long arm of chromosome IX and in subterminal position on the long arm of chromosome XI) correspond to the nucleolus organizers (Fig. 8).

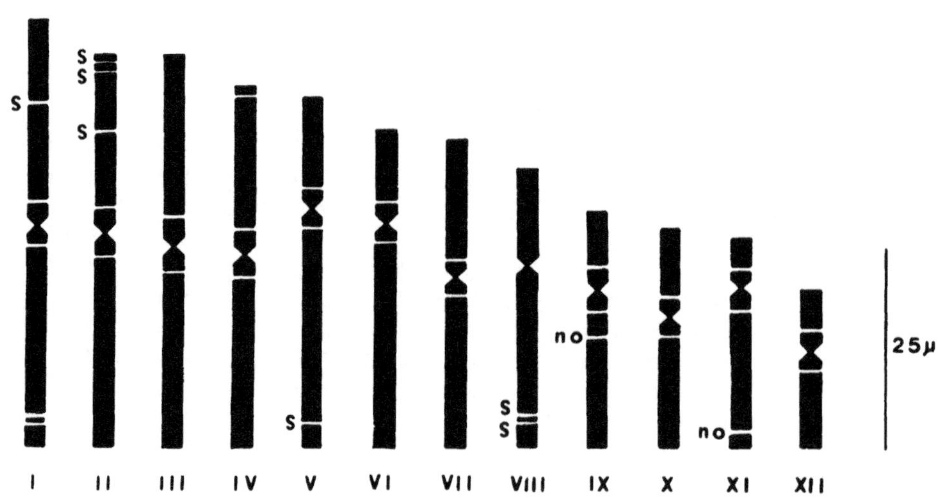

Fig. 8. Idiogram of the 12 mitotic chromosomes of the haploid set from an embryonic cell of T. a. apuanus. S = sphere site, no = nucleolus organizer.

After C-staining, the karyotype of T. a. apuanus demonstrates the following characteristics. Centromeres generally appear as dark granules and pericentric bands are evident close to the centromere region in all the chromosomes (Fig. 9). C-bands are found at the nucleolus organizer level, although, in some preparations, they can appear as light regions delimited by slightly stained edges. Subterminal C-bands are present on some chromosomes; among them, the bands on chromosomes II and VIII which are double and the one on chromosome V, which is single, are particularly interesting. These subterminal C-bands seem, in fact, to coincide with the heterochromatic regions believed to correspond to the sphere sites of lampbrush chromosomes. Finally, the terminal regions of chromosome IV appears as a darkly stained cap.

As a whole, the banding pattern of T. a. apuanus mitotic chromosomes appears very similar to that of T. a. alpestris recently described by Schmid and Krone (1976).

The lampbrush chromosomes. At present, lampbrush maps are known only for T. a. apuanus (Mancino and Barsacchi 1965). Their description has been completed (Ragghianti et al. 1972); this allows us to establish a possible correspondence between the differentiation of mitotic chromosomes and the structures inserted on

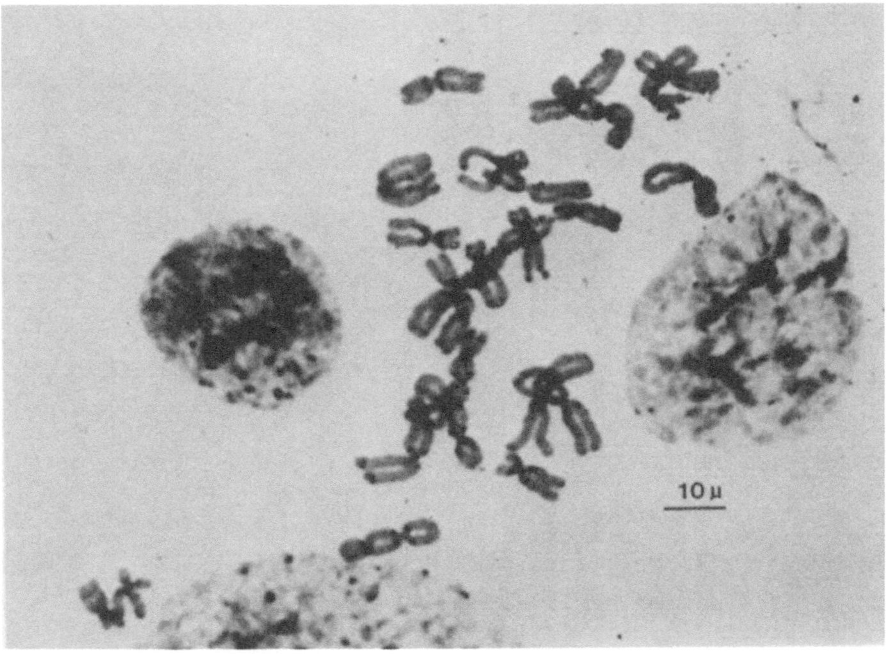

Fig. 9. The 24 C-stained mitotic chromosomes from a larva of T. a. apuanus.

lampbrush chromosomes (i.e., nucleolus organizer regions and spheres). Centromeres have been identified on each chromosome; they appear as enlarged chromomeres, generally devoid of loops, situated between the two pericentric regions in which chiasmata preferentially form.

Localization of the Ribosomal Genes on Salamander Chromosomes by Means of *In Situ* Hybridization

Callan and Lloyd's (1975) report concerning "working maps" of the lampbrush chromosomes of several amphibian species shows that much attention has been devoted to the individualization and exact localization of nucleolus organizer regions, since they represent, in each species, one of the most significant cytogenetic landmarks. Table 1 shows the localization of such regions in some salamandrid species.

Table 1. Number and localization of the nucleolus organizers on salamandrid lampbrush chromosomes.

Species	Chromosome pair(s)	Arm	Position	Refer.*
Notophthalmus viridescens	III	short	subterm.	1
T. c. carnifex	VI	short	subterm.	2
	IX	short	interc.	
T. marmoratus	X	long	subterm.	3
T. a. apuanus	IX	long	interc.	4
	XI	long	subterm.	
T. v. meridionalis	XI**	long	subterm.	5
Pleurodeles waltl	III	short	subterm.	6
	XI	long	subterm.	
Pleurodeles poireti	III	short	subterm.	7
	XI	long	subterm.	

*References: (1) Gall 1954; (2) Mancino *et al*. 1972a; (3) Nardi *et al*. 1972; (4) Ragghianti *et al*. 1972; and present work; (5) Barsacchi *et al*. 1970; (6) and (7) Lacroix 1968.

**According to statistical length value, this chromosome should be chromosome X.

The introduction of the technique of in situ molecular hybridization of a homogeneous RNA species to denatured eukaryotic chromosomes (Pardue and Gall 1970) has confirmed the precision of the cytological findings; for instance, the nucleolus organizer regions on chromosome X of T. marmoratus, on chromosomes VI and IX of T. cristatus carnifex and on chromosome XI of T. vulgaris meridionalis are constantly labeled with ^{3}H 18S + 28S rRNA by RNA/DNA in situ hybridization (Nardi et al. 1974; Barsacchi Pilone et al. 1974a, 1974b). The nucleolus organizer region is also labeled in the American newt Notophthalmus viridescens (Hutchison and Pardue 1975). The molecular hybridization technique thus is useful in localizing nucleolus organizer regions in salamandrid species, such as Taricha granulosa (León 1976), T. italicus, T. helveticus and Salamandra salamandra (Barsacchi Pilone, Nardi and their associates, unpublished) in which they have not been observed on mitotic or lampbrush chromosomes (Table 2).

In most individuals of T. vulgaris meridionalis the in situ hybridization has revealed, besides the locus constantly found on both homologs XI and already known to be the nucleolus organizer, the presence of other genetic loci containing 18S and 28S ribosomal

Table 2. Chromosome localization of the genes for 28S and 18S ribosomal RNA.

Species	Chromosome pair(s)	Arm	Position	Refer.*
T. c. carnifex	VI	short	subterm.	1
	IX	short	interc.	
T. marmoratus	X	long	subterm.	2
T. v. meridionalis	XI**	long	subterm.	3
T. italicus	III	short	subterm.	4
Salamandra salamandra	XI	long	interc.	5
Taricha granulosa	II	short	telom.	6
	IX	short	centrom.	

*References: (1) Barsacchi Pilone et al. 1974a; (2) Barsacchi Pilone et al. 1974b; (3) Nardi et al. 1974; (4) and (5) Barsacchi Pilone, Nardi and associates, unpublished; (6) León 1976.

**According to statistical length value, this chromosome should be chromosome X.

RNA cistrons (Nardi et al. 1974). Although these loci are constant for number and localization in both germinal and somatic cells of each individual, they vary in number and position from one individual to another; moreover, they have, until now, appeared only in a heterozygous condition (Barsacchi Pilone et al. 1976). The cross between individuals with a known distribution pattern of ribosomal genes has shown that these loci follow a Mendelian inheritance; this means that they are stably integrated into the chromosome structure, as an integral part of the genome of each individual (Barsacchi Pilone et al. 1976).

Hutchison and Pardue (1975) found that, in Notophthalmus viridescens, the regions which hybridize in situ with ^{3}H 18S and 28S rRNA are on three chromosome pairs. Doubts arise, however, concerning their exact chromosomal localization, for, as the authors point out, they did not construct an exact mitotic karyotype of the species. The regions are characterized by inequality of the hybridization between homologous chromosomes. The locus corresponding to the nucleolus organizer region identified on lampbrush chromosome III (Gall 1954; Callan and Lloyd 1975) was labeled in all the animals examined, even though it did not appear to be the major locus of hybridization in any of them. Problems concerning the exact number of nucleolus organizer regions and the overall number of chromosomal loci that bind labeled rDNA also arise from analysis of the in situ hybridization experiments performed on the chromosomes of American plethodontids by Macgregor and Mizuno (1976).

The 5S gene sequences cannot be revealed without resorting to the in situ molecular hybridization technique. Utilizing ^{3}H 5S rRNA, Barsacchi Pilone et al. (1974a, 1974b) and Nardi et al. (1974) have found the locus for 5S rRNA in an intercalary position on chromosome X of T. cristatus and T. marmoratus; T. v. meridionalis has only one 5S locus, presumably on the same chromosome and localization. In T. marmoratus this locus is near the chromomere of the dense matrix loop mapped on lampbrush chromosome X at 75 units or on a few chromomeres next to this loop. In Notophthalmus viridescens, Hutchison and Pardue (1975) and Pukkila (1975) have localized the genes which code for 5S rRNA near the centromere of chromosomes I, II, VI and VII. Pukkila (1975) has shown that the lampbrush loop, which extends from one of the extremities of each centromere region, is a carrier of the 5S sequence. In the other Nearctic species, Taricha granulosa, only one locus for 5S rRNA is present, close to or on the centromere region of chromosome VII (León 1976). Table 3 shows the results concerning number and localization of the genes for 5S rRNA in several salamandrids.

At present, in situ molecular hybridization with homogeneous RNA classes other than ribosomal RNA classes is increasing in scope gradually improving our knowledge of the gene arrangement along the

Table 3. Chromosome localization of the genes for 5S ribosomal RNA

Species	Chromosome pair(s)	Arm	Position	Refer.*
T. c. carnifex	X	long	interc.	1
T. marmoratus	X	long	interc.	2
T. v. meridionalis	XI**	long	interc.	3
T. italicus	several locations			4
Notophthalmus viridescens	I, II, VI, VII		peri-centric	5
Taricha granulosa	VII		centrom.	6

*References: (1) Barsacchi Pilone et al. 1974a; (2) Barsacchi Pilone et al. 1974b; (3) Nardi et al. 1974; (4) Barsacchi Pilone, Nardi and associates, unpublished; (5) Hutchison and Pardue 1975; Pukkila 1975; (6) León 1976.

**According to statistical length value, this chromosome should be chromosome X.

single chromosomes of the complement. Within urodeles, tritiated or iodinated 4S RNS has been hybridized to the chromosomes of Taricha among the salamandrids and of Batrachoseps among the plethodontids (León 1976). León (1976) has shown that, similar to the ribosomal RNA genes, the transfer RNA genes are also clustered in the genome of these two amphibians.

Sex Chromosomes

The identification of differentiated sex chromosomes in urodele amphibians represented for a long time a promising field of research, although the results were unsatisfactory (cf. Spurway 1945; White 1946; Makino 1947; Humphrey 1948). An XY sex determining mechanism characterizes plethodontid salamanders of the subfamily Bolitoglossinae; J. Kezer (unpublished) found differentiated XY sex chromosomes in Oedipina, Thorius and Chriopterotriton including C. bromeliacea, C. rabbi and C. cuchumatanos from Guatemala but probably not the Mexican Chiropterotriton species (see also Mancino and Barsacchi 1966b). Since the structure of the sex chromosomes of Thorius and Chiropterotriton differs from that of Oedipina, Kezer believes that there have been independent lines of XX/XY evolution within the family Plethodontidae.

In salamandrids, there are some cytological observations which may be interpreted to indicate male heterogamety in some newt species; the male of T. italicus, for instance, possesses a terminal dark cap on only one homolog of pair II (Fig. 6) and bivalent II from metaphase first spermatocytes does not form chiasmata at this region (Fig. 10). These findings, however, cannot represent sufficient evidence for assuming male heterogamety. In T. cristatus and T. marmoratus the morphological asymmetry of lampbrush bivalent I together with a difference in length of the two partner chromosomes and the lack of chiasmata between the two "heteromorphic" segments were interpreted as cytological evidence for female heterogamety (Callan and Lloyd 1956, 1960b, 1975; Mancino and Nardi 1971; Mancino et al. 1972c). However, the difference in length between the two partners of lampbrush bivalent I is equal to the difference in length between the two mitotic chromosomes I seen in both sexes. Moreover, male and female karyotypes, after C-staining, have an

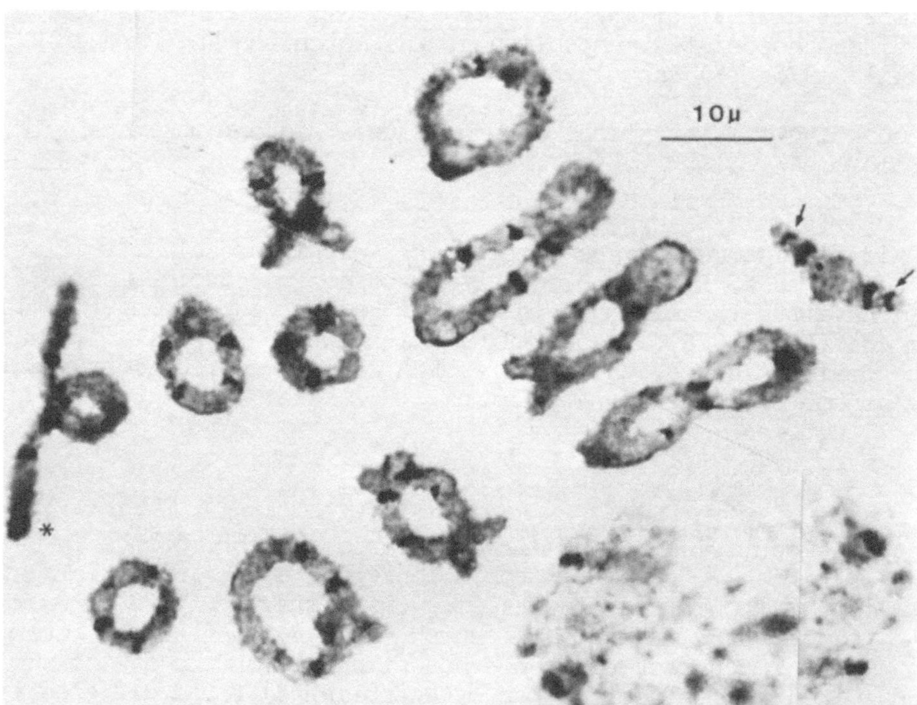

Fig. 10. The 12 C-stained bivalents from a metaphase spermatocyte I of T. italicus. Bivalent II (left side) is open. The terminal dark cap is indicated with *. Centromeres of bivalent XII are indicated by arrows.

intensely Giemsa positive segment on the long arm which corresponds to the "heteromorphic" and achiasmate region on lampbrush bivalent I (Mancino et al. 1973). Finally, a study of spermatogenesis of T. c. dobrogicus revealed that the two partners of bivalent I from metaphase I spermatocytes are joined to each other by chiasmata only between the short arms and, occasionally, between the end regions of the long arms. Chiasmata have never been observed inside the darkly stained Giemsa positive segments that correspond to the "heteromorphic" region of lampbrush bivalent I (Fig. 11). Chromosome pair I in both sexes, therefore, has the same general morphology and structure as well as behavior at meiosis. It is necessary to perform experimental genetic tests for ascertaining the type of digamety in every salamandrid species, as done in Ambystoma mexicanum (Humphrey 1942, 1945) and in Pleurodeles waltl (Gallien 1954a, 1954b).

Lacking more precise findings, the male of European newts of the genus Triturus is still regarded by some authors as the heterogametic sex on the basis of experimental hybridization (cf. Haldane 1922; Spurway 1945; Benazzi 1956) and of gonadic structure in specimens characterized by genome mutation (cf. Fankhauser 1938; Fischberg 1945; Kawamura 1951).

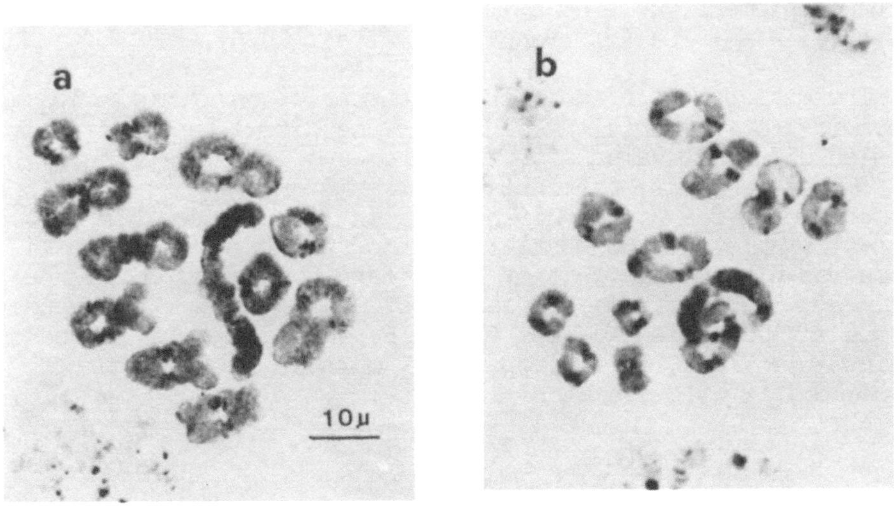

Fig. 11. Metaphase I spermatocytes of T. c. dobrogicus (a) and of T. marmoratus (b). In both cells, a darkly stained segment on the long arm of bivalent I is clearly visible.

Chromosome Constitution of Interspecific Hybrids and Specific Nuclear DNA Values

The C-staining method is particularly useful for cytotaxonomic investigations because it induces specific banding patterns on salamander chromosomes. We therefore used this method to study the cytogenetics of interspecific newt hybrids. Our aims were: 1) to ascertain the chromosome constitution of interspecific F_1 hybrids and the exact contribution from each parental species to the hybrid complements; 2) to correlate possible differences in chromosome lengths between the two parental sets of a diploid hybrid to the different nuclear DNA values found in the parent species. Table 4 shows that the most common class among the hybrid F_1 is diploidy with an identical contribution from each parental species (2n, n♀ + n♂; Figs. 12-14). Cases of eutriploidy and of aneutriploidy occur; they arise from fertilization of a diploid oocyte and a haploid sperm (3n, 2n♀ + n♂). Aneutetraploid hybrids have also been recognized; these individuals probably originate from a chromosome doubling in the zygote or during very early cleavage (4n, 2n♀ + 2n♂). Finally, one case of trisomy was constant in all cells and compatible with regular embryonic and larval development. Since the extra chromosome present in the single cells belongs to group C, which includes the four shortest elements of the complement, this constitution can be indicated as trisomy C (2n, n ♀ + n ♂ + 1C) (cf. Ragghianti *et al.* 1975).

The measurement of chromosome lengths of reciprocal diploid F_1 hybrids (*T. v. meridionalis* x *T. c. carnifex*) has constantly revealed different values for the genomes of the two parental species. When we refer, for instance, to the cross *T. v. meridionalis* ♀ x *T. c. carnifex* ♂, the differences in lengths were highly significant (Table 5): the total length of the *T. v. meridionalis* genome was higher than that of *T. c. carnifex* in all hybrid cells examined. Such difference, therefore, can be related to the difference in nuclear DNA values (Table 6) characterizing the parental species (Olmo 1973, 1974 and personal communication; Olmo and Morescalchi 1975; Morescalchi 1975).

CONCLUSIONS

Karyological studies have provided useful information on the phyletic relationships among and within the three amphibian orders of Gymnophiona (cf. Wake and Case 1975), Urodela (cf. Morescalchi 1975) and Anura (cf. Goin and Goin 1968; Bogart 1973). According to a hypothesis put forward by Matthey (1951) and, more recently, by Morescalchi (1970), chromosome evolution occurred by means of a series of modifications which have caused the passage from a high chromosome number formula, including meta- and acrocentric elements

Table 4. Chromosome constitution of interspecific hybrid larvae

Hybrids*	Numbers of Larvae	Ploidy				
		2n (n♀ + n♂)	3n (2n♀ + n♂)	Aneutri-ploid	Aneutetra-ploid (4n, 2n♀ + 2n♂)	Trisomic (2n, n♀ + n♂ + 1C)
T. v. meridionalis x x T. c. carnifex	33	28	2	1	1	1
T. c. carnifex x x T. v. meridionalis	36	35		1		
T. a. apuanus x x T. v. meridionalis	5	5				
T. a. apuanus x x T. c. carnifex	5	5				
T. v. meridionalis x x T. a. apuanus	4	3			1	
T. v. meridionalis x x T. m. marmoratus	4	4				
T. italicus x x T. v. meridionalis	1	1				

* The first species of the cross represents the female parent.

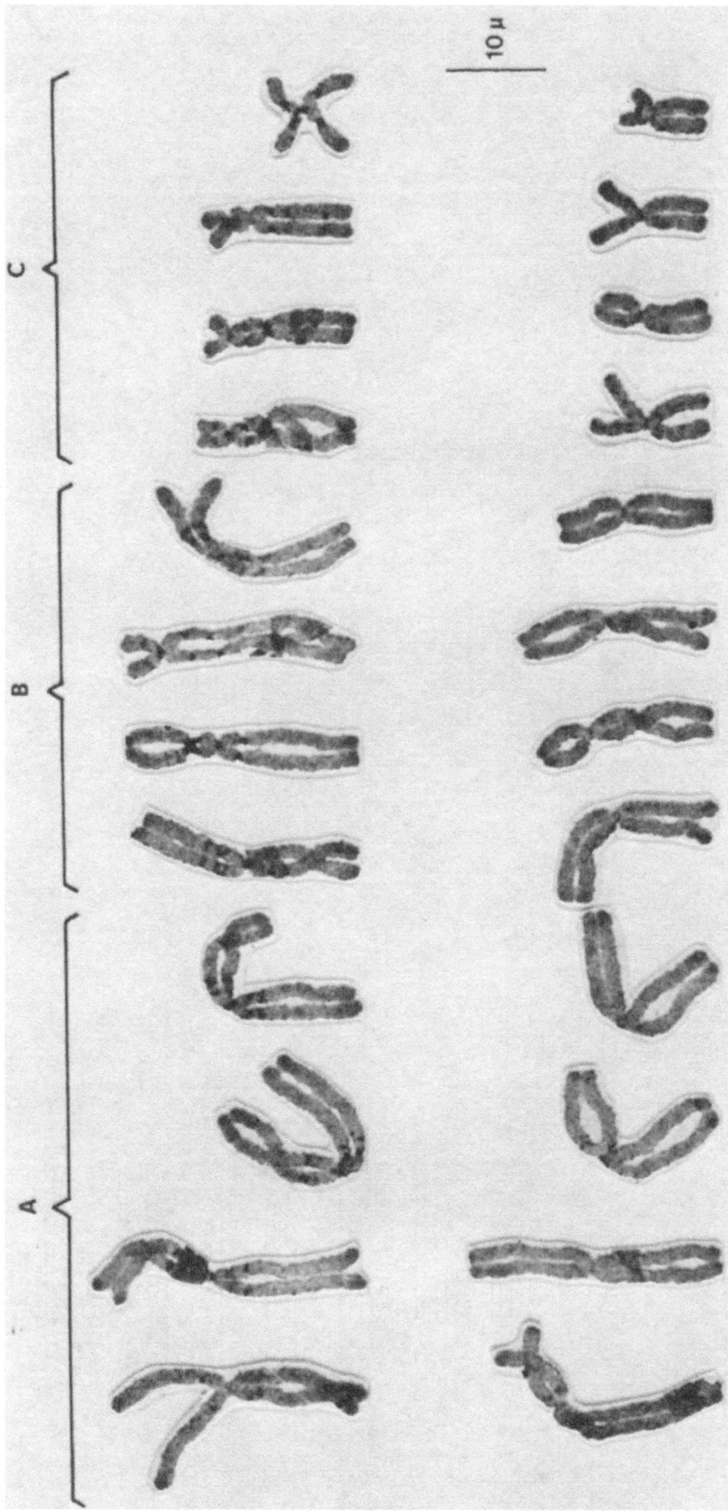

Fig. 12. The karyotype of the 24 C-stained mitotic chromosomes of a diploid hybrid larva (T.V. meridionalis ♀ x T. c. carnifex ♂). The maternal set, upper alignment; the paternal set, lower alignment.

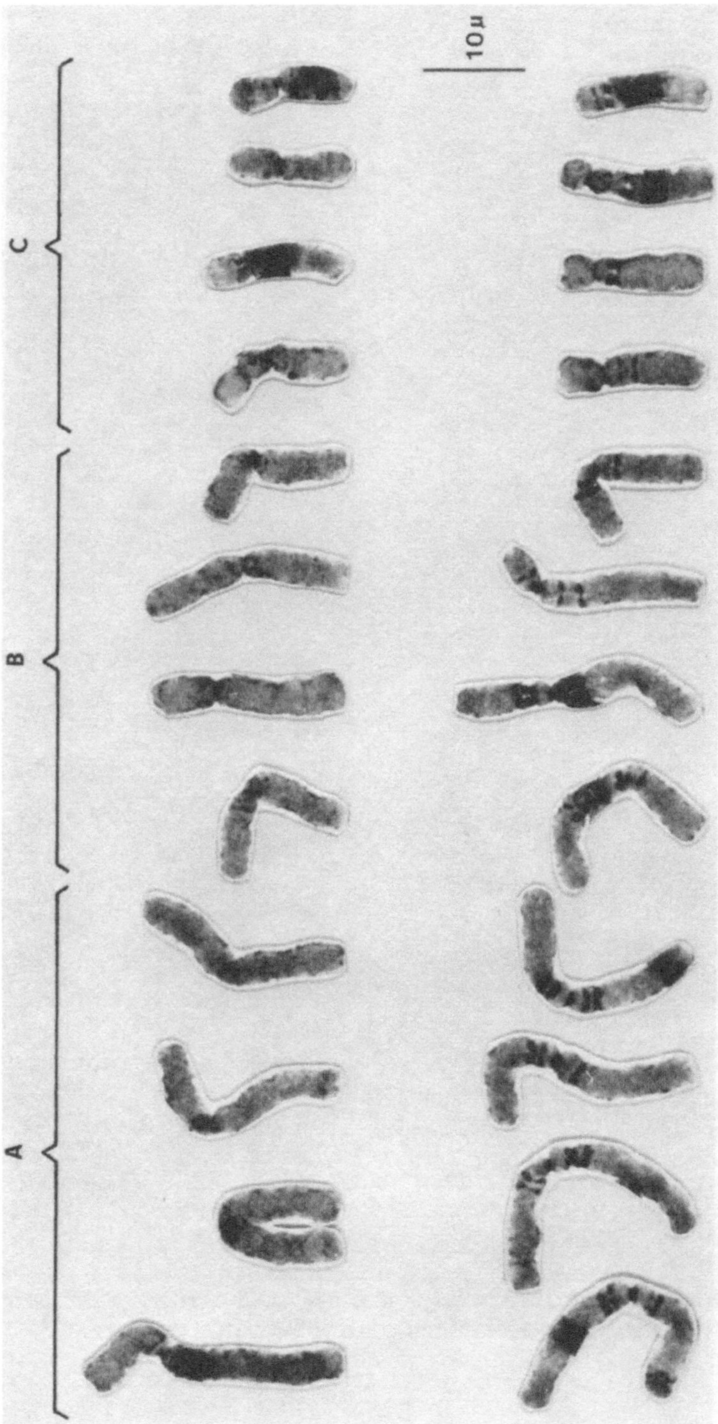

Fig. 13. The karyotype of the 24 C-stained chromosomes of a diploid hybrid larva (*T. c. carnifex* ♀ x *T. v. meridionalis* ♂). The maternal set, upper alignment; the paternal set, lower alignment.

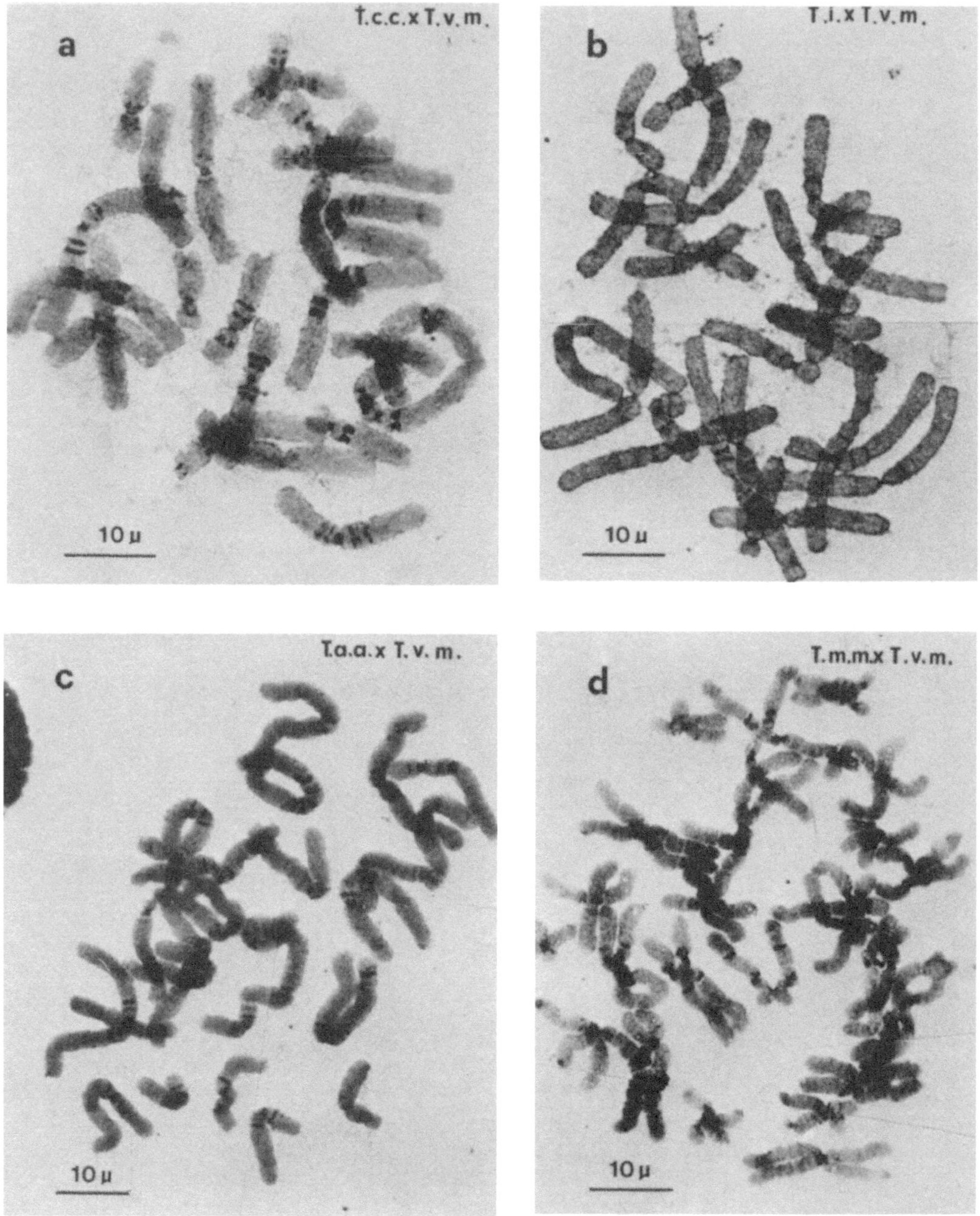

Fig. 14. Diploid F_1 hybrid larvae: the two C-stained parental genomes are recognizable by their specific banding patterns.

Table 5. Statistical analysis on 8 metaphase cells of diploid larvae *T. v. meridionalis* ♀ x x *T. c. carnifex* ♂.

Hybrids	Total length of hybrid genome (μ)	Length of genome *T. v. meridionalis* (μ)	%	Length of genome *T. c. carnifex* (μ)	%
1	791	412	52.09	379	47.91
	989	515	52.07	474	47.93
2	792	432	54.54	360	45.46
3	1452	785	54.06	667	45.94
	562	308	54.80	254	45.19
4	481	253	52.60	228	47.40
5	593	318	53.63	275	46.37
6	868	444	51.15	424	48.85

According to the method of paired comparisons, $t = 2.3521$; $P = 0.025$ (one-tailed test).

Table 6. Nuclear DNA values of some European salamandrids.

Species	DNA (pg/N)
Triturus cristatus carnifex	44
Triturus cristatus cristatus	38
Triturus cristatus dobrogicus	44
Triturus marmoratus marmoratus	42
Triturus vulgaris meridionalis	48
Triturus helveticus helveticus	41
Triturus italicus	44
Triturus alpestris apuanus	48
Salamandra atra	65
Salamandra salamandra	66
Salamandrina terdigitata	41
Euproctus montanus	46

Data kindly supplied by Dr. E. Olmo, Napoli.

as well as microchromosomes, to a lower chromosome number formula containing neither microchromosomes nor acrocentric elements. Referring to urodeles, a generalization of such a pattern is still considered premature because only about 35% of the species have been karyologically studied (Bogart 1973). Nevertheless, two different mechanisms can be assumed to have contributed to the chromosome number reduction verified during the evolution of urodeles: 1) Robertsonian mechanisms of centric fusion which caused the disappearance of all or nearly all the acrocentrics, and 2) a series of translocations of the genetically active traits onto the largest chromosomes which caused the loss of microchromosomes. According to modern cytogenetics, the first mechanism is recognized as valid in itself. The second is in agreement with White (1973b) and Matthey (1975) who seem to share the opinion that the disappearance of microchromosomes in the Amphibia presumably occurred by fusion rather than by actual loss; this agrees with the so-called principle of karyotypic orthoselection (White 1973c).

Cytogeneticists, therefore, wonder what the structure and the function of amphibian microchromosomes actually are; they also want to understand the significance of their presence in certain families and of their absence in others. In fact, we know only that microchromosomes are very small elements (less than two microns in length) and possibly are all acrocentric (Morescalchi 1973). More information is available in other animal groups, mainly birds; avian microchromosomes contain relatively fewer Mendelian genes per unit length than macrochromosomes, but they have a large quantity of heterochromatin. They could be associated with the formation of nucleoli and carry rDNA cistrons (cf. Ohno et al. 1962; Schmid 1962; Donnelly and Newcomer 1963; Comings and Mattoccia 1970; Brown and Jones 1972; Ray-Chaudhuri 1973).

The results of the experiments on urodeles using in situ molecular hybridization with ^3H 18S + 28S rRNA suggest that the loss of at least a part of the microchromosomes occured as a result of translocations. In fact, in these experiments, the major nucleolus organizers are always labeled, even though with a diverse intensity on both homologs. In contrast, the regions which are preferentially localized at the centromere and telomere regions or at C-bands and which hybridize in heterozygous condition with the same type of RNA may correspond to regions primitively belonging to microchromosomes but later translocated onto the larger elements. This hypothesis would be supported if such results were obtained using in situ hybridization in many species and genera of salamandrids and of other urodeles. It would also be useful to determine whether at least a part of the microchromosomes of the living primitive families of urodeles (Hynobiidae and Cryptobranchidae) would hybridize with 28S + 18S rRNA. If so, we would, perhaps, have further evidence favoring the translocation hypothesis. The utility of such a mechanism which, while conserving other possible Mendelian genes besides the cistrons for the 28S + 18S rRNA, has determined, in the course of karyological evolution, a numerical and structural reorganization which is advantageous to the general cell economy and has possibly influenced the mechanism of speciation, is extremely great. At present, this hypothesis concerns only karyotypic events in amphibian phylogeny; it could, however, also provide chromosomal information to integrate recent molecular hypotheses such as that concerning the evolution of rDNA and 5S DNA sequences (cf. Brown and Sugimoto 1973; Smith 1973; Pardue 1973; Wellauer et al. 1976a; Wellauer et al. 1976b; Reeder et al. 1976).

The results on the number and distribution of 5S RNA genes are still too scanty to allow karyological deductions about phylogeny and speciation processes. However, in several if not all amphibian species, these genes seem to be located in heterochromatic regions, such as telomeres and centromeres, which have a notable significance in chromosomal rearrangements (see also Pardue et al. 1973). Moreover, genera of Nearctic and Paleartic salamandrids appear to vary greatly in the number and arrangement of ribosomal 18S + 28S and 5S DNAs. This also occurs within the genus *Triturus*. Two taxonomically distant species, *T. marmoratus* and *T. v. meridionalis*, have both the 5S DNA sequences and the nucleolar organizer on one chromosome pair; in contrast, *T. c. carnifex*, which is closely related to *T. marmoratus*, shares with it only the localization of the locus for 5S genes; *T. italicus*, which is closely related to *T. v. meridionalis*, has a different number and localization of the ribosomal 18S + 28S and 5S DNAs.

The karyotypes of all the genera of European salamandrids are morphologically similar or only slightly dissimilar if studied with the traditional methods; this suggests that the karyotypic pattern is conserved by natural selection. A more detailed analysis performed by means of the most advanced cytological techniques reveals, however, the presence within each genus and even within each species of chromosomal differentiations which can be related to mechanisms of speciation. For instance, all the species and subspecies of *Triturus* studied up to now show a more or less extensive chromosome repatterning, as shown by 1) the specific DNA nuclear content values, 2) the quantity of centromeric heterochromatin and also by the amount and distribution of constitutive heterochromatin in other chromosome regions, 3) the different number, morphology, texture and distribution pattern of the lampbrush loops and of other important cytogenetic markers, including nucleolus organizer regions and sphere sites, and 4) the different patterns of gene arrangement found by *in situ* hybridization. "Homosequentiality" thus does not occur in those subspecies that have karyotypes that are nearly identical in number, lengths, and arm lengths. Considering both the chromosomal and the molecular data, the four subspecies of *T. cristatus*, which have long been geographically and reproductively isolated, appear to be at an advanced stage of speciation. Similarly, *T. italicus*, still considered as a subspecies of *T. vulgaris* (cf. Mertens and Wermuth 1960; Steward 1969), like *T. marmoratus* within the *T. cristatus* group (cf. Spurway and Callan 1960), must be considered to have already reached the species level (cf. Mancino 1968). The above facts stress the great importance of chromosome repatterning as a mechanism of specific and subspecific differentiation. The correlation between chromosomal events and evolutionary changes in regulatory systems, that is, changes in the patterns of gene expression, can therefore contribute to support some of the most recent theories concerning speciation mechanisms (cf. Britten and Davison 1969; Wilson *et al*. 1974; Mayr 1975).

REFERENCES

Ahrenfeldt, R. H. 1960. Mating behaviour of *Euproctus asper* in captivity. *Brit. J. Herp.* 2:194-197.

Arrighi, F. E. and T. C. Hsu. 1971. Localization of heterochromatin in human chromosomes. Cytogenetics 10:81-86.

Atchley, W. R. 1972. The chromosome karyotype in estimation of lineage relationships. Syst. Zool. 21:199-209.

Avery, R. A. 1968. Food and feeding relations of three species of Triturus (Amphibia Urodela) during the aquatic phases. Oikos 19:408-412.

Avery, R. A. 1971. The ecology of newt tadpoles: food consumption, assimilation efficiency and growth. Freshwater Biol. 1:129-134.

Ayala, F. 1975. Genetic differentiation during the speciation process. IN: T. Dobzhansky, M. K. Hecht and W. C. Steere (eds.), Evolutionary biology VIII. Plenum Press, New York, pp. 1-78.

Barsacchi, G., L. Bussotti and G. Mancino. 1970. The maps of the lampbrush chromosomes of Triturus (Amphibia Urodela). IV. Triturus vulgaris meridionalis. Chromosoma (Berl.) 31:255-279.

Barsacchi Pilone, G., F. Andronico, R. Batistoni, I. Nardi and G. Lisanti. 1976. Localization of the ribosomal cistrons in lampbrush and mitotic chromosomes of various individuals of Triturus vulgaris meridionalis (Amphibia Urodela). Atti Ass. Genet. Ital. 20:110-111.

Barsacchi Pilone, G., I. Nardi, R. Batistoni, F. Andronico and E. Beccari. 1974a. Chromosome location of 28S, 18S, and 5S ribosomal RNA genes in Triturus (Amphibia Urodela). Boll. Zool. 41:456.

Barsacchi Pilone, G., I. Nardi, R. Batistoni, F. Andronico and E. Beccari. 1974b. Chromosome location of the genes for 28S, 18S and 5S ribosomal RNA in Triturus marmoratus (Amphibia Urodela). Chromosoma (Berl.) 49:135-153.

Batistoni, R., I. Nardi and G. Barsacchi Pilone. 1974. Banding patterns on lampbrush chromosomes of Triturus marmoratus (Amphibia Urodela) by the Giemsa stain. Chromosoma (Berl.) 49:121-134.

Bell, G. A. C. 1974. The reduction of morphological variation in natural populations of smooth newt larvae. J. Anim. Ecol. 43: 115-128.

Bell, G. and J. H. Lawton 1975. The ecology of the eggs and larvae of the smooth newt (Triturus vulgaris (Linn.)). J. Anim. Ecol. 44:393-423.

Benazzi, M. 1956. Il sesso negli ibridi di tritoni e la questione della digametia sessuale negli urodeli. Ricerca Scient., Suppl. 26:3-11.

Bogart, J. P. 1973. Evolution of anuran karyotypes. IN: J. L. Vial (ed.), Evolutionary biology of the anurans. Univ. Missouri Press, Columbia, pp. 337-349.

Britten, R. J. and E. H. Davidson. 1969. Gene regulation for higher cells: a theory. Science 165:349-357.

Brown, D. D. and K. Sugimoto. 1973. The structure and evolution of ribosomal and 5S DNAs in Xenopus laevis and Xenopus mulleri. Cold Spring Harb. Symp. Quant. Biol. 38:501-505.

Brown, J. E. and K. W. Jones. 1972. Localization of satellite DNA in the microchromosomes of the Japanese Quail by in situ hybridization. Chromosoma (Berl.) 38:313-318.

Callan, H. G. 1942. Heterochromatin in Triton. Proc. Roy. Soc. Lond. B 130:324-335.

Callan, H. G. 1955. Recent work on the structure of cell nuclei. IN: Symp. on fine structure of cells. I. U. B. S. Publ., Ser. B 21:89-109. Noordhoff, Groningen.

Callan, H. G. and L. Lloyd. 1956. Visual demonstration of allelic differences within cell nuclei. Nature (Lond.) 178:355-357.

Callan, H. G. and L. Lloyd. 1960a. Lampbrush chromosomes of crested newts Triturus cristatus (Laurenti). Phil. Trans. R. Soc. Lond. B 243:135-219.

Callan, H. G. and L. Lloyd. 1960b. Lampbrush chromosomes. IN: P. M. B. Walker (ed.), New approaches in cell biology. Acad. Press, New York, pp. 23-46.

Callan, H. G. and L. Lloyd. 1975. Working maps of the lampbrush chromosomes of amphibia. IN: R. C. King (ed.), Handbook of genetics IV. Plenum Press, New York, pp. 57-77.

Carson, H. L., F. E. Clayton and H. D. Stalker. 1967. Karyotypic stability and speciation in Hawaiian Drosophila. Proc. Natl. Acad. Sci. U. S. A. 57:1280-1285.

Comings, D. E. and E. Mattoccia. 1970. Studies of microchromosomes and a G-C rich DNA satellite in the quail. Chromosoma (Berl.) 30:202-214.

Donnelly, G. M. and E. H. Newcomer. 1963. Autoradiographic patterns in cultured leukocytes of the domestic fowl. Exp. Cell Res. 30: 363-368.

Fankhauser, G. 1938. Sex differentiation in a haploid salamander, Triton taeniatus Laur. J. Exp. Zool. 79:35-49.

Fischberg, M. 1945. Über die Ausbildung der Geschlechts bei triploiden und einem haploiden Triton alpestris. Rev. Suisse Zool. 52:407-414.

Gall, J. G. 1954. Lampbrush chromosomes from oocyte nuclei of the newt. J. Morph. 94:283-353.

Gallien, L. 1954a. Démonstration de l'homogamétie du sexe mâle chez le Triton Pleurodeles waltlii Michah par l'étude de la descendance d'animaux à sexe physiologique inversé, après un traitement hormonal gynogène (benzoate d'oestradiol . C. R. Ac. Sci. Paris 238:402-404.

Gallien, L. 1954b. Inversion expérimentale du sexe, sous l'action des hormones sexuelles, chez le Triton Pleurodeles waltlii Michah. Analyse des conséquences génétiques. Bull. Biol. Fr. Belg. 88:1-51.

Galton, M. and P. R. Bredbury. 1966. DNA replication patterns of the sex chromosomes of the pigeon (Columba livia domestica). Cytogenetics 5:295-306.

Geormaneanu, C. 1975. The karyotype of urodele amphibian Triturus montandoni. Rev. Roum. Biol. 20:275-279.

Goin, O. B. and C. J. Goin. 1968. DNA and the evolution of vertebrates. Am. Midl. Nat. 80:289-298.

Haldane, J. B. S. 1922. Sex ratio and unisexual sterility in hybrid animals. J. Genet. 12:101-109.

Halliday, T. R. 1974a. The profligate private life of the newt. New Scient. 62:14-15.

Halliday, T. R. 1974b. Sexual behaviour of the smooth newt, Triturus vulgaris (Urodela, Salamandridae). J. Herp. 8:277-292.

Halliday, T. R. 1975. On the biological significance of certain morphological characters in males of the smooth newt Triturus vulgaris and of the palmate newt Triturus helveticus (Urodela: Salamandridae). Zool. J. Linn. Soc. 56:291-300.

Hedgecock, D. and F. J. Ayala. 1974. Evolutionary divergence in the genus Taricha (Salamandridae). Copeia 1974:738-747.

Humphrey, R. R. 1942. Sex of the offspring fathered by two Amblystoma females experimentally converted into males. Anat. Rec. 82, suppl.:77.

Humphrey, R. R. 1945. Sex determination in ambystomid salamanders: a study of the progeny of females experimentally converted into males. Am. J. Anat. 76:33-66.

Humphrey, R. R. 1948. Reversal of sex in females of genotype WW in the axolotl (Siredon or Amblystoma mexicanum) and its bearing upon the role of the Z chromosomes in the development of testes. J. Exp. Zool. 109:171-186.

Hutchison, N. and M. L. Pardue. 1975. The mitotic chromosomes of Notopnthalmus (= Triturus) viridescens: localization of C banding regions and DNA sequences complementary to 18S, 28S and 5S ribosomal RNA. Chromosoma (Berl.) 53:51-69.

Kawamura, T. 1951. Reproductive ability of triploid newts, with remarks on their offspring. J. Sci. Hiroshima Univ., Ser. B 12:1-10.

Lacroix, J. C. 1968. Étude descriptive des chromosomes en écouvillon dans le genre Pleurodeles (Amphibien, Urodèle). Ann. Embryol. Morphogen. 1:179-202.

Lantz, L. A. 1947. Hybrids between Triturus cristatus Laur. and Triturus marmoratus Latr. Proc. Zool. Soc. Lond. 117:247-258.

León, P. 1976. Hybridization of 4S, 5S, and 18S + 28S RNA to salamander chromosomes. J. Cell. Biol. 69:287-300.

León, P. and J. Kezer. 1974. The chromosomes of Siren intermedia nettingi (Goin) and their significance to comparative salamander karyology. Herpetologica 30:1-11.

Macgregor, H. C. and S. Mizuno. 1976. In situ hybridization of "nick-translated" ^{3}H-ribosomal DNA to chromosomes from salamanders. Chromosoma (Berl.) 54:15-25.

Makino, S. 1947. A study of chromosomes in the two sexes of Hynobius retardatus (an Urodelan), with a consideration on the chromosomes and sex. J. Fac. Sci. Hokkaido Univ. Ser. 6 Zool. 9: 261-265.

Mancino, G. 1968. Sulla validità specifica di Triturus italicus (Peracca 1898) (Anfibi Urodeli). Rend. Acc. Naz. Lincei (Roma) 44:697-700.

Mancino, G. and G. Barsacchi. 1965. Le mappe dei cromosomi "lampbrush" di Triturus (Anfibi Urodeli). I. Triturus alpestris apuanus. Caryologia (Firenze) 18:637-665.

Mancino, G. and G. Barsacchi. 1966a. The maps of the lampbrush chromosomes of Triturus (Amphibia Urodela). II. Triturus helveticus. Riv. Biol. 59:339-351.

Mancino, G. and G. Barsacchi. 1966b. Cariologia di Salamandrina perspicillata (Anfibi Urodeli). Boll. Zool. 33:251-267.

Mancino, G. and G. Barsacchi. 1969. The maps of the lampbrush chromosomes of Triturus (Amphibia Urodela). III. Triturus italicus. Ann. Embryol. Morphogen. 2:355-377.

Mancino, G. and I. Nardi. 1971. Chromosomal heteromorphism and female heterogamety in the marbled newt Triturus marmoratus (Latreille, 1800). Experientia (Basel) 27:821-822.

Mancino, G., I. Nardi and G. Barsacchi. 1970. Spontaneous aberrations in lampbrush chromosome XI from a specimen of Triturus vulgaris meridionalis (Amphibia, Urodela). Cytogenetics 9:260-271.

Mancino, G., I. Nardi and M. Ragghianti. 1972a. Lampbrush chromosomes from semi-albino crested newts, Triturus cristatus carnifex (Laurenti). Experientia (Basel) 28:856-860.

Mancino, G., I. Nardi and M. Ragghianti. 1972b. Structural correspondence between nucleolus- and sphere-organizing regions of the lampbrush chromosomes and secondary constrictions of the mitotic chromosomes. Experientia (Basel) 28:586-588.

Mancino, G., M. Ragghianti and S. Bucci-Innocenti. 1973. I cariotipi di Triturus marmoratus e T. cristatus studiati con il "C-staining method." Rend. Acc. Naz. Lincei (Roma) 55:559-564

Mancino, G., M. Ragghianti, I. Nardi and P. Andreuccetti. 1972c. Sex chromosomes in newts. Boll. Zool. 39:639.

Matthey, R. 1951. The chromosomes of the vertebrates. Adv. Genet. 4:159-180.

Matthey, R. 1975. Caryotypes de Mammifères et d'Oiseaux. La question des microchromosomes. Quelques réflexions sur l'évolution chromosomique. Arch. Genetick 48:12-26.

Mayr, E. 1975. The unity of the genotype. Biol. Zbl. 94:377-388.

Mertens, R. and H. Wermuth. 1960. Die Amphibien und Reptilien Europas. Verlag Waldemar Kramer, Frankfurt am Main.

Morescalchi, A. 1970. Karyology and vertebrate phylogeny. Boll. Zool. 37:1-28.

Morescalchi, A. 1973. Amphibia. IN: A. B. Chiarelli and E. Capanna (eds.), Cytotaxonomy and vertebrate evolution. Acad. Press, New York, pp. 233-348.

Morescalchi, A. 1975. Chromosome evolution in the caudate amphibia. IN: T. Dobzhansky, M. K. Hecht and W. C. Steere (eds.), Evolutionary biology. VIII. Plenum Press, New York, pp. 339-387.

Namur, P. 1969. Étude des chromosomes de la blastula de Triturus vulgaris Linné. Bull. Soc. Linn. Normandie 10:67-72.

Namur, P. and J. Signoret. 1967. Étude comparée du caryotype de quelques espèces de Tritons communes en Normandie. Bull. Soc. Linn. Normandie 8:183-194.

Nardi, I., R. Batistoni, G. Barsacchi Pilone, M. Bartoli and F. Andronico. 1974. The genes for 28S, 18S and 5S ribosomal RNA in the chromosomes of Triturus vulgaris meridionalis (Amphibia Urodela). Boll. Zool. 41:509.

Nardi, I. and G. Mancino. 1971. Mitotic karyotype and nucleoli of the marbled newt Triturus marmoratus (Latreille). Experientia (Basel) 27:424-427.

Nardi, I., M. Ragghianti and G. Mancino. 1972a. Characterization of the lampbrush chromosomes of the marbled newt Triturus marmoratus (Latreille, 1800). Chromosoma (Berl.) 37:1-22.

Nardi, I., M. Ragghianti and G. Mancino. 1972b. Morphology of the mitotic chromosomes of embryos and of adults of Italian alpine newt Triturus alpestris apuanus (Bonaparte). Experientia (Basel) 28:591-592.

Nardi, I., M. Ragghianti and G. Mancino. 1973. Banding patterns in newt chromosomes by the Giemsa stain. Chromosoma (Berl.) 40:321-331.

Ohno, S., L. C. Christian and C. Stenius. 1962. Nucleolus-organizing microchromosomes of Gallus domesticus. Exp. Cell Res. 27:612-614.

Olmo, E. 1973. Quantitative variations in the nuclear DNA and phylogenesis of the Amphibia. Caryologia 26:43-68.

Olmo, E. 1974. Further data on the genome size in the urodeles. Boll. Zool. 41:29-33.

Olmo, E. and A. Morescalchi. 1975. Evolution of the genome and cell size in salamanders. Experientia (Basel) 31:804-806.

Özeti, N. 1967. The morphology of the salamander Mertensiella luschani (Steindachner) and the relationships of Mertensiella and Salamandra. Copeia 1967:287-298.

Özeti, N. and D. B. Wake. 1969. The morphology and evolution of the tongue and associated structures in salamanders and newts (family Salamandridae). Copeia 1969:91-123.

Pardue, M. L. 1973. Localization of repeated DNA sequences in Xenopus chromosomes. Cold Spring Harb. Symp. Quant. Biol. 38: 475-482.

Pardue, M. L., D. D. Brown and M. L. Birnstiel. 1973. Localization of the genes for 5S ribosomal RNA in Xenopus laevis. Chromosoma (Berl.) 42:191-203.

Pardue, M. L. and J. C. Gall. 1970. Chromosomal localization of mouse satellite DNA. Science 168:1356-1358.

Pukkila, P. J. 1975. Identification of the lampbrush chromosome loops which transcribe 5S ribosomal RNA in Notophthalmus (Triturus) viridescens. Chromosoma (Berl.) 53:71-89.

Ragghianti, M., S. Bucci-Innocenti and G. Mancino. 1975. Indagine citogenetica in ibridi interspecifici di Triturus (Anfibi Urodeli). Boll. Zool. 42:482.

Ragghianti, M., I. Nardi and G. Mancino. 1972. Completion of the morphology of the lampbrush chromosomes of the Italian alpine newt Triturus alpestris apuanus Bonaparte. Experientia (Basel) 28:588-590.

Ray-Chaudhuri, R. 1973. Cytotaxonomy and chromosome evolution in birds. IN: A. B. Chiarelli and E. Capanna (eds.), Cytotaxonomy and vertebrate evolution. Acad. Press, New York, pp. 425-483.

Reeder, R. H., D. D. Brown, P. K. Wellauer and I. B. Dawid. 1976. Patterns of ribosomal DNA spacer lengths are inherited. J. Molec. Biol. 105:507-516.

Rensch, B. 1929. Das Prinzip geographischer Rassenkreise und das Problem der Artbildung. Borntraeger, Berlin.

Rudak, E. and H. G. Callan. 1976. Differential staining and chromatin packing of the mitotic chromosomes of the newt Triturus cristatus. Chromosoma (Berl.) 56:349-362.

Salthe, S. N. 1967. Courtship patterns and the phylogeny of the urodeles. Copeia 1967:100-117.

Schmid, M. and W. Krone. 1976. The relationship of a specific chromosomal region to the development of the acrosome. Chromosoma (Berl.) 56:327-347.

Schmid, W. 1962. DNA replication patterns of the heterochromosomes in Gallus domesticus. Cytogenetics 1:344-352.

Smith, G. 1973. Unequal crossover and the evolution of multigene families. Cold Spring Harb. Symp. Quant. Biol. 38:507-513.

Spurway, H. 1945. Sex determination in Triturus vulgaris Linn. (taeniatus Schneid.). Am. Nat. 79:377-380.

Spurway, H. and H. G. Callan. 1960. The vigor and male sterility of hybrids between the species Triturus vulgaris and T. helveticus. J. Genet. 57:84-117.

Steward, J. W. 1969. The tailed amphibians of Europe. David and Charles, Newton Abbott.

Thireau, M. 1975a. L'allométrie pondérale encéphalo-somatique chez les urodèles. I. Relations intraspécifiques. Bull. Mus. Natn. Hist. Nat. (Paris) 297:467-482.

Thireau, M. 1975b. L'allométrie pondérale encéphalo-somatique chez les urodèles. II. Relations interspécifiques. Bull. Mus. Natn. Hist. Nat. (Paris) 297:483-501.

Thorn, R. 1968. Les salamandres. P. Lechevalier, Paris.

Tortonese, E. and B. Lanza. 1968. Pesci, anfibi e rettili. IN: Piccola Fauna Italiana. A. Martello, Milan.

Ullerich, F. 1970. DNS-gehalt und Chromosomenstruktur bei Amphibien. Chromosoma (Berl.) 30:1-37.

Vallée, L. 1959. Recherches sur Triturus blasii de l'Isle, hybride naturel de Triturus cristatus Laur. x Triturus marmoratus Latr. Impr. M. Declume, Lons-le-Saunier.

Wake, D. B. and N. Özeti. 1969. Evolutionary relationships in the family Salamandridae. Copeia 1969:124-137.

Wake, M. H. and S. M. Case. 1975. The chromosomes of Caecilians (Amphibia: Gymnophiona). Copeia 1975:510-516.

Wellauer, P. K., I. B. Dawid, D. D. Brown and R. H. Reeder. 1976a. The molecular basis for length heterogeneity in ribosomal DNA from Xenopus laevis. J. Molec. Biol. 105:461-486.

Wellauer, P. K., R. H. Reeder, I. B. Dawid and D. Brown. 1976b. The arrangement of length heterogeneity in repeating units of amplified and chromosomal ribosomal DNA from Xenopus laevis. J. Molec. Biol. 105:487-505.

White, M. J. D. 1946. The spermatogenesis of hybrids between Triturus cristatus and T. marmoratus (Urodela). J. Exp. Zool. 102: 179-207.

White, M. J. D. 1954. Animal cytology and evolution, 2nd ed. Cambridge Univ. Press, London.

White, M. J. D. 1973a. Animal cytology and evolution, 3rd ed. Cambridge Univ. Press, London.

White, M. J. D. 1973b. Chromosomal rearrangement in mammalian population polymorphism and speciation. IN: A. B. Chiarelli and E. Capanna (eds.), Cytotaxonomy and vertebrate evolution. Acad. Press, New York, pp. 95-128.

White, M. J. D. 1973c. The chromosomes. Chapman and Hall, London.

Wickbom, T. 1945. Cytological studies on Dipnoi, Urodela, Anura and Emys. Hereditas (Lund) 31:241-346.

Wilson, A. C., V. M. Sarich and L. R. Maxson. 1974. The importance of gene rearrangement in evolution: evidence from studies on rates of chromosomal, protein, and anatomical evolution. Proc. Natl. Acad. Sci. U. S. A. 71:3028-3030.

CONCLUDING REMARKS

Thomas Uzzell

The Academy of Natural Sciences
Philadelphia, Pennsylvania 19103

Successful reproduction by any biparentally reproducing metazoan involves not only production of gametes and finding appropriate mates to fertilize them, but also the development of those fertilized gametes to reproductive adults, at least to the extent that the individual replaces itself in the next generation. Clearly, the scope of possible studies on amphibian reproduction is enormous, and any symposium on the subject must be highly selective in the topics covered. Members of the SSAR are largely interested in the comparative aspects of reproduction rather than the general aspects, in the natural history rather than the biochemistry. Presentations at the 1976 symposium share this bias towards certain aspects of amphibian reproduction.

Within these reasonable biases, however, there are other limitations. The three groups of amphibians are not equal in numbers of taxa, have non-comparable geographic distributions, and are not equally easy to observe within their ranges and habitats. Our knowledge of the three groups is enormously different. Virtually all that is known about caecilian reproduction can be reviewed by a single person, as in Wake's contribution; whereas many different speakers addressed different problems in the reproductive biology of frogs and salamanders. These differences reflect the amount of research effort devoted to the different groups, a function primarily of their differing accessibility. North temperate groups have been investigated by many students for many years; tropical groups are much less well known. The caecilians, a tropical group many of which are also burrowers, are least well understood.

The different groups of organisms pose different problems. Frogs are often vocal, caecilians never and salamanders, rarely so;

the sophistication of studies of vocalization and sound reception among the frogs is quite remarkable. The different groups lend themselves to different kinds of studies, even though the problems may be general. Salamanders have especially large chromosomes, and the detailed analyses of chromosomes of the genus Triturus reported by Mancino and his colleagues are among the best studies yet made of vertebrate chromosomes.

Despite the necessary heterogeneity in our level of knowledge about various groups of amphibians, or various aspects of their reproductive biology, certain themes run through the papers presented at this symposium. An obvious theme is that of obtaining the most propitious mate. A second theme, less strongly expressed, is that of reproductive schedules; the production of enough gametes (usually ova) at appropriate times to permit populations to persist. A third theme is exploration of the taxonomic implications of life history or other comparative data.

Development of detailed knowledge about amphibian reproduction is nowhere better illustrated in these papers than in the series on acoustical behavior of frogs. Selective advantages that may shape the spectrographic characters of salientian calls are discussed by Littlejohn, who analyzes his material in the framework of a communication system, consisting of an emitter, a communication channel, a receiver, and a shared repertoire of signs. Schneider reviews extensive research on the vocalization of European frogs, and using Hyla arborea as a model, explored in depth the features of the emitter system. At the other end of the line, Capranica has probed ever more deeply into the sound receiving apparatus of frogs. The finding of geographic variation in the auditory sensitivity of the receiver that is correlated with geographic variation in the sound emitted, and of differing sexual sensitivities and responses to male vocalizations were noted by several of the speakers. The extent to which the neural basis for Littlejohn's "communication system" is being explored is thus quite remarkable. Some of the same patterning features in salientian sounds that Littlejohn discusses are also discussed by Wells but more from an organismal than from a spectral viewpoint.

These elaborately studied vocal behaviors and the responses to them serve many functions but many of these same functions may be served, even in frogs, without so much use of sounds. Wells' review of courtship patterns in dendrobatid frogs emphasizes their differences from the more familiar temperate zone Bufo-Rana-Hyla pattern. Similar elaborate patterns in courtship are described in this symposium by Halliday (salamandrid salamanders) and by Arnold (plethodontid and ambystomatid salamanders); again, many of the functions served by vocal behavior in frogs are served by other behaviors in these animals. Halliday emphasized the selective potential of slow female response to courtship in shaping features of male courtship, both in terms of competition between conspecific males and in terms of avoidance of hybridization. Arnold emphasized comparative aspects of male behavior, and pointed out the roles of sexual interference and sexual defense as responses to intermale competition.

What behavioral or sensory features serve these functions in caecilians is almost unknown.

In all of these studies, whether of frogs or salamanders, the patterns, vocal or otherwise, can be interpreted as serving to maximize successful reproduction. Both Wells (for dendrobatid frogs) and Halliday (for newts of the genus *Triturus*) consider the courtship patterns of the species that they describe in terms of mate assessment. Each develops plausible interpretations of the behavior described that would associate success in mating with greater fitness. Success in mating may itself be part of such fitness, but more than this is implied, especially by Halliday, who argues that those male newts that can stay submerged longest, that can deposit more spermatophores, that can support a larger tail crest, and that have reached larger body size and presumably greater age are likely to be genetically more fit in other respects as well.

Differential success at reproducing successfully (natural selection) is perhaps most clearly seen in interspecies hybridization. Many of the elaborate studies of interspecies crosses reported by Kawamura and Nishioka reveal a gradation in the developmental stage at which failure of such hybridizations occurs. Some of the crosses do not even produce activated and cleaving zygotes. Others fail at later stages of development, and others yet result in partial sterility of one or both sexes. Studies by the group at the Laboratory for Amphibian Biology in Hiroshima have resulted in recognition that many of the taxa of Japanese frogs are distinct biological species; additional progress in this direction will occur. It seems certain for example that the Japanese *Hyla* cannot be conspecific with *Hyla arborea* of Europe on the basis of their results.

As demonstrated in great variety by Kawamura and Nishioka, crosses between species of amphibians usually fail to some degree in one way or another, although the degree and timing vary from case to case. For this reason, among others, stable hybrid populations such as *Rana esculenta* in its various guises are particularly interesting. One can find, in Berger's review of many years' work in Poland, parts of an explanation for this anomaly. The hybridogenetic pattern of inheritance results in progeny for generation after generation with essentially F_1 genotypes; having essentially intact genomes from two species, these individuals are highly heterozygous, with whatever advantages that may bestow. For example, *R.*

esculenta has to some degree the strong odor of R. ridibunda, which may protect it against predation, and can hibernate both underground like R. lessonae and under water. On the other hand, females, which do most of the breeding in most populations, produce the higher numbers of eggs associated with R. ridibunda, rather than some intermediate number or the lower number of R. lessonae. These observations, together with the behavioral patterns reported by Blankenhorn, in which small males of Rana lessonae mate preferentially with larger conspecifics, while the larger males of R. lessonae more frequently mate with the yet larger female hybrids, provide the framework for an adequate explanation of the persistence of such hybrid lineages without appeal to group selection. The hybrids are somatically vigorous; fertility is impaired in males but high in females; fecundity is high in females and they are favored in matings by large R. lessonae males; F_2 recombinants are mostly absent.

Stable hybrid populations such as Rana esculenta highlight some of the difficulty of "fitness" as a concept. We all seem to know what fitness is, or at least, we all talk about it. One way that it has been defined is as the relative ability of individuals or of genotypes to leave fertile adult offspring. Males of Rana lessonae that mate with females of Rana ridibunda appear to leave more fertile adult offspring than those that mate with females of Rana lessonae, although they leave fewer Rana lessonae progeny. Fitness has also been defined as the probability of leaving descendants after a given long period of time; I suppose that these two definitions span the range of those proposed. Since the hybridogenesis in R. esculenta generally results in loss of the R. lessonae chromosomes, we can be sure that, in terms of grandprogeny, hybridizing R. lessonae males are leaving few progeny. My point has been made many times by others: if we determine fitness by the first definition, we often have little information about fitness by the second.

The problem surely is much more complex. Using biochemical markers, eventually we probably can obtain enough genetic information about individual adult amphibians in each of two successive generations, at least locally, to estimate the contribution of individuals of one generation to the second. With enormous effort we probably could do this for two generations. Finite population sizes, however, compound the difficulty of showing that observed differentials are selective rather than random.

Given these difficulties, both in reconciling long and short term fitness and in determining contributions of individuals to successive generations, another approach, analyzing "components of fitness," is often used. Natural selection is differential reproduction by different phenotypes; reproduction here subsumes surviving to reproduce. One can suppose that fitness has these two components: survival and reproduction. Individual longevities and reproductive schedules, if known, should provide a measure of fitness of individuals. Failing that, we can use such measurable quantities as the number of ova produced by females or the number of matings by males as indices of differential reproduction. It is plausible that these indices of fitness are positively correlated with fitness by whatever definition. I even suppose that they usually are. Trade-offs between quality and quantity of offspring, however, or between risking reproduction and longevity could give some of these indices negative correlations with fitness. To me, the correlation between "components of fitness" and "fitness" is a question requiring data rather than assumption. The use of the concept is a prime example of what Arthur Cain has called explaining phenomena away, thus avoiding the hard work of investigating them. In any case, our data showing genetic variance in any of these components of fitness is, as pointed out by Tilley, woefully limited, especially in amphibians.

I suppose, in passing, that a general function, incorporating effort and risk in surviving and reproducing, will show that the average male and the average female are likely to have equal success in reproducing.

Several of the presentations, especially those by Littlejohn, Schneider, Wells, and Blankenhorn on frogs and those by Arnold and Halliday on salamanders, emphasized behavioral features of courtship. Several other papers were more indirectly associated with courtship, including Capranica's and Schneider's contributions on the neural and muscular basis for vocal aspects of the courtship behavior of frogs. A contribution to our understanding of the hormonal control of courtship behavior in frogs was offered by Obert, who pointed out that European frogs fall into two groups, Type I with a single reproductive period each season, the commoner pattern; and Type II, with several distinct periods of spawning and calling each season. It seems likely to me at least that an analysis of the hormonal relationships in the dendrobatids described by Wells would reveal that they belong among the Type II frogs defined by Obert.

Two careful studies, that by Tilley on the reproductive patterns shown by individuals of two populations of the salamander _Desmognathus ochrophaeus_, and that by Houck comparing clutch size and the relation of climate to reproduction in plethodontid salamanders from the eastern and western United States and from Guatemala, scarcely discussed courtship at all. Tilley's study forces into our consciousness something that many of us have not wanted to acknowledge: that in addition to differences between populations, there are differences between individuals in life history features such as age or size at maturity, frequency of oviposition by females, and growth rates. While studying these individual variations requires long term commitment, without knowledge of them, differences between populations are more difficult to interpret correctly, although easy to interpret, since hypotheses are available to cover all contingencies.

Houck's studies on the reproductive patterns of neotropical plethodontids, together with her survey of the literature, reveal three sets of patterns among the Plethodontidae that can be associated in general with climatic features. Exceptions to the patterns are usually interpretable within the general context, and often seem at second glance to support the associations between pattern and climate that Houck invokes. One of her findings, that many tropical plethodontids have relatively larger rather than relatively smaller clutches, is a contrast to the pattern seen in other tetrapods. This is a start, and an excellent one; comparisons among these three groups in other aspects of their biology will probably be equally informative about the bases for the adaptive radiation and subradiations in this remarkable family of salamanders.

If our knowledge of tropical plethodontids is just beginning to develop, our knowledge of the caecilians is even more skimpy. Quite a bit can be said about them, however, on the basis of dissections of the material available coupled with careful thought. That caecilians are remarkable animals has long been sensed; I can hardly say known, since they still are so poorly known. Wake, in her review of what is known, which incorporates many of her unpublished observations, particularly emphasized that, just as salamanders and frogs are unlike in reproductive biology, so too are caecilians different; there is no general pattern of amphibian reproduction. Some generalities across the three groups probably will be made eventually, of course, and Wake carefully points out what some will be, and what some probably will not be.

The studies on newt chromosomes reported by Mancino and his associates, while perhaps not involved in reproduction in the same way that many of the contributions are, draws on studies of the reproductive cells themselves for most of the data. To me, the astounding feature is that the karyotypes of these newts, while so similar in terms of chromosome length and arm length ratio, have markedly different arrangements of genes, at least for those markers so far studied. What selective pressures maintain the overall similarity of the karyotype are unknown. Such evidence of rearrangement within apparent karyotypic homogeneity highlights even more sharply the suggestion that rates of karyotypic evolution and of morphological evolution are more highly correlated with each other than either is with the biochemical evolution examined by electrophoresis or immunology.

I cannot pass by the opportunity to look into the future of studies of amphibian reproductive biology. A first trend is the extension of current research efforts, applying them to new groups. While there is a school of thought that maintains that science consists of falsifying or at least attempting to falsify hypotheses, without observations, it is not possible to propose hypotheses, and without other observations, impossible to reject them. We have in this symposium an excellent body of observations, sometimes with hypotheses suggested, sometimes awaiting the appropriate hypothesis. Given the diversity of the Amphibia, the usefulness of acquiring comparative data to suggest hypotheses for testing is obvious, and the imagination displayed by the contributors in using such data points out the strength of this comparative approach.

A second trend, then, is to find appropriate hypotheses, or currencies, to mediate generalizations across the diversity that constitute the Amphibia. We see some of these, for example, in interpretations of courtship patterns as permitting "assessment" of fitness of mates. The hard work of testing this hypothesis, of showing that males that inseminate females have greater fitness in other than their insemination of the females has scarcely begun; it's an easy assumption, but tough to demonstrate, although the detailed studies reported by Tilley are surely a step in this direction.

Some other areas in which I expect new data to suggest hypotheses, or to serve in testing hypotheses, will include the following:

1. Rates of evolution. One of the most interesting areas for exploration will be in rates of evolution, an idea that is in the air, and that was explicitly suggested at this symposium for behavior by Arnold. As our knowledge of relationships of major and lesser taxa becomes more refined, through traditional morphological studies as well as through newer, mostly biochemical studies, and as the times associated with divergences become more convincing, again using traditional paleontological and biogeographic methods as well as newer biochemical techniques, the explorations of rates in biology will become considerably more widespread. I foresee not only rates of behavioral evolution, but general studies: rates of phenetic divergence, of detailed chromosomal rearrangements, of ecological adaptations. The possibilities of developing convincing cladograms with believable times associated with the branch points will spur many workers to quantify divergence in behavior, in ecology, in morphology, etc., so that rates of evolution can be estimated, compared, and perhaps understood.

2. Speciation. Amphibians ought also to be especially good for extending studies of speciation. In my view, speciation is

the acquisition of differences between populations such that crosses between populations have considerably less likelihood of leaving grandprogeny than crosses within. Differences between populations in sympatry that reduce the likelihood of cross mating between the populations or that reduce the competition between members of the two populations may be established during speciation or may be sequelae to speciation itself, or, more plausibly, may start to evolve synchronously with speciation but may be reinforced during sympatry.

We know at present almost nothing about the genetics of speciation as I define it. A considerable amount is known about electrophoretically detectable differences between pairs of species, but the consensus is that such differences are not the differences that cause problems for hybrids. It is easy to retreat into the near-mysticism of regulatory evolution as the basis of problems in interspecies hybrids, but this is not explanation. Surely there must be regulatory genetic material, and surely it differs between species, but so far appealing to regulatory differences only says that the critical differences are not in the places we're looking (a plausible but largely undemonstrated assumption).

Problems that hybrids have can be those discussed by Kawamura and Nishioka, and may occur at various developmental stages. They may occur even later, in phenomena such as adaptation to natural environments or in ability to find mates (one can't help wondering about the vocalizations and cochlear tuning in the complex of hybrids called Rana esculenta). We have as yet, however, almost no clue as to what exactly goes wrong in any interspecies hybrids, and whether many or few genetic differences lie at the basis of these hybrid incompatibilities. Amphibians, because they can be experimentally manipulated, should offer many opportunities for exploring these problems, for developing a genetics of speciation.

What we do know is that the problems of hybrids are in fact a negative selection on individuals that hybridize. What we cannot demonstrate at present, is whether or not the various phenotypic differences that reduce hybridization among sympatric taxa arose in sympatry or in allopatry, in small, perhaps peripherally isolated populations, or in large ones. It is not easy to see how such a demonstration could be made. I believe that it may be possible to find something of an answer by studying the genetics of the problems that hybrids between cognate species have. If these problems arise from a few important genetic differences, such differences seem more likely to have arisen in small isolated populations in which the deleterious effects of being heterozygous for such differences might not prevent such traits from becoming fixed. If many small genetic differences are contributing to such problems, however, these seem more likely, to me at least, to have arisen in larger populations, perhaps somewhat subdivided, than in small ones.

It is in such areas as this, for example, in which the large chromosomes of many amphibians perhaps should be useful. It is often stated that chromosomal differences are the basis of speciation, and this may well be so. The correlation of species-specific differences of any sort with speciation does not, however, demonstrate a causal relationship between them, in either direction. The demonstration that chromosome heterozygosity of hybrids reduces their vigor or fertility, or that it at least does so in backcrosses, for example, is a step towards showing such a relationship. We can then proceed to answer the questions of how the karyotypic differences have their effects, and more importantly, whether all or only some do so, and whether particular sets of loci are involved or not; in short, to move from general assertions to detailed explanations.

3. The origin of tetraploidy. What are the paths to tetraploidy in vertebrates? Tetraploids are now known in several families of frogs, and are often thought to have been important in vertebrate evolution. Kawamura and Nishioka report that all the Japanese frogs that they studied, including members of a variety of families, have male heterogamety. Probably this is the general pattern in the Salientia although visibly detectable sex chromosomes are not usual. One of the difficulties in tetraploidy, at least in autotetraploidy, is in the simultaneous occurrence of diploid male and female gametes. It is usually assumed that these would be XXXX and XXYY in any case, and that the progeny would not produce a stable tetraploid. For these reasons, progression through diploid hybrids (such as Rana esculenta, for example), to triploid, clonal taxa, to tetraploids with restored biparental reproduction is often invoked. If two taxa are hybridizing, however, there may be many F_1 diploids. If these include males and females, and have any increased tendency to produce diploid gametes (X_1X_2 ova and X_1Y_2 sperm), progeny may be $X_1X_1X_2Y_2$ males. When back crossed to X_1X_2 females producing unreduced gametes, the back cross progeny might well contain $X_1X_1X_2X_2$ females and $X_1X_1X_2Y_2$ males, which could lead readily to a stable tetraploid population, without invoking a clonal triploid intermediate. Such a system probably requires preferential hybridization between males of one species and females of the other, such as rarely occurs in the origin of Rana esculenta. Amphibia should be particularly favorable material for such studies, but we do not know whether or not any tetraploid frog has in fact a hybrid origin.

Participants in the Symposium on the Reproductive Biology of Amphibians. First row, left to right: Sheldon I. Guttman, Symposium Co-chairman (Miami University, Ohio); Stephen G. Tilley (Smith College, Massachusetts); Midori Nishioka (Hiroshima University, Japan); Toshijiro Kawamura (Hiroshima University, Japan); Lynne D. Houck (University of California, Berkeley); T. R. Halliday (Oxford University, England). Second row, left to right: H. J. Blankenhorn (University of Zurich, Switzerland); Kentwood Wells (University of Connecticut, Storrs); Robert R. Capranica (Cornell University, New York); Giorgio Mancino (University of Pisa, Italy); Leszek Berger (Polish Academy of Sciences, Poland); Murray J. Littlejohn (University of Melbourne, Australia). Third row, left to right: Hans Schneider (University of Bonn, Germany); Thomas Uzzell (The Academy of Natural Sciences, Philadelphia); Max Nickerson (Milwaukee Public Museum, Wisconsin); Stevan J. Arnold (University of Chicago, Illinois); Douglas H. Taylor, Co-chairman of Symposium (Miami University, Ohio). Not pictured, Marvalee H. Wake (University of California, Berkeley).

SYSTEMATIC INDEX

SUBJECT INDEX

GPSR Compliance

The European Union's (EU) General Product Safety Regulation (GPSR) is a set of rules that requires consumer products to be safe and our obligations to ensure this.

If you have any concerns about our products, you can contact us on ProductSafety@springernature.com

In case Publisher is established outside the EU, the EU authorized representative is:

Springer Nature Customer Service Center GmbH
Europaplatz 3
69115 Heidelberg, Germany

Batch number: 08228323

Printed by Printforce, the Netherlands